WORKS ISSUED BY
THE HAKLUYT SOCIETY

Series Editors
W. F. Ryan
Michael Brennan

THE GUIANA TRAVELS OF ROBERT SCHOMBURGK
1835–1844

Volume II: The Boundary Survey 1840–1844

THIRD SERIES
NO. 17
(Issued for 2006)

INTERNATIONAL REPRESENTATIVES OF THE HAKLUYT SOCIETY

Australia	Dr Martin Wood, Curator of Maps, National Library of Australia, Canberra, ACT 2601
Canada	Dr Joyce Lorimer, Department of History, Wilfred Laurier University, Waterloo, Ontario, N2L 3C5
France	Contre-amiral François Bellec, 1 place Henri Barbusse, F92300 Levallois, France
Germany	Thomas Tack, Ziegelbergstr. 21, D-63739 Aschaffenburg
Japan	Dr Derek Massarella, Faculty of Economics, Chuo University, Higashinakano 742–1, Hachioji-shi, Tokyo 192–03
New Zealand	John C. Robson, Map Librarian, University of Waikato Library, Private Bag 3105, Hamilton
Portugal	Dr Manuel Ramos, Av. Elias Garcia 187, 3Dt, 1050 Lisbon
Russia	Professor Alexei V. Postnikov, Institute of the History of Science and Technology, Russian Academy of Sciences, 1/5 Staropanskii per., Moscow 103012
Spain	Carlos Novi, Calle de Vera 26, E-17255 Begur (Girona), Spain *and* 39 Hazelmere Road, Petts Wood, Kent BR5 1PA, England, UK
USA	Dr Norman Fiering, The John Carter Brown Library, P.O. Box 1894, Providence, Rhode Island 02912 and Professor Norman Thrower, Department of Geography, UCLA, 405 Hilgard Avenue, Los Angeles, California 90024–1698

Figure 1: Sir Robert Schomburgk in about 1857. Photograph in the editor's possession.

THE GUIANA TRAVELS
OF
ROBERT SCHOMBURGK
1835–1844

VOLUME II: THE BOUNDARY SURVEY
1840–1844

Edited by
Peter Rivière

Published by
Ashgate
for
THE HAKLUYT SOCIETY
LONDON
2006

Published for The Hakluyt Society by

Ashgate Publishing Limited
Gower House
Croft Road
Aldershot
Hants GU11 3HR
England

Ashgate Publishing Company
Suite 420
101 Cherry Street
Burlington
VT 05401-4405
USA

Ashgate website: http://www.ashgate.com

British Library Cataloguing in Publication Data
Schomburgk, Robert H. (Robert Hermann), Sir, 1804–1865
The Guiana travels of Robert Schomburgk 1835–1844
Volume II: The boundary survey 1840–1844. –
(Hakluyt Society. Third series; no. 17)
1. Schomburgk, Robert H. (Robert Hermann), Sir, 1804–1865 – Travel – Guiana 2. Guiana – Discovery and exploration – British 3. Guiana – Discovery and exploration – German
I. Title II. Rivière, Peter III. Hakluyt Society
918.8'042

Library of Congress Cataloging-in-Publication Data
The Guiana travels of Robert Schomburgk 1835–1844 / edited by Peter Rivière.
p. cm. – (Works issued by the Hakluyt Society, ISSN 0072–9396 ; third ser., no. 17)
Includes bibliographical references.
ISBN 0–904180–88–3 (alk. paper)
1. Guyana – Discovery and exploration – British. 2. Guyana – Description and travel. 3. Royal Geographical Society (Great Britain). 4. Schomburgk, Robert H. (Robert Hermann), Sir, 1804–1865 – Travel – Guyana. I. Rivière, Peter. II. Series.
F2384.G85 2006
918.81042–dc22
2005024025

ISBN–13: 978–0–904180–88–6
ISBN–10: 0–904180–88–3
ISSN 0072 9396

Typeset by Waveney Typesetters, Wymondham, Norfolk.
Printed and bound in Great Britain by
the University Press, Cambridge.

CONTENTS

ILLUSTRATIONS

MAPS

PREFACE

Robert Schomburgk had arrived in London in September 1839 having completed three journeys in the interior of Guiana[1] between 1835 and 1839, under the sponsorship of the Royal Geographical Society. The first of these, up the Essequibo and Rupununi rivers, had done little to add to the sum knowledge of the geography of the interior but it did result in the first accurate survey of these rivers and it provided important experience for Schomburgk himself. While on the second expedition, up the Corentyne and Berbice rivers, Schomburgk failed to achieve his main aim, to reach the mountains of the interior, once again a great deal of previously unknown geographical knowledge about the interior, including the length and course of the rivers, was obtained. This journey is best known for Schomburgk's discovery, in the upper reaches of the Berbice, of the giant water lily, *Victoria amazonica*.[2] It was the success of his third journey, undertaken in 1838–9, that meant that he was received in London with Europe-wide acclaim and honours. This journey took him first to the headwaters of the Essequibo, and then via Mount Roraima to Esmeralda on the Orinoco to join up his survey with that of Alexander von Humboldt, who had been there forty years earlier. His reports of these journeys are the subject of Volume I of *The Guiana Travels of Robert Schomburgk 1835–1844, Explorations on behalf of the Royal Geographical Society 1835–1839*.

This second volume of the travels of Robert Schomburgk contains his accounts of the surveys he made as Her Majesty's Boundary Commissioner for British Guiana. These were undertaken in 1841–43 and were composed of three distinct journeys. The first, to the north-west, traced the boundary with Venezuela southward from the mouth of the Orinoco to the confluence of the Cuyuni and Akarabisi rivers, between April and July 1841. The second expedition, from December 1841 until January 1843, surveyed the boundary with Brazil, first southward to the sources of the Takutu River and then northward to Mount Roraima, from where the survey of the boundary with Venezuela was joined to the point on the Cuyuni River reached the previous year. The final survey, that of the eastern frontier with Dutch Guiana, occupied the months February to October 1843. The survey team then remained in Georgetown, completing calculations and maps, before returning to London in May 1844.

[1] As explained in the Introduction to Volume I, Guiana is used to describe a wide geographical area that includes parts of Brazil and Venezuela as well as the three political entities, Guyana, Surinam and French Guiana. British Guiana is used when reference is to a period prior to the independence of the modern republic of Guyana.

[2] Schomburgk was not the first European botanist to discover this plant. At least four others, Thaddäus Hänke, Aimé Bonpland, Alcide d'Orbigny and Eduard Pöppig, had previously come across it, but, for various reasons, had never publicized their find. Schomburgk, however, by requesting that the lily be dedicated to the young Queen Victoria (it was first called *Victoria regia*) and through delivering a painting and an almost intact specimen to London, ensured his place in its botanical history.

As he now held a government appointment, Schomburgk reported on his activities to the Governor of British Guiana, Sir Henry Light, who forwarded his reports to the Colonial Office in London; although, as explained in more detail in the editorial introduction to each survey, edited versions of these reports were then forwarded to the Royal Geographical Society. In the case of two of the reports, the Colonial Office does not appear to have retained a copy. No map accompanied the report on the first survey, that of the boundary with Venezuela, and a new map of the journey has been prepared for inclusion in this volume. As in Volume I, the maps that accompanied the original published versions in the *Journal of the Royal Geographical Society* of the other two surveys, those of the boundaries with Brazil and Dutch Guiana, have been reproduced here.

There are a number of differences between Schomburgk's earlier accounts, those contained in Volume I, and those reproduced here. In the first place, the impact of Schomburgk's contact with the scientific world of Europe is obvious; for example, he shows himself familiar with a much wider range of scientific authorities, some of whose works he seems likely to have had with him in the interior. He himself also writes with more authority and in better English; the editor of the *Journal of the Royal Geographical Society* made far fewer editorial amendments to his words for the published versions. His accounts, however, remain equally austere but it should be remembered that he was writing official reports. For the reader who wants a more human and readable, if less reliable, story of what went on, there is the account by his younger brother, Richard Schomburgk, *Travels in British Guiana*, which, while not that readily available, is not that rare either.

The editorial conventions employed in Volume I are maintained here, and the text has been transcribed exactly as appears in the manuscripts without attention being drawn to misspellings, or grammatical and other mistakes corrected. Occasionally, where it is called for in order to clarify the sense, punctuation has been added in square brackets. Once again an attempt has been made in the provision of footnotes to steer the correct course between the over- and underestimation of the reader's knowledge.

Since Volume I went to press, a new book has appeared which may be of interest to readers.[1] It was published in commemoration of the 200th anniversary of Schomburgk's birth in 1804, and contains twelve chapters covering many aspects of his life and work. Usefully it includes an English translation of his autobiographical essay which appeared in *Leopoldina* in 1859.[2]

As the editorial work for both Volumes I and II was mainly completed at the same time, all those whose help I acknowledge in the Preface to Volume I are equally to be thanked for assistance with this Volume. There are a few exceptions to this and the later help of specific individuals is recorded at the relevant place. Finally, as a result of the publication of Volume I, John Moore, a photograph by whom was included as Figure 9 on page 277 of that volume, but whom I had been unable to trace to obtain permission for its reproduction, contacted me, having been alerted by a member of the Hakluyt Society. He was pleased to grant retrospective permission.

[1] *Sir Robert H. Schomburgk. Essays in Honour of an Explorer and Natural Scientist*, ed. James Rose, Georgetown, University of Guyana and German Technical Cooperation, 2004. Whereas the contributors to this volume are drawn from a number of different national backgrounds, the accompanying celebrations seem to have been a purely German and Guyanese affair.

[2] See 'Autobiographie', *Leopoldina*, 1, 1859, pp. 34–9, and also Volume I, p. 1, n. 1.

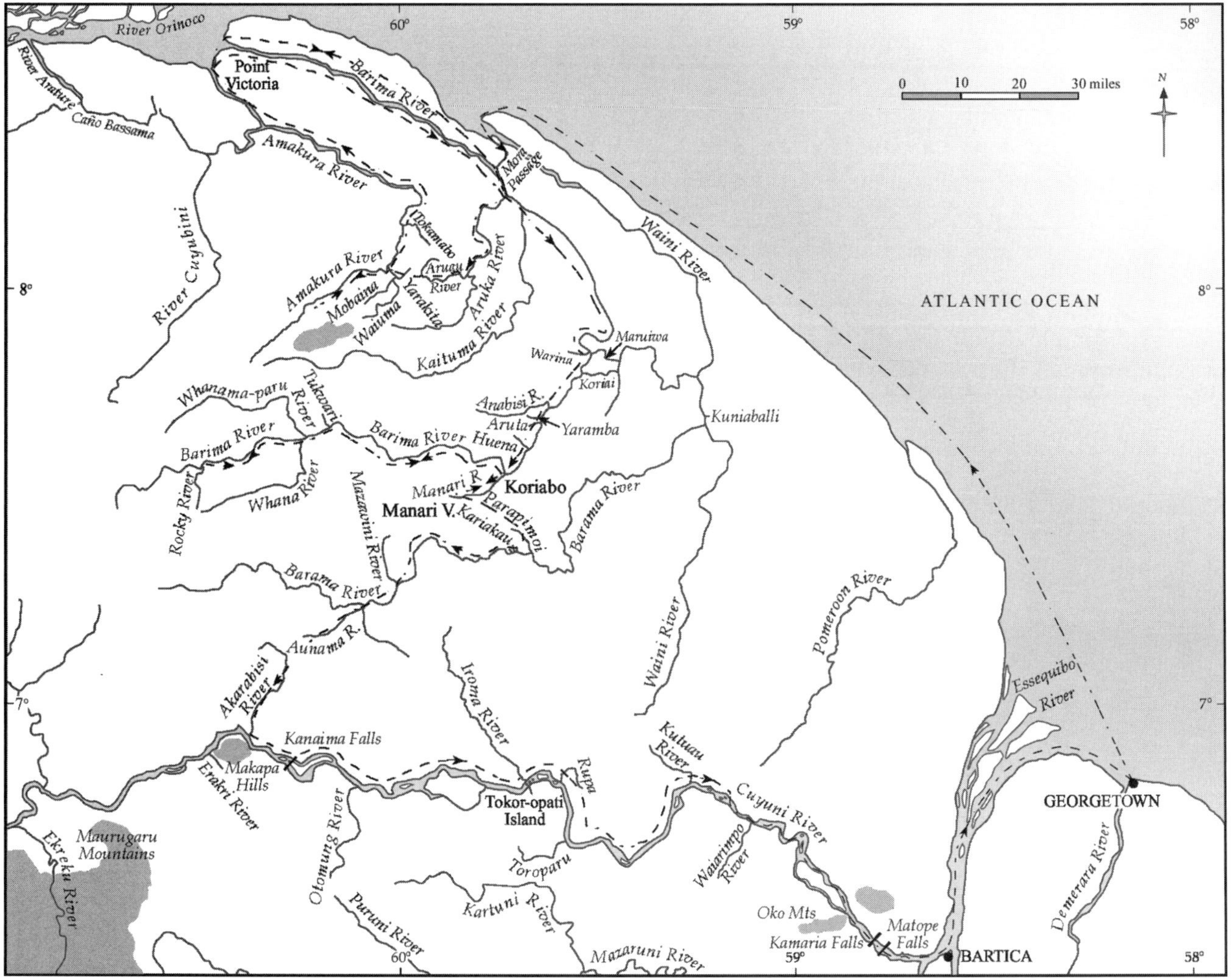

Map 1: Sketch map of Schomburgk's route while surveying the north-west boundary of British Guiana. Drawn by Julie Snook.

CHAPTER 4

Venezuela and the North-west Frontier: 1841

Robert Schomburgk had arrived in England in September 1839 and he was to be there for the next fifteen months with occasional visits to the continent to see his family in Germany or to meet Alexander von Humboldt in Paris. He had busied himself with writing and had involved himself in the scientific world of mid-nineteenth-century London. What, however, was probably of most concern to him was whether he might be able to continue his South American explorations as boundary commissioner for British Guiana. He had high hopes of this appointment.

In July 1839, on his return from Esmeralda, Schomburgk had presented to Governor Light a memorandum on the line of the boundaries of British Guiana, to become famously known as the 'Schomburgk line'. He proposed that the boundary with Brazil should be formed by the Takutu and then run north along the Surumu and Cotingo to Roraima. The boundary with Venezuela would be along the watershed between the Mazaruni, Cuyuni, Barama and Barima Rivers and the tributaries of the Orinoco until it reached the sources of the Amakura River whence it would follow the course of that river to the Orinoco Delta. Unfortunately no one had any idea just where this watershed was, and it turned out to be much further west than was realized. This was particularly true of the Cuyuni which rises far west, not far from the Caroni River. To the south and east the boundary was to run along the Serra Acarai from the sources of the Takutu to those of the Corentyne and then down that river to the sea. These claims have given rise to endless and continuing trouble, of which a brief account is to be found in Appendix 1, but it might be noted here that Schomburgk, at least as far as the Surumu/Cotingo line was concerned, realized that the claim was probably excessive and already appreciated that the Mahu or Ireng would form an acceptable fall-back position. Governor Light transmitted Schomburgk's views on the boundaries to the Colonial Office on 15 July 1839 and had strongly urged that some immediate action be taken on the question of defining British Guiana's frontiers; he also recommended that Robert Schomburgk be employed as boundary commissioner.[1]

It was not until February 1840 that the Colonial Office acted, partly, it would appear, because the memorandum prepared by Schomburgk had been mislaid. An internal minute of the 11 February expressed the view that there was no longer any reason to delay consideration of the boundary question and went on:

> Motives of humanity & the obligations which this country may be considered to have contracted towards the Aborigines, w[d] seem to urge very strongly the duty of

[1] The National Archives (hereafter TNA), CO 111/164. Letter with enclosures from Light to Colonial Office, 15 July 1839.

> extending to the Indians as far as we have any right to extend it, the protection of British territory.[1]

Schomburgk had originally held the view that it would be better for the Indians to move downstream, closer to the colony, where they could be afforded protection. In an 1837 draft of what was to be published in 1840 as *Description of British Guiana* he had recommended that the Macushi and Wapishiana be induced to settle among the colonists for whom they could provide labour while learning to reap the benefits from contact with 'civilization and religion'.[2] However, by the time *Description of British Guiana* appeared his views on this matter had changed. There is no longer a question of persuading the Indians to settle among the colonists. In fact, his thinking was now the reverse, and he argued that at present there would be no hope of so persuading them and that it would only be the introduction of 'religion, civilization and industrious habits' that would induce future generations of Indians to come and settle among the colonists. In keeping with this new advice the Colonial Office now discarded its earlier policy as unrealistic for nothing would induce Indians to leave their homes. It was concluded:

> [Accordingly it would be] extremely desirable for motives of humanity & also with a view to support the efforts of the missionary ... [to] give them all the protection we can where they are ... The Indians are the only parties who have any just or natural claim to them [territories] ...

However, their lack of power made such a claim nugatory and therefore it fell to Great Britain, in whom the Indians had confidence, to protect them. Pirara lay a long way inland but it was thought that Brazil may not be 'tenacious of this wilderness'.[3]

A memorandum along these lines was sent to the Foreign Office and the response of the Foreign Secretary, Lord Palmerston, on 18 March, was to propose that Britain should proceed in the following high-handed manner: that a map be drawn according to the boundaries proposed by Robert Schomburgk and copies of it be sent to the governments of the three countries adjoining British Guiana (that is, to Brazil, The Netherlands and Venezuela) as a statement of the British claim, and if any of the three governments objected:

> It would then rest with each of the three governments above mentioned to make any objections which they might have to bring forward against these boundaries and to state the reasons upon which such objections might be founded, and Her Majesty's Government would then give such answers thereto as might appear proper and just.

In the meantime, commissioners should proceed with the erection of *permanent* markers along boundaries claimed by Britain. Palmerston further considered that 'it would be expedient that the Brazilian detachment should be required to withdraw from Pirara'.[4] This remark was to lead to a military expedition being sent to retake Pirara from the Brazilians.

In April Robert Schomburgk was duly appointed boundary commissioner and he began making plans for the survey. The summer was spent in negotiating with the

[1] TNA, CO 111/162. Letter with enclosures from Light to Colonial Office, 9 January 1839.

[2] TNA, CO 111/150. Letter with enclosures from Carmichael Smyth to Colonial Office, 30 August 1837.

[3] TNA, CO 111/162. Note of 11 February 1840 on letter of Light to Colonial Office, 20 February 1839.

[4] TNA, CO 111/164. Memorandum of Foreign Office to Colonial Office, 18 March 1840, enclosed with letter of Light to Colonial Office, 15 July 1839.

Colonial Office on the matter of expenses, Schomburgk insisting that it was essential to have high-quality survey instruments and assistants. The Colonial Office immediately started getting cold feet about the cost of the whole business and a minute to one of Schomburgk's letters reads:

> It was originally supposed that on the writer's intended journey he would be able to accomplish this business with but little trouble to himself and little expence to the Public. But the fact is obviously otherwise.

It was also suggested that the cost of the boundary survey should be equally divided between the home country and British Guiana, and this proposal was conveyed to Light.[1]

Inevitably the preparations took longer than expected and it was not until 19 December that Schomburgk left London for Georgetown. He was on this occasion accompanied by one of his younger brothers, Richard Schomburgk, who was to make collections on behalf of the Royal Prussian Museum and the Botanical Garden in Berlin.[2] Three other members of the commission recruited in London were Lieutenant Adam Glascott RN[3] as assistant surveyor, Thomas Hancock as secretary,[4] and W. L. Walton as draughtsman.[5]

Schomburgk arrived back in British Guiana in January 1841, but it was not until mid-April that the work of the boundary commission got underway. There were various reasons for the delay, most of them local obstructions of a more and less serious nature. On the petty level was the refusal of Georgetown city council to give permission for Schomburgk to erect on town land a temporary wooden observatory from which to take observations to provide a baseline for his survey. More serious was the position adopted by the Combined Court, dominated by planters, who insisted that the colony's boundaries and the cost of fixing them were solely the responsibility of London and that

[1] TNA, CO 111/175. Letters from Schomburgk to Colonial Office, 29 May 1840, 10 July 1840; TNA, CO 112/21. Letter from Colonial Office to Light, 17 August 1840.

[2] For a life of Moritz Richard Schomburgk, see the PhD thesis by his great-granddaughter Pauline Payne, 'Dr Richard Schomburgk and the Adelaide Botanic Garden', University of Adelaide, 1992. Richard is probably as well known as his brother and they are often confused. He is best known for his three-volume account of his travels, originally published in German in 1847 as *Reisen in Britisch-Guiana in den Jahren 1840–1844*, and then in English, in two volumes, in 1922 as *Travels in British Guiana*. Richard's work, although it contains natural historical and ethnographic information, belongs more to the genre of travel writing than the rather drier scientific reports of Robert. It suffers from inaccuracies of various kinds and Richard was not above suggesting that he witnessed certain events which he knew only from hearsay. In 1848 he migrated, together with his brothers and sisters, other than Robert, to Australia. He became the successful director of the Adelaide Botanic Garden from 1865 until his death in 1891.

[3] Adam Gifford Glascott had already had experience of surveying in the region of Kurdistan and the Black Sea. He was given leave by the Admiralty to join the British Guiana Boundary Commission.

[4] It is not clear who this Hancock is, but it would be surprising if he were unrelated to the Dr J. Hancock who lived in British Guiana in the early decades of the nineteenth century and whose name has already been referred to on several occasions.

[5] We know little about Walton. Schomburgk had some difficulty in finding a draughtsman and he had to be taken on at rather the last minute without entirely satisfactory references. This was clearly a mistake as Walton took fright on arrival in Georgetown, first at the possibility of contracting yellow fever and then at the dangers of the expedition, and demanded to return home, which he was allowed to do (Richard Schomburgk, *Travels in British Guiana*, I, p. 60). He does not, however, seem to have departed in a hurry as he was still in Georgetown at the end of August (see Goodall, 'The Diary of Edward Goodall Esq. during his Sojourn in Georgetown from 28th July to 11th December 1841', *Journal of the British Guiana Museum and Zoo*, 35, 1962, p. 39).

the colony was not going to contribute to the expense. As we have seen the Colonial Office had assumed that the colony would bear half the cost and had written to Governor Light in those terms. When confronted with refusal, it was unable to understand the position of the Combined Court and considered cancelling the whole exercise. In fact, the Combined Court never did provide any money and the survey only went ahead because Governor Light, despite objections from the Combined Court, drew on a contingency fund at his disposal. The people who suffered most were the members of the commission as Light was unable to meet the half-share of their salary from this source, and they were on half-pay until the matter was sorted out after the commission had finished its work.

Schomburgk submitted two reports on the Boundary Commission's first survey, that to the north-western region of British Guiana and the frontier with Venezuela. The originals of both of them are to be found among Colonial Office papers in The National Archives. The first of these, in the form of a communication to Governor Light, was written on 22 June 1841 at the River Manari, and was forwarded by Light to the Colonial Office.[1] The main part of Schomburgk's communication is a letter of forty-three pages, not all of which are numbered, together with certain enclosures to which reference is made as appropriate.

The second part was written in Georgetown and takes the form of an official report.[2] It consists of thirty-five sheets, written on both sides, and numbered 1–70. The Colonial Office sent copies of both parts to the Royal Geographical Society, but before doing so deleted certain passages. Virtually all references to the frontier, to the setting-up of boundary markers and to reports on the Venezuelan treatment of the native population were excised. Attention to examples of these is drawn at the appropriate places in the text.

The Royal Geographical Society published the reports consecutively in 1842, in Volume 12 of its *Journal*, as 'Expedition to the Lower Parts of the Barima and Guiania Rivers', and 'Excursion up the Barima and Cuyuni Rivers'. It made few changes to the manuscript received from the Colonial Office. Those it did are mainly of an editorial nature, and even if, for example, the editor changed 'negligence' to 'sluttishness' as a description of the state of the Warao Indians, on the whole Schomburgk's own prose survives remarkably intact compared with what it had previously suffered. The one major change is the removal of a lengthy passage on the Bartica mission (pp. 36–8), which had also been sent by the Colonial Office to the Church Missionary Society.

[FIRST REPORT]

River Manari,
a tributary of the Barima:
June 22nd. 1841

Sir

In conformance with the plan which I had the honor to place before Your Excellency and which received Your Excellency's approbation, the Boundary Expedition under my

[1] TNA CO 111/179. Letter from Schomburgk to Light, 22 June 1841, enclosed with letter of Light to Colonial Office, 19 July 1841.

[2] TNA, CO 111/!80. Enclosure in letter from Light to Colonial Office, 21 October 1841.

command, composed of the individuals mentioned in the accompanying document,[1] left Georgetown on the afternoon of the nineteenth of April in the schooner Home, which had been chartered for the purpose of conveying us to the Waini, or Guiania. After a stormy passage, which the vessel and her crew appeared to be but ill calculated to meet, we arrived in the afternoon of the twenty-first of April at the mouth of the Waini, where I resolved on disembarking our baggage; and selected a bank composed of sand and shells, heaved up by the sea, as the site of our camp. With the exception of some provisions, which were damaged, all our baggage was disembarked in good order.

I resolved on remaining at the mouth of the Waini a sufficient length of time to enable me to fix the geographical situation of that point with some precision, and also for the purpose of ascertaining to what extent the entrance of the river was navigable. I accordingly commenced a survey, and, with the assistance of Mr Glascott, completed it by the thirty-first of April.[2] I have the honor to send herewith for transmission to the Right Honorable Her Majesty's Secretary of State for the Colonies,[3] a copy of the original Survey.[4] It will be observed that, although shallows and sand-banks do not qualify that river as a resort for larger vessels, nevertheless if, at a future period that part of British Guiana should become of importance, it may serve for vessels of smaller size, as during high water it affords a navigable channel of from twelve to eighteen feet at the bar, and a greater depth in the basin. It labours, however, like all tidal rivers along this coast, under the disadvantage that fresh water can only be procured within the distance that can be made in a boat with one tide in its favour. During our sojourn at the shell-bank, I had to send a boat's crew to the river Aruka,[5] a tributary of the Barima, in order to procure drinkable water, which was connected with the delay of a day and a half. The scarcity of water induced me to despatch, on the twenty-seventh of April, part of our Expedition who were not indispensable for the survey to Cumaka, a settlement of Warrau Indians on the banks of the Aruka; and Mr King, the Superintendant of Rivers and Creeks,[6] kindly took upon himself to command them. The remainder of the party followed on the first of May, after the survey had been completed. – On the twenty-eighth of April we received the visit of a Warrau chieftain from the Canyaballi,[7] a tributary of the Waini, and about two days' journey from

[1] The accompanying document is 'Enclosure A'. The individuals were the three officers of the expedition; Schomburgk himself, Adam Gifford Glascott, RN, as assistant surveyor, and William Leahy Echlin, a medical doctor who volunteered to act as well as artist/draughtsman, in order to replace Walton who, as we have seen, abandoned the commission before it set out.

The rest of the crew consisted of two coxswains, Hermann Peterson and Class van Cornelinsen, both of whom had accompanied Schomburgk on previous expeditions, four chief boatmen, an interpreter and nine canoe men including Hamlet Clenan, the cook and clown of the expedition. Attached to the expedition, but not officially part of it, were Schomburgk's brother, Richard, and Thomas Hancock, who is described as a secretary and had accompanied the commission from London. There may well have been others since we know Richard Schomburgk hired two German assistants, a young man, Florenz, and a Wurtemberger called Gottlob Stöckle, although there is some doubt about the status of the latter as Robert Schomburgk lists him as a crew member (Richard Schomburgk, *Travels*, I, p. 79).

[2] An unknown hand has written in the margin 'A new day'.

[3] This was Lord John Russell.

[4] The map resulting from the survey is to be found in TNA, CO 700/20.

[5] The Aruka River is a left-bank tributary of the Barima, the mouth of which is a few miles upstream from the Mora Passage.

[6] Robert King, Superintendent of Rivers and Creeks for Demerara.

[7] Possibly Kuniaballi, a right-bank tributary just below the mouth of the Barama River.

its mouth; who, having heard of our arrival, came with part of his men to visit us,[1] and appeared rejoiced that at last it should be decided whether the Waini was in the British or in the Venezuelan territory, as at present they did not consider themselves secure against being carried away by the Venezuelans, and forced to work at low wages at Angostura, or in other parts of the Venezuelan territory. The captain is known among the colonists of this part under the name of Sam Peter, and appeared a very intelligent old man. – During the interval, the weather had changed, and it became now apparent that the short rainy season had set in. We ascended the Waini to the remarkable passage which connects that river with the Barima, and although not navigable for sailing vessels, affords a ready communication in boats and canoes between the two rivers. This natural channel, which may be compared in some respects to the Cassiquiare, which connects the upper Orinoco with the Rio Negro, and is known in the colony under the name of the Mora creek.[2] The Warrau Indians, who inhabit these rivers, call it Morawan. Where we entered it from the Waini, I estimated its width one hundred and ten feet; and near the entrance we found a depth of sixteen feet.

During the flow of the tide, the current sets from the Waini to the Barima, and with such a velocity that the Steersman has to use precautions not to be swept against trees, which in one or two places obstruct the bed of the river, and which become the more dangerous, since the passage is so very winding – for this reason, though the depth would permit vessels of six to eight feet draught to navigate the Mora, its numerous windings and rapid tide render it only fit for boats and canoes. The ebb tide sweeps with equal velocity through this natural channel, from the Barima to the Waini. The Barima offered, where we entered it from the Mora, the sight of a much larger river than I would have expected it to be. I estimated its breadth seven hundred feet; its water, still subjected to the influence of the tides, was of a dark colour, and its depth from eighteen to twenty-four feet. About five miles distant from the Mora, flows the river Aruka into the Barima on its left bank. – Before the conjunction the two rivers are nearly of equal breadth, namely about four hundred feet. – The Aruka has, however, yellowish muddy water. A few houses, inhabited by Warrau Indians, are within a short distance of the confluence of the Aruka with the Barima. They, with others who inhabit the lower Aruka, acknowledge a Warrau by the name of William as their Chieftain, who resides at the small brook Atopani. We followed Mr King to the Warrau settlement Cumaka, within a short distance of Atopani, where we landed in the evening and found a large assemblage of Warraus, with their Chieftain William, all of whom confessed that they had always considered themselves under British protection; and, as proof thereof, their Chieftain bore one of the sticks which are given as a badge of Chieftainship by the authorities of British Guiana, and which he is said to have received as early as seven years ago.[3]

The Indians assembled offered a distressing sight of suffering under Opthalmia.[4] My former travels have made me acquainted with numerous tribes who inhabit British Guiana or the adjacent territories, and though that disease is by no means unusual

[1] The rest of this sentence was deleted from the copy sent to the Royal Geographical Society.

[2] Mora Passage.

[3] The practice of recognizing Amerindian chieftains by presenting them with a staff as a badge of office has a long history in the Guianas and was certainly practised by the Dutch in the 17th century.

[4] I.e., ophthalmia, an inflammation of the eye.

among them, I no-where saw it so frightfully exhibited as here, where at least fifty per cent of the inhabitants are suffering under the disease; or, in consequence of it, have their eye-sight impaired. I ascribe it to their inhabiting the low marshy grounds, where it appears they are more subjected to colds than in the open savannahs or on the high mountains, and to inexcusable neglect.

Cumaka is situated on rising ground. These hillocks, which are the first high ground from the sea inland, form a small chain that extends in a western direction; they are composed of indurated clay, highly ochreous; and, to judge from their vegetation, and the provision grounds of the Indians on their declivities, I consider the soil fertile. It is only here that the vegetation of the banks of the rivers commences to change. Hitherto it consisted of Curida,[1] and Mangrove trees, and numerous Truli and Manicole palms; but when we had reached the rising ground, we observed noble forest trees; as, per example, the Crab-nut tree, useful for building materials, Locust, Curahara, Siruaballia, Souari and others.

From the Curahara, the Warraus prepare canoes and corials; and, from the size of these, I judge of the height of the trees from which they are made.

Several of the crew were indisposed, and the first Coxswain dangerously ill. It was therefore necessary to make a stay of some days at Cumaka, to restore the health of those who suffered. The skill and usefulness of Mr Echlin, who accompanied the Expedition as Artist, but who by his study and experience on attending the Colonial Hospitals, possesses medical knowledge, were therefore in constant requisition. I employed the interval to determine the geographical situation of Cumaka, as a point in the interior on which to rest our pending operations, and to calculate and draw the plan of the river Waini.[2] A native Warrau, who spoke somewhat of the English language, was engaged as Interpreter, and through him we gave the Indians who continued to visit us, plainly to understand, that it was the wish of Her Majesty's Government to afford every protection to those who inhabited the regions within the limits of British Guiana, and that the object of our present Expedition was to ascertain how far Her Britannic Majesty had the right to claim these parts. Many of these Indians had to relate acts of cruelty committed by the Venezuelans, and in some instances they accused their persecutors even of murder.

I resolved, as soon as the general health of my crew was restored, to proceed to the mouth of the Barima for the purpose of examining that part of the river, and to plant a Boundary Post at its eastern point as a testimonial of Her Majesty's right of possession, and another at the western point of the river Amacura,[3] as a testimonial of Her Majesty's claim to the right bank of that river, as the western limit of Her colony of British Guiana. I engaged six Warrau Indians, under the command of the Chieftain's son, to accompany us, and we set out on our journey on the tenth of May; and having paddled through the greater part of the night, we landed the following day at the mouth of the Barima, where we encamped not far from Point Barima on the river's right bank.

The survey of the river was commenced on the twelfth, and after having inspected the localities in the neighbourhood, I fixed upon a small sandy bay at a short distance

[1] See Appendix 2, 'Glossary of Vernacular Plant Names', for the identification of these plants.

[2] This is the plan referred to earlier and is to be found in TNA, CO 700/20. The passage from here to the end of the paragraph was deleted from the Royal Geographical Society's copy.

[3] Amakura River or Río Amacuro.

south from Point Barima to plant the first Post. This took place on the thirteenth of May, with such ceremony as circumstances would permit. From thence we proceeded to the river Amacura, where we planted on the same day a Post, as a testimonial of Her Majesty's claim to its right bank as the boundary of British Guiana. The two original documents which accompany this, will serve as an attestation of our proceedings. We took the liberty to name the point of the Amacura, where the post was planted, after the Her Most Gracious Majesty, Point Victoria.[1]

The situation of the river Barima, near its mouth, offered various difficulties to fix on a base-line for its survey. I resolved, therefore, to determine the respective distances of some of its chief points from each other, by intervals noted by chronometer between the flashes and reports of Guns fired from three stations. Mr Superintendant King offered his services to the Assistant Surveyor Mr Glascott in firing the Guns on the 18th. of May, when, I am sorry to say, he experienced much temporary injury by the explosion of one of them. I was at first apprehensive for his sight; but am now happy that my fears on that score are entirely removed.[2] Our survey of the Barima was finished by the nineteenth of May; and I have the pleasure to send herewith for Your Excellency's transmission to Her Majesty's Secretary of State for the Colonies, a copy of the original plan.[3]

It will become evident by an inspection of this plan, that the Barima near its mouth labours under similar disadvantages with the Waini; but if once entered, it offers an uninterrupted navigation to vessels of two hundred and fifty to three hundred tons burthen from its mouth to the junction of the Aruka. Indeed, a finer river for steamers could not be desired. Its banks are, however, marshy to its junction with the Aruka, and so much subjected to the tide that we could not find any spot fit for our night-quarters. If the lower tracts were to be put into cultivation, it would cost the same labour and expense which were required to render the coast land of Demerara arable and productive. This has no reference to the upper regions, which have not been visited as yet by me.

If the difficulty of procuring at all times fresh water could be overcome by building tanks, &c. the Barima and the Waini would offer excellent fishing stations, and the easy communication, either by sea or by the Mora passage, between these rivers, enhances their importance as such. The fish known under the name of Querriman[4] in the Colony, abounds in these estuaries, and its value is acknowledged, as in its dry state it brings in the market of Georgetown five to six bits (one shilling and nine-pence to two-shillings

[1] The two documents referred to are to be found as 'Enclosures B and C'. Enclosure B records the erection of the boundary marker in these words: 'Post branded with Her Majesty's initials, as a testimonial of Her Majesty's right of possession to the River Barima and its tributaries, and all the land through which they flow'. Enclosure C claims the right bank of the Amakura River to its sources.

It was the erection of these two boundary posts which was to cause great offence to the Venezuelan authorities. Everything in this and the previous paragraph was deleted from the copy sent to the Royal Geographical Society. This account, as well as others, certainly belie any claim that Schomburgk and the British government made, when the Venezuelan government protested, to the effect that the marks were made purely for scientific and surveying purposes.

[2] One of the mortars went off prematurely as a result of pouring gunpowder into its barrel while it was still hot from the previous firing.

[3] This plan is to be found in TNA, CO 700/21.

[4] Queriman: 'A large grey fish said to be exactly like a large mullet; it is much prized for the richness of its flesh; *Mugil braziliensis*' (Richard Allsopp, *Dictionary of Caribbean English Usage*, Oxford, 1996; hereafter *DCEU*).

and three pence) each. Of equal, if not greater value is the Morocotto,[1] which frequents the rivers that fall into the Orinoco, and which weigh in their natural state from ten to twelve pounds. I consider it of importance to point out every resource that the country possesses. These fisheries, if followed up in a proper manner, would no doubt become a useful branch of internal commerce.[2]

I scarcely need observe to Your Excellency, that during our operations at the Barima, we met no obstructions from the Venezuelans; of whom we saw none, although the Commandant at Coriabo,[3] which is the first Post in Orinoco, must have been aware of our proximity, as several Indians who had visited us in our camp, went from thence to Coriabo. It was at first my intention to pay a visit to the Commandant, and to assure him, as being the nearest Venezuelan authority, of the friendly intention of her Majesty's Government, and that the present demarcation was merely a preliminary measure, open to future discussion between the respective Governments; but after having proceeded a considerable distance, an uncommonly rough sea, such as our corial was not at all calculated to encounter, obliged us to bear away for the nearest beach, and there remain till the following morning, when finer weather enabled us to return to our camp.

In a memorial on the boundaries of British Guiana which I had the honor to address to Your Excellency, I observed that the Dutch, when in possession of these Colonies, were in actual occupation of the mouth of the Barima, and some Merchants of Middleburg, subjects of the States General, had a Colony in that river. Colonel Moody[4] of the Royal Engineers, who was sent in the earlier part of this century to report on the Military situation of the Orinoco, observed at the mouth of the Barima the remains of the former Post. I report this circumstance, as the site of our camp, at the mouth of the Barima, gave evident proofs that the ground had been under cultivation, and the environs showed vestiges of trenches. I noted some straggling Cassada plants, and a few shrubs of Arnotto, which does not grow wild on grounds subjected to tidal influence. These circumstances, as simple as they appear, contribute to attest the undoubted right of Her Majesty to the Barima, with all the tributary streams which flow into it. But as in the demarcation of a territory it is of great importance to fix upon a line of boundary which is permanent and fixed in nature, and which cannot be destroyed by human hands, I though it advisable to claim the eastern or right bank of the river Amacura, preserving for Her Majesty, or for such of Her subjects as may deem it advantageous for their purposes, the same rights to the navigation and fisheries of that stream as the Venezuelans may claim hereafter.

The pale or post at the mouth of the Barima was planted as an attestation of Her Majesty's undoubted right of possession to that river. This point in the possession of Great Britain, is of great value in a military respect. The peculiar configuration of the only channel (Boca de Navios)[5] which admits vessels of some draught to the Orinoco,

[1] Morocot: 'A freshwater fish with a flat body, averaging 2–2½ ft in length and sometimes growing to weigh 30 lbs; it is much prized for its firm salmon-like flesh and is often salted and dried; *Colossoma spp.*' (*DCEU*).

[2] The next four paragraphs, all on the subject of the boundary, were deleted from the copy sent to the Royal Geographical Society.

[3] Curiapo.

[4] Thomas Moody. He was commissioned into the Royal Engineers in 1806, and saw active service in the Caribbean between 1809 and 1815. He retired in 1846 and died in 1849.

[5] Also Boca Grande or Great Mouth.

passes near Point Barima; so that if hereafter it became of advantage to command the entrance to the Orinoco, this might be easily effected from that point. This assertion is supported by Colonel Moody's evidence, who visited the spot in his military capacity in the commencement of this century.

When the limits of British Guiana are established, it will be highly advisable that some person of authority should be placed at this point, not only for the protection of the native tribes, but likewise to command from the neighbouring States that respect, to which a British Colony like Guiana has full right. Venezuela has a Post, and a Commandant within a short distance from the mouth of the Orinoco; the Post nearest to the western boundary of British Guiana is in the river Pomeroon, a distance of one hundred and twenty miles from the Amacura; and it follows, consequently, that the Postholder of the Pomeroon can never exercise his influence, or protection over the Indians who are settled on the Barima, or its tributaries; and which, as I have been assured by the Superintendant of Rivers and Creeks in this district, amount to a large number.

The unsettled state of the weather during the period we encamped on the Barima, made our astronomical observations very precarious. Mr Glascott and myself, however, succeeded in fixing the situation of our camp to our satisfaction; but, as much as I should have liked to extend the survey of the mouth of the Barima to the Boca de Navios of the Orinoco, the unfavourable weather, the ill state of health of my crew, and the delay, which would have been connected with it, prevented me from executing a work, which although my instructions did not point out such an undertaking, would have found every excuse by its general usefulness to navigation, if circumstances had been more favourable.

We left the mouth of the river Barima on the 20th. of May, and arrived at Cumaka, which we had selected as our depot, the following day.

The exposure to the heavy rains which had set in, did not fail to show its influence on the crew, and five were reported on the sick list. The twenty-seventh of May arrived, therefore, before we could start for the Amacura. Mr Glascott, the Assistant Surveyor, being indisposed, he remained at Cumaka; and I was only accompanied by Mr Echlin.

Thirteen miles from Cumaka, in a southern direction, the Aruka is joined by the Aruau,[1] by means of which the portage is reached, which facilitates the communication between the rivers Aruka and Amacura. I resolved however, to follow the Aruka some distance beyond the junction, in order to visit a Warrau settlement, and to become acquainted with the nature of that river at its upper course. It lessens materially in size, being scarcely more than thirty yards across. The banks, still swampy, are studded with Manicole and Truli palms, along the stems of which we saw the aromatic Vanilla trailing in large quantities, forming natural festoons, and its numerous white flowers diffusing a delicious perfume. The water of the river was of a jet black, and so clear, that it proved difficult to discern where the reflected image, which the trees and shrubs that bordered its banks cast into the river, separated from the real object. It was late in the evening before we reached the Warrau settlement, which consisted of eighteen individuals. Another village of fifteen inhabitants was higher up, which it appears is the last inhabited place in the Aruka; that river having its source about fifteen miles farther south. The incompetency of my crew for the pending journey, in consequence of several

[1] The Aruau River is a left-bank tributary of the Aruka.

having been left sick at Cumaka, made it necessary that I should engage some Indians, to assist in transporting the corial across the portage and through the smaller creeks; and three Warraus were accordingly engaged for that purpose.

We returned next morning to the junction of the Aruau with the Amacura,[1] and, following the former river upwards, reached in the evening the portage, whence we had to transport the corial to one of the rivulets which flow into the Amacura. The ground rises here to about forty or fifty feet; and, extending from N.W. to S.E., forms the separation between the small streams which flow into the Amacura and the Barima. The portage is somewhat more than a mile long, in a south west direction. The size of our boat, and the narrowness of the path, were such, nevertheless, that our crew were occupied nearly two days ere they had got the corial across to the river Yarikita,[2] which falls into the Amacura. The soil consisted of rich loam; and I observed several trees useful for Naval and Civil Architecture, as the Crabwood, Siruaballi, Suari, Mora and many others. One of the Mora trees astonished me by its gigantic size. If required, and a thicker population and increased industry were to render it expedient, there would be little or no difficulty in connecting, by means of a permanent water-course, the river Barima with the Amacura: this might be effected by cutting a canal across the portage. The soil, as already observed, is an ochreous clay, and, with the exception of a few blocks of granite, which no doubt had been transported by water, there was no rock *in situ*, that appeared to offer obstructions to such an undertaking. The course of the Yarikita was W.N.W., towards its junction with the Amacura. After having been joined at its right bank by the small rivers Waina, and Wayuma,[3] it increases considerably in size. – The Botanist would have been here much delighted, in a diversified and interesting Flora. Orchideous plants, the *Peristeria*,[4] or flower of the holy spirit; several *Epidendra*, with scarlet blossoms, and many others of equal interest, adorned the trees. A *Crinum*[5] with white flowers and a delicious perfume bordered the banks; *Bignoniaceae* trailed along the trees; and the *Brownea racemosa*,[6] which has been compared to our rose, added to the variety by its bright scarlet colour, especially when contrasted with the green of the surrounding shrubs and trees. The river is subjected to the influence of the tide, which, it appears, rises here about two feet. A short distance from its junction with the Amacura, rise on its right bank some hills to the height of about five hundred feet. They are called Manibari, and were the highest we had seen since we left Demerara. On the left bank, and close to the confluence, is the hillock Arikita, of less size than the former.

We entered the Amacura at two o'clock in the afternoon, and, following its course downwards were at five o'clock in the evening at the mouth of the rivulet Otucamabo,[7] flowing into the Amacura from the right or eastern bank. We ascended it, in order to pay a visit to Assecuru, a settlement of Arawaaks and Warraus, under the Arawaak

[1] This is clearly a mistake for Aruka. It has been corrected in the published version, p. 173.

[2] Yarakita.

[3] The Wayuma is the Waiuma, and the Waina may be the Mobaina. They are left-bank tributaries. This has been corrected in the published version, p. 174.

[4] A genus of Vandoid orchid.

[5] A genus of the family Liliaceae.

[6] *Brownea coccinea.*

[7] Probably the Tokamabo.

chieftain Jan. We were received by him in a very friendly manner; and found in him an intelligent man, who spoke the Creol Dutch perfectly. The settlement consisted mostly of Arawaaks, and only a few Warraus. The greater cleanliness in person of the former, compared with the latter, was striking. We did not observe among any of the Arawaaks, (whether children or adults) those tumours which are caused by an accumulation of chigoes, and which, being neglected to be extracted in time, render many of the Warrau children lame: indeed, as the chigoes penetrate other parts as well as the feet, these poor children suffer, by the neglect of their parents, not only the greatest pain, but are rendered in their appearance absolutely offensive. This was not the case with the Arawaaks, among whom the filthy state of the Warraus is proverbial; nor did they suffer from those opthalmic complaints which I have mentioned as being so common to the Warraus of these rivers, and of which the extent has been underrated in the statement that even fifty per cent of them suffer under it.[1]

The complaints of Captain Jan of the cruelty of the Venezuelans or Spaniards, as they are generally styled, were numerous. He related that they frequently came to his place and took from him and his people, plantains, cassava, hammocks, paddles &c. without paying for them at all, or at the best very inadequate prices. One of the women showed me an ell[2] of salempores, of the value of one shilling, for which they had taken from her four bunches of plantains. But as hard as these proceedings must fall on the poor people, who frequently by brutal force are obliged to give the produce of their fields to their oppressors, and suffer in the sequel want themselves, it would be well if there it stopped. But, alas! the system of carrying Indians from their habitations to distant parts of the Orinoco and Venezuelan Guiana, and there oblige them to work in bondage and subject them to chastisements, is frequently perpetrated upon these poor beings. In the small river Otucamabo, which Jan inhabits, there was, at a short distance from the mouth, a settlement of Warrau Indians, called Awarra, who, a few months ago, were surprised by a number of Venezuelans, led, as they told me, by the Commandant of the lower Orinoco, and three of them were carried to the Venezuelan post Coriabo. Some time after, however, they found means to get away at night, in a small canoe; and, as they are now staying at Assecuru, I conversed with them through our interpreter. Even supposing that much of what they related is exaggerated, still, if any part of it be true, the conduct of the Venezuelans towards the Natives is stamped with tyrannic cruelty. One of the Indians, who had been forced to work at an estate called Carussima, as I understood, said that those who, by age or infirmity, were not able to finish their tasks, were flogged with a four-tongue whip of ox-hide or they were drawn up with their hands to a beam in the workhouses, and, when thus hanging above the ground, were unmercifully beaten. Their nourishment, during the period they were forced to work, was scanty, and of the coarsest description; and, as to a reward for their labours, if they should be allowed to return to their homes, this was out of the question. I will not relate any more of the cruelties which were mentioned as having been inflicted upon them or others; but the truth of these was attested by all the Indians who were present; and they observed, likewise, that it frequently had occurred, that Indians who travelled with their

[1] The whole of the next two long paragraphs was omitted from the copy transmitted to the Royal Geographical Society. Except for major excisions no further examples of such deletions will be given.

[2] An 'ell' was a measurement of length that varied from country to country. In England it was 45 inches, and in Scotland 37.2 inches.

families in canoes, had been overtaken by the Venezuelans, who, after having tied the men, had violated in their presence their wives and daughters. – I cannot think for a moment that the higher authorities of the province are acquainted with these diabolical proceedings, to which, no doubt they would put a stop. But the poor Indian, who, in consequence of the distance of the seat of the Provincial Government can never bring forward his wrongs or expect any redress, must not suffer; much less he, who, according to the right of possession, or the claim of Her Britannic Majesty, considers himself under her protection.

As I possess, myself, some knowledge of the Creol language, and as the chief Boatmen, Prentice and George Albert, speak it perfectly, I examined Captain Jan whether he believed in a supreme being, a future life, or was aware of the nature of an oath, and the punishment which awaits those who perjure themselves; to all which he answered in the affirmative, having acquired some knowledge of the Christian religion during his stay at the Essequibo, where in former times he worked for wages; and he asserted that, if required, he would confirm by an oath the truth of what he had told me. I desired him to assemble next morning his people, and I found that their number amounts to fifty-nine; namely, nineteen men from the age of fifteen years upwards, fourteen boys, thirteen women, from fifteen upwards and thirteen girls. I told them, through our interpreter, the object of our coming here, and that it was not the wish of Her Britannic Majesty's Government, since it claimed the right bank of the Amacura as the limit of the British Colony of Guiana, that they should be molested; and that I should make it my duty to bring their complaint, through Your Excellency, to the knowledge of the Right Honorable Her Majesty's Secretary of State for the Colonies. Meanwhile, I advised them, if these oppressions of the Venezuelans continued, rather to leave their present habitation, although it might be dear to them; and to draw nearer to the cultivated parts of the British Colony, where the law would secure them against such oppression; which proposal they promised to consider. I drew up the accompanying document,[1] of which I left a copy in the hands of the Chieftain Jan; and under the supposition that the Venezuelan authorities of the Orinoco have been informed through their Government of the intentions of Her Majesty's Government, I hope that, by presenting this document, it may prevent the repetition of similar atrocities. At the same time, I beg leave, submissively, to suggest to Your Excellency, to transmit a copy of this protest to the authorities at Angostura, and to desire them to stop these proceedings of their inferior servants, as the Indians of the Amacura accuse the Commandant of the Orinoco, who resides at Coriabo, of having led the party who surprised the Warrau Indians at the Amacura, and carried them into bondage.

With Captain Jan of Assecuru as a guide, and our crew increased by several of his followers, we left the settlement on the second of June, and followed now the Amacura upwards. After having passed the Yarikita, which we had descended two days previously, we found that the Amacura decreased materially in size, lessening in the course of the day to a stream. We ascended, at five o'clock in the evening, the stream Curriyabo, which joins the Amacura from its western bank, on the Venezuelan territory, where we intended to halt during night at a Warrau settlement, especially as it had rained almost continuously and in torrents during the whole day.

[1] This is 'Enclosure D'.

The Indians have all withdrawn from the banks of the Amacura, and selected small streams for establishing their settlements upon. They suppose that by doing so they are less subject to the visits of the so-called Spaniards; and, in order to increase their security, these streams are almost allowed to be grown over, so that one only who is well acquainted with their navigation, would suppose them to be inhabited by human beings, or be able to reach their abodes.

The intricate navigation rendered it very difficult to make any progress in our large boat. The settlement consisted of only twelve individuals; but there are several other settlements in the vicinity, and about forty Caribs live in the neighbourhood. The whole population, including the Caribs, amounts nearly to ninety persons; but as these natives are, according to the limit at present claimed by Her Majesty, in the Venezuelan territory, I did not consider myself authorised to hear, or interfere in their complaints against Venezuelan oppression; nor would I give them any encouragement to settle in British Guiana, in order to prevent the Venezuelan Government from accusing me, hereafter, of having enticed inhabitants of their territory to settle in the British Colony.

The succeeding day (June third) proved so rainy that we were obliged to remain stationary. We started, however, on the fourth of June, to continue the survey of the Amacura to its falls or rapids, which are caused by a ledge of granitic rocks that cross the river, and throw an impediment to its farther navigation. It had dwindled the previous day to a stream; but the torrents of rain which had fallen lately, rendered it impetuous in its course. Near the mouth, the Curriyabo is only divided by a short neck of land from the Amacura, which latter river has still low banks, and is quite serpentine. As we advanced, I found its banks to increase in height, and studded with noble forest trees. The gorgeous flowers of the *Brownea racemosa* and *Gustavia angustifolia*,[1] were so abundant, that they added considerably to the beauty of the sylvan scenery. A few miles above the junction of the Curriyabo with the Amacura, the stream Tusa joins the latter river from its right bank. It appeared to be of the same size as the Amacura. The course of the Amacura is much farther westward, in ascending, than laid down on extant maps; our course was today generally west-south-west to the fall Cuyurara. This fall is about twelve feet in perpendicular height; two others are higher up, and the whole descent may amount to about thirty feet. The small size of the river renders the aspect of the falls by no means imposing, and it may be said, that the Amacura above its junction with the Yarakita, is only fit for the navigation of the small boats of the Indians.

We did not proceed farther, which in our large boat would have been impracticable; nor did it appear to me that I was so far warranted in risking the health of the individuals who accompanied me, as to prosecute the stream's course in small boats, where protection against the inclemency of the rain proved impossible. Astronomical observations were so precarious, that, since we departed from Cumaka, we had not seen either sun or stars. There are no more inhabitants at the banks of the Amacura, or its tributaries beyond the junction of the Curriyabo; and, according to the evidence of the Indians, who pretended to have been at the source of the Amacura, it is about two days' journey, in their small boats, from the fall Cuyurara.

The fifth of June saw us on our return to Assecuru. Arrived at the junction of the Yarikita with the Amacura, I selected one of the trees on its left bank to engrave on it

[1] Also *Japarandiba angustifolia*.

Her Majesty's initials, as a boundary mark. This tree is situated on the northern foot of the hill Arikita, and about five hundred yards distant from the junction of the rivers, which bears North 37° West.

On leaving Cumaka, and considering the present journey as a pioneering expedition, I had only provided myself with a chronometer, a sextant, an artificial horizon and prismatic compass. The unfavourable state of the weather enabled me only to procure observations of the sun for the chronometer, on the morning of the sixth of June; and ten days having elapsed without any intermediate observations, I could not depend upon its rate. However, I had desired Mr Glascott, who, in consequence of indisposition had remained at Cumaka, to fire at six o'clock on the evening of the sixth of June, three guns, which we distinctly heard at Assecuru. We thus procured the direct compass bearing of Cumaka; and, combined with my observations for latitude, I received as the result, the difference of longitude between Cumaka and Assecuru.

I was fortunate enough to procure here, and at the upper Amacura, a large supply of Indian provisions, for which we paid, to the full satisfaction of the Indians, in such articles as they most desired; namely, cutlasses, knives, calico, salempores, beads, &c. The provisions which we had brought with us from Georgetown, being nearly exhausted, this supply was very welcome; and, as I had received information from Mr Glascott and his party at Cumaka, that they were short of provisions, I despatched a large supply by two small canoes, across the portage of Yarikita.

We left on the seventh of June on our farther descent to the mouth of Amacura. The Arawaak Captain Jan who went with us to the upper Amacura, and who proved himself very useful and intelligent, accompanied us farther; as his knowledge of the localities and the names of streams which fall into the Amacura, rendered his services valuable. The streams which join the river from the eastern or right bank are very numerous, and it increases materially in breadth. I state its average depth at its lower course as eighteen feet, though there are places which much exceed that depth. A peculiar feature in this river are large patches, consisting of matted grass; the splendid blue water-lily (*Ponthederia azurea*)[1] and several other water-plants, which, torn off by the increased stream during the rainy season, come floating down with the current, and, reaching that part of the river where it is subjected to the tides, they are carried to and fro, as the tide may be flowing or falling. We might have numbered thousands of these little floating islands. We reached, in the afternoon at three o'clock the Coyuni,[2] which like the Mora from the Waini to the Barima, and *vice versa*, offers an uninterrupted passage in canoes from the Amacura to the Araturi.[3] The Coyuni connects the Amacura with the Waicaicaru or Bassama,[4] which falls into the Araturi. This river flows, opposite the island Imataca into the Orinoco, and is another instance of a remarkable connexion between the tidal rivers of this coast. A short distance above the mouth of the river Araturi, is the Venezuelan post Coriabo. The importance of the natural canal in a military, or a commercial point of view, is undeniable; but its importance to Venezuela (if a denser population should make it such) is rendered abortive in a military aspect, if Great Britain possesses the right or eastern bank of the Amacura.

[1] *Pontederia azurea = Eichhornia azurea.*

[2] Río Cuyubini.

[3] Caño Arature.

[4] Caño Basama.

There is no doubt that the Amacura is navigable for smaller vessels and steamers to the Yarikita; the bar at its mouth, and the inconsiderable breadth, which seldom amounts to more than 300 yards, render it unfit for larger vessels. It abounds in that delicious fish, the Morococo.

It was late in the evening before we reached Victoria Point, at the mouth of the Amacura; and we were happy to observe that the Boundary Post which we had planted here on the 13th of May, was still standing. The same refers to the Post which we had planted at Point Barima, and which we visited next morning on our ascent of the Barima.

We arrived on the tenth of June at Cumaka, where to my great pleasure I found the invalids mostly restored, and Mr Superintendant King rejoicing in his recovered eyesight.[1] We prepared the succeeding day for our departure, when a murder, which had been committed two months ago upon an Indian from the Orinoco, obliged Mr Superintendant King to take cognizance of it, and to enter into judicial proceedings. As this case comes before Your Excellency in a more detailed form, I shall only here allude to it in general terms, and make such remarks as my knowledge of Indians & customs calls forth.

It appears that an Indian from the Orinoco, by the name of Waihahi, frequently visited the Indians at the Aruka; among whom he was much feared as a Pi-aiman or Conjuror, who, by his malpractices or charms, was enabled, like the Obëah man of the Africans, to injure his fellow creatures. This Waihahi was accused of having killed by his charms or by secretly administering poison, the family of a young Indian boy, named Maicarawari; his mother, whom he loved affectionately, being the last victim: and when he, who now remained the only member of the family, reproached the Pi-aiman for his deed, he was laughed at in derision, and was told that a similar fate awaited him. – The Warraus of these rivers have not the slightest knowledge of Religion; they know nothing of a God, or a future life; but the principle of revenge, based upon "blood for blood, life for life," is implanted in their breasts from the time they are able to understand their maternal language. This Indian boy considered himself unalterably appointed to revenge the death of his family upon him, who did not deny that he was the cause of his bereavement; and when Waihahi came again to the Aruka, an opportunity offered itself to execute his revenge. At a drinking feast which Waihahi gave to the Indians in the neighbourhood, and to which Maicarawari accompanied his Chieftain, a dispute arose; when the Pi-aiman said in anger, that he would leave the place next morning, but that the Chieftain William and his followers should die shortly after in the bush, and that there should be no person to bury them, but the carrion crows. The boy, who had taken no share in the revels, had been sitting apart; and this threat no doubt confirmed him in his resolution of executing his intentions. When he saw the Pi-aiman asleep in his

[1] Approximately the next five pages, dealing with a case of Amerindian homicide and the problems of bringing to trial under British law Amerindians unacquainted with its principles, were omitted from the copy sent to the Royal Geographical Society; that is until the end of the first paragraph on p. 18 here. The reliability of Amerindian witnesses was a topical question at the time as in a recent trial of a colonist who had shot an Amerindian, the case was thrown out because the Amerindian witnesses possessed no religious beliefs. In 1841 a law was passed allowing Amerindians to give evidence, but with this proviso: 'that the admissibility of such evidence shall be subject to such deduction from its weight as any court may deem proper to attach to the fact of such Testimony being unsworn' (see M. N. Menezes, *British Policy towards the Amerindians in British Guiana 1803–1873*, Oxford, 1977, pp. 136–7).

hammock he rushed towards him, and taking his war club in both his hands, he killed him with one stroke, by completely fracturing his skull.

As the deed was committed within the assumed limits of British Guiana, namely east of the Amacura, and in a river which falls into the Barima, it would come under the jurisdiction of the Colony; but a serious question arises, – whether the Indian, who has no knowledge of the Christian Religion, and does not acknowledge our Laws, can be punished for an act which civilized nations consider a capital crime, but which, according to the manners and customs he has been brought up in, is a meritorious deed. And this persuasion has not left him: he himself went to the Indian Chieftain Cabaralli, who bears the highest authority in these rivers, and informed him of what he had done; and since the Superintendant thought it his duty to proceed with him and the witnesses to Georgetown, he has followed voluntarily and without restraint, to be tried by a Court of Justice, and adjudged by laws, of even the existence of which he has no knowledge. His judges are not able to enter into his feelings; nor do they see that, by the maxims of his tribe, he was, as it were, ordained to commit an act, which any other Indian under similar circumstances, and equally unacquainted with the Christian Religion, would have considered it his sacred duty to perform. But this tragical event, which is no doubt one among many that have come to our knowledge, gives rise to the mournful reflection, that there are in this Colony, and comparatively within so short a distance from its capital, thousands of Indians who walk in perfect darkness with regard to the Christian Religion. Should the moment arrive when religious teachers shall be sent amongst them, and they shall be converted, such cases must cease entirely, or the perpetrators will be answerable to the Courts of Law for their misdeeds. In the present case, between Indian and Indian, both of whom are uncivilized, it is my opinion, based upon my knowledge of the Indian character – of their customs and manners – that Maicarawari is not amenable to the Courts of Law of this Colony for the deed which he has committed.

I have to apologize to Your Excellency for the remarks and the opinions which I have advanced; but another opportunity might not occur, to show how much it is required to tender civilization to the native tribes who inhabit British Guiana. If Great Britain, by its commercial connections and shipping, derives any benefit from the possession of this Colony, it must be recollected that the territory belonged once to those tribes from whom European nations have wrested it.

The interest in the welfare of the natives of this Colony which Your Excellency has shown, and the wise ordinances which have been framed for their protection,[1] render it impossible that such cruel acts as they are subjected to in the neighbouring territory, could be committed upon those in British Guiana; but, as long as those tribes are not converted to Christianity, they labour under a disability, which I am sure Your Excellency will give your consideration to remove: otherwise the Indian is liable to be oppressed by every unprincipled and designing man. The first question which is put in a Court of Justice, to a person who appears as Plaintiff or as Witness, is, whether he is acquainted with the nature of an oath – of a God and a future life. A satisfactory answer

[1] Light had made various attempts to introduce measures aimed at civilizing by evangelization the Amerindian population, but had been consistently blocked by the Combined Court. He did manage to raise, together with matching funds from the bishop of the diocese, some money for this purpose (see Menezes, *British Policy towards the Amerindians*, pp. 244–6).

cannot be expected from him, who has never been instructed in the Christian Religion; but, as sacred as is his affirmation to the Quaker, equally so is to the Indian his assertion, that he tells "the truth and no lie"; by which strong expression only, I can convey the meaning of the Indian sentence. Before such an assertion, however, is taken in lieu of an oath, the unprincipled Colonist may subject the native to every oppression, without running the risk of punishment for his misdeeds. I am informed it is not so in Her Majesty's East India Colonies, where a law is in existence, by which the evidence of the natives, who are heathens, is nevertheless regarded valid in the Courts of Justice. The Indian of Guiana is no idolater: he either believes in a good spirit, or walks in perfect darkness, without giving a thought on the existence of a God; and, as he cannot swear by his idols, some other binding form must be substituted.[1]

Although the rainy reason has for some time past set in, and although our stores are materially reduced, and we have been deprived of many comforts, I yet deem it my duty to persevere, and continue the survey to the Cuyuni. The two large corials, which we are not able to transport across the land, have received orders to proceed round the coast to the Essequibo, where they are to remain at Bartika point; while the Coxswain Cornelius is to meet us with some small canoes and a supply of the most necessary provisions on the Cuyuni. As far as I can foresee, three to four weeks may elapse before the expedition can return, for refitting, to Georgetown.[2]

The map which I am constructing will point out more clearly the route which we have taken, and those points where Boundary marks have been planted. I shall lose no time, on my return to the Colony, to lay this map before Your Excellency; the incompleteness of which at this moment, where my investigations are unfinished, and unprovided as I am with the necessary materials for its construction, prevents me from enclosing it herewith. I must not, however, omit to observe that more unfavourable weather for astronomical observations we could not have had, than we have experienced during our Expedition.

I cannot close this report without bringing to Your Excellency's notice the alacrity and good conduct of the Officers belonging to the Expedition. It gives me, likewise, pleasure to observe, that the men who compose the crew have performed their duty to my satisfaction; and I have only to wish that they may continue in their good behaviour.

The Expedition is highly indebted to Mr King, the Superintendant of Rivers and Creeks of the County; who, by his active co-operation in farthering the ends of the Expedition, and his acquaintance with the Indians of these parts, has made his assistance the more valuable.

I have the honor to be, Sir, Your Excellency's most obedient and humble Servant,
Robert H. Schomburgk,
H.M. Commissioner of Survey

[1] The fundamental problem is not so much one of truth derived through a sworn statement, but the fundamentally different ideas concerning cause and effect in European and Amerindian cosmologies.

[2] As will have been noted this letter was written at Manari on 22 June. It will be found that the period from 13 to 22 June is covered at the beginning of the second report where the return of the two large canoes to Georgetown is reported again.

SECOND REPORT OF ROBERT H. SCHOMBURGK, HER MAJESTY'S COMMISSIONER FOR SURVEYING AND MARKING OUT THE BOUNDARIES OF BRITISH GUIANA.

Demerara:
August, 1841.

The party under my command left Cumaka, where we had sojourned for some time, as detailed in my former report, on the thirteenth of June; and having arrived at the junction of the Aruka with the Barima, we continued the ascent of the latter river in an east-south-eastern direction. We reached, on the next day, its junction with the Kaituma, which falls in on the left bank from the south, and is at its mouth about two hundred feet wide. The Kaituma is inhabited by Warrau and Waika Indians; and is connected with the upper Barima by several intermediate brooks.[1]

Numerous rivulets join the Barima at both its banks; some of them inhabited by Warraus. It has still, however, the appearance of a tidal river, being margined by Mangrove and Curida bushes, over which Manicole and Truli palms raise their heads. Its banks form continued swamps, which, only through the industry of man, could be made arable.

We encamped on the night of the fifteenth – sixteenth of June, at a Warrau settlement; the Chieftain of which called himself Marawari. The noise of the Indian drum and songs, on our approach, announced that the Inhabitants were revelling; and on landing we had sufficient evidence that Marawari was intoxicated. One of his wives was in the same state; and thus we witnessed, for the first time since we had left the Colony, the effects of that horrible vice, drunkenness. The settlement consisted of five huts, surrounded by rich provision grounds; and I observed with pleasure some Lemon and Lime trees near their houses. Their provision fields abounded in cashew nuts (*Anacardium occidentale*).

We passed on the following day the small river Maruiwa, or Hohanana, which, by the interlacing of numerous other rivulets, affords a passage in boats from the Barima to the Waini: a journey which the Indian effects, generally, in two days.[2] At a short distance above this river, rise some hillocks from both its banks.[3] They are the first which are met with in the Barima. The Warrau chieftain, Clementi, had selected one of them to build on its summit a large hut, which, by its construction and neatness, distinguished itself from the generality of Indian houses. The erection of this house with its gallery, in imitation of a second story, was the more gratifying, as he did it to afford accommodation to the Superintendant of the district on his periodical visits; the good effects of which, as exhibited in the character and better conduct of the Indians in their social intercourse, I had several times opportunity to witness. By studious attention, I seized here a favourable moment to determine the position of the place; a circumstance which the unfavourable weather had not afforded us, since we left Cumaka. Warina[4] is, according to my observations, in latitude 7°50′15″ North, and longitude 59°43′30″ West; and the height of the house we had selected for our night's quarters, was about seventy feet above the Barima. The

[1] The sources of the Amakura, Kaituma and Barima rivers are relatively close to one another.

[2] The reference map shows riverine communication possible between the Barima and Waini by both the Maruiwa and the Koriai, the latter a few miles further upstream.

[3] That on the right bank is called Mount Everard, and that on the left Mount Terminus.

[4] The Warina is marked as a small left-bank tributary of the Barima on the reference map.

chain of hillocks, on one of which the house was built, extends in longitudinal ridges in a north and south direction, N12°E, and S12°W.

I always considered it my duty, wherever an opportunity offered for illustrating, however slightly, the geology of the Colony, to observe how far the structure of the country might be favourable to cultivation; it being undeniable, that, the quality of the soil depends, generally, upon the rocks which form the strata below the arable land. The super-strata at the hills of Warina consist of ochreous clay, intermixed with mould, pebbles, and that due proportion of sand which would particularly qualify it for the cultivation of coffee. The large blocks of ferruginous clay which lie dispersed on the surface, insure the necessary moisture for the cultivation of that plant: for it is well known to the Agriculturist how beneficially such blocks operate on the soil on which they lie, contributing not only to the retention of the moisture, which would otherwise evaporate, but to the precipitation of atmospheric vapours.

The rivulet Curiye[1] offers another medium of communication with the Maruiwa and the Waini; but it can only be made use of by small boats. We passed, at nine o'clock on the eighteenth of June, the river Amissi,[2] which joins the Barima on its left bank. It is of considerable size; and at the junction of the two rivers, it would seem as if the Amissi were the larger one. The Indians, however, informed me that its course has not the length of the Barima, and that its banks are mostly swampy: – the current appears insignificant. During the rainy season the influence of the tide is felt to this point; in the summer months, it is felt still higher up. The swampy banks of the Amissi render it unfit for habitations. Even the Warraus, whom the earlier authors of travels described as living on the tops of trees, but who, in reality, raised only a platform just above the level of the water, and rested their miserable dwellings on stumps of Ita trees, prefer now higher ground to build their huts upon.[3] The Amissi affords, by natural canals, a communication with the river Kaituma.

Since we had left Warina, the Barima, in ascending, had adopted a more south-western course; its banks, also, became higher; while the Palms and Mangrove bushes, which till now had been so numerous, became less frequent, and were replaced by a more varied vegetation. Our Indian Guides informed us, that, by ascending the rivulet Yaramuku[4] half a day, we would reach high hills and savannahs. We continued, however, the ascent of the Barima, and passed the rivulets Aruta and Pegua; the latter inhabited by Warraus. The Barima narrows, above this creek to forty yards, and flows with a strong current, which impeded our progress; its depth was still from three to four fathoms. The banks (it being now the middle of the rainy season) were full to overflowing, and rose scarcely a foot above the water's edge. In lieu of Palms, the most stately Mora trees overshadowed the river. In all my former travels in Guiana I have nowhere

[1] Koriai.

[2] Anabisi River. There is also a right-bank tributary of the Kaituma called Anabisi, and although both rivers have their sources in the same general area, they appear to be too far apart for there to be any riverine connexion between them.

[3] The story that the Warao lived in the tops of trees goes back in the earliest Spanish sources (see Walter Ralegh, *The Discoverie of the Large, Rich and Bewtiful Empyre of Guiana*, ed. Neil Whitehead, Manchester, 1997, p. 158). Humboldt (*Personal Narrative of Travels to the Equinoctial Regions of the New Continent during the Years 1799–1804*, transl. Thomasina Ross, 3 vols, London, 1907, III, p. 8) repeated it, but by the 1830s it was recognized that the Warao lived in pile dwellings over the water.

[4] Probably the Yaramba.

seen trees of this description so gigantic, as on the land adjoining the Barima at its upper course. Indeed, frequently, when our boat rounded some point which the river made in its course, and a long reach was before us, these majestic trees appeared in the background as hillocks clothed with vegetation, until a nearer approached showed our mistake; and we found that, what we considered to have been a hillock, was a single tree, rising to the enormous height of one hundred and thirty to one hundred and fifty feet; forming by itself, as it were, a forest of vegetation. The importance of the Mora in Naval Architecture, is now fully recognised in Great Britain, and a new export trade has been opened to the Colony. At the upper Barima this tree is so abundant, and grows to such a size, that the whole British navy might be reconstructed merely from the trees which line its banks: a circumstance well worth consideration, especially as, being near a river which is navigable to vessels of twelve feet draught, the craft intended for the transport of the timber might load at the very spot where the trees are cut down.

It is only lately that the timber of Guiana has come into notice in England; but, so superior is the Mora and the Green Heart for objects of Naval Architecture, that a higher price is given for them in sea-port towns, than for any other wood imported into England.

It appears that, at the commencement of this century, a white man – very likely a Dutch settler – had advanced so far as the Herena[1] river. The Indians showed us the place where he had cultivated sugar; and they told us, that he had possessed a schooner, and several punts, with which he carried on a timber trade. The Indian, in his expressive language, called the former settlement "the last place of the white man".

We entered, on the 19th. of June the Caruwavu or Caruawa river,[2] a tributary of the Barima, and halted at a settlement of Warrau Indians. While among the Warraus I had heard much of one of their games, which they exhibit during festivities; and I had the satisfaction of seeing it here performed. It is played in parties, two against two; and the champions, painted and dressed according to the taste of their tribe, show their athletic skill by attempting to push each other from a space of ground by means of the *Haha*,[3] which I might resemble to a shield. It appeared to us an innocent pastime, which gave agility to their limbs, and displayed to the greatest advantage their muscular power and fine proportions.

There are several Warrau settlements on the banks of the Caruwavu. I estimate their number at two hundred individuals. The Manari, a river with a stronger current than the Caruwavu, joins the latter on its left bank, at the distance of about a mile from its confluence with the Barima.[4] The Manari is mostly inhabited by Warraus; but there is a settlement of Waikas[5] about five miles up, where we intended to stay for a few days. I had understood from some Indians, who were well acquainted with the Cuyuni, that there had been once a Dutch post at an island called Tokoro,[6] which was much further

[1] Huena, which is the way it is spelt in the published version (p. 189).

[2] Koriabo River.

[3] This game is still played among the Warao, and is known as 'isahi', although it is also called after the name of the shields, 'nahakara' (which suggests that 'Haha' in the text should have been 'Haka'). In the past it was not a game but a fight to settle disputes (personal communication Dieter Heinen).

[4] The reference map shows the Manari as a right-bank tributary of the Barima, joining the latter river a little above the Koriabo.

[5] Akawaio

[6] See below, p. 35.

west than that part of the Cuyuni where, from the information I had received previously to my submitting the memorial on the Boundaries of British Guiana, I conceived the boundary line ought to cross to the river Cuyuni. The path overland led from this settlement to the river Barama, and from thence to the Cuyuni; and it became, hence, necessary to select it as a starting point. Our larger canoes, being much too bulky, were now of no further use; and as the official duties of Mr King, the Superintendant of the County, required his speedy return to Georgetown, I resolved on sending the two canoes, with such of the crew as I considered not qualified for the fatigues of an overland journey, under his command, to the coast.

We landed at the settlement Manari in the afternoon. It consisted of five Indian houses, the largest of which was given up to us to reside in. In my former Report, I alluded to the negligence, in person and in the houses of the Warraus; and mentioned how superior the Arawaak Indians were, in that regard, to the latter. Equally superior, if not surpassing the Arawaaks, are the Waikas. Their cleanliness, both in person and domestic arrangements, was a gratifying picture, after having travelled for months among the Warraus. The Waikas are of much fairer complexion than the other Indians who inhabit the coast regions, whom they surpass in athletic form and regularity of features. The Warraus indulge in bigamy: I met even several instances of polygamy among that tribe. It is different with the Waikas, whom I found, uniformly to possess but one wife.

The land adjoining Manari is wonderfully productive. We saw sugar-canes vying with the best on the coast; Indian corn or Maize far surpassing any ever produced at the coast regions; and bunches of Bananas weighing from eighty to a hundred pounds. The super-stratum is a rich loam, intermixed with vegetable earth and sand; and, as it lies upon clay, a sufficient moisture to advance vegetation is always kept up, thus affording ever capability for the cultivation of the staple produce of the Colony.

Mr King, and those of the party who were to return with him, left Manari on the 22nd of June. Our provisions were at that time much reduced, and the period of the year did not warrant me in supposing that I would be able to get much from the Indians. I desired, therefore, that the boat of the expedition should proceed up the Essequibo, and await the land party at Bartika point, from whence a supply of the most necessary provisions should be sent up the Cuyuni, in a smaller corial, to meet us. This service was confided to the second Coxswain, Claas van Cornelinsen.

I was anxious to examine the Barima beyond its falls. I started, accordingly, on the twenty-fourth of June in a small canoe, accompanied by Mr Glascott the Assistant surveyor, and Mr Echlin the Artist of the Expedition; and, descending the Manari for a short distance, we reached the Barima by two of those natural canals (the Taima and Ataima) which so frequently connect rivers having a parallel course in these swampy regions. The almost continual torrents of rain which we had had for some weeks, had caused the Barima to overflow its banks, and we found the current running at the rate of from four to four and a half miles an hour; our progress was consequently slow. A short distance above the off-flow which connects the Barima and Manari, we visited a Warrau settlement called Emu, where we admired a gigantic bamboo, several hundred yards in circumference. Two of the Indians were occupied finishing a native canoe, which they had cut out of cedar (*Icica altissima*): a species of wood uncommonly well qualified for that purpose, and resembling, in its durability, odour, and reddish colour, the famed Bermuda cedar, although a genus

Figure 2: Waika house and dance. This dance was held at the Akawaio village of Manari and is fully described by Richard Schomburgk, *Travels in British Guiana*, I, p. 160. The picture is taken from a copy of this work in the editor's possession.

quite distinct from the Icica. As the cedar tree of Guiana is by no means scarce, it deserves more attention.[1]

The Warraus are famed for their skill in finishing canoes out of the single trunk of a tree. They formerly furnished the Colonists, as well as the tribe of Indians inhabiting the coast regions, with canoes and corials which, for durability and speed, far surpassed any boats ever introduced from Europe. Of late years their industry has much relaxed; and they are loud in their complaints that the Spaniards of the Orinoko take away all their largest craft, and destroy them; and that the smaller only escape, by their being

[1] Richard Schomburgk in the index to *Travels*, I, has under *Icica altissima*, see *Protium altissimum*. According to the NMNH checklist *Icica altissima* = *Tetragastris altissima*. There are no species of *Icica* listed in Mennega et al. (*Check-list of Woody Plants of Guyana: based on D. B. Fanshawe's 'Check-list of the Indigenous Woody Plants of British Guiana'*, Ede, The Tropenbos Foundation, 1988), but numerous species of *Protium*, most of which are used for making canoes. However, in the same work, the tree known locally as 'red cedar' is listed as a species of *Cedrela* and the 'white cedar' as a species of *Tabebuia*. Meanwhile Johannes Wilbert ('To Become a Maker of Canoes: An Essay in Warao Enculturation', in *Enculturation in Latin America. An Anthology*, ed. Johannes Wilbert, Los Angeles, University of California Press, 1976, p. 307) states that the most popular tree among the Warao for manufacturing canoes is the 'red cedar' (*Colophyllum lucidum* Benth.), and then the 'white cedar'(*C. lucidum*). All of which illustrates some of the problems encountered in identifying botanical species.

able to hide them. The famed Spanish *launches*, employed during the revolutionary war of Venezuela, were made by the Warraus. Some of these were roomy enough for from fifty to seventy people. They refuse, now, to make any of so large a size: not for want of the trees fit for the purpose, but that, they say, if the Spaniards hear of their making any large craft, they send a party of men to take them away, or cut them in pieces, in order to prevent them from being sold, and used for smuggling by the people at the mouth of the Orinoco. Such cruel acts cannot be practised upon the Indians who live within the British boundary, if that boundary is once politically recognised.

We passed the small rivers Ararisi, Yabritin, Buruparu, Mariawaballi;[1] and landed on the evening of the 25th. of June at the Warrau village Simuita. We measured here the breadth of the river, and found it fifty-one yards. The river Kaituma runs hence about nine miles in a North-east direction. The Barometer stood, at 6 hours A.M. 30°0′20″ English inches; the thermometer at 70°5′ Fahrenheit.

We were accompanied by a number of Indians from Simuita and the neighbouring settlements, who intended to ascend the river to the falls, to shoot the delicious fish called Marocotto or Ossibu; which, at the time these waters are full, migrate beyond the falls for the purpose of depositing their spawn. We formed a flotilla of small boats, our canoe being the leading frigate. Several fish were procured on the first day. In order to attract them to the shore, a number of the seeds of the Carapa or crab-nut are pounded, and having been enclosed in a netting made up of withes, they are put in the water, and soon attract the greedy Maracotto: an Indian stands ready with a light spear, which he lances into them, one after another, with unerring skill. I have in my former Report, alluded to the importance of following up the Fisheries, as an additional resource for the Colony. I here observed that the Maracotto reaches frequently a length of thirty inches, and is twenty-six inches in girth; while its delicious flavour recommends it to attention, as an article of trade.

We observed on the 27th. of June a tract of sandstone which was heaped up in numerous blocks. It is fine-grained, and much used by the Warraus in lieu of grindstones, to sharpen their tools for the manufacture of their boats. We arrived in the afternoon at the fall Mecoro-vussu, which threw the first impediment to the navigation of larger vessels on the river. A few miles below the falls, we found a depth of three fathoms. The Barima is therefore navigable so far for steamers of considerable size, although it might prove tedious to sailing vessels to reach that distance, in consequence of the serpentine course and strong contrary current.

It is not known to the Indians inhabiting these regions, that white men had ever penetrated so far before. We might have stopped here, and commenced our return, the more especially since the weather was so unfavourable, but I found the course of the Barima so different from what it is laid down on maps, that I considered it of importance to trace it higher up, as, by its western course, on its ascent, every mile would add to the British territory. This course, differing so much from the Barima of theoretical Geographers, will, I presume, be deemed sufficient evidence of the importance of the measure which Her Majesty's Government have resolved upon; namely, that an actual survey should prelude the definitive negotiations with the Governments interested in the determination of these boundaries.

[1] None of these names appears on the reference map.

Only the rainy season could have afforded us the opportunity of ascending the Barima any further; and I resolved, therefore, to continue until we could make no further progress in our corial.

The first series of falls were ascended without any accident, and we halted the same evening, near some temporary huts, which certain Waikas from Manari had erected opposite to a place where they intended to found a new settlement. Although months had doubtless elapsed since any human beings had sojourned in these huts, we found them swarming with Fleas and Tshigoes which made us soon relinquish the idea of using them for our night's quarters.

We passed, in the course of the next day (June twenty-eight) numerous rapids, of which one called Uropocari was the largest. The river kept its breadth, but was studded with rocks. We passed in the afternoon a large rivulet with black water, called Duquari:[1] it comes from the west-north-west. I afterwards observed stratified quartz, and could not but admire some huge blocks of granite which rose above the level of the river, and are called, by the Indians, Arauta. Their shining surface and symmetrical form were equally remarkable.

The river Wanama[2] (so named from a species of Bamboo which grows at its mouth, and which the Warraus call Wanama) joins from the left bank, and is one of the largest tributaries of the upper Barima. About half a mile farther south-west, the river Mehokawaina[3] unites with the Barima: both tributary and recipient are, previously to their junction, of the same breadth. The Mehokawaina comes from the south-east, and the Barima proper from the south-west.[4]

I found it now advisable to discontinue the ascent in corials, as numerous trees which had fallen across the Barima, would have thrown the greatest difficulties in the way of any further attempt to advance with the boats.

After having marked three trees with Her Majesty's initials, I left Mr Glascott in charge of the camp which we had formed at the junction of the two rivers; and having armed the most effective of the crew with cutlasses and axes, we pathed ourselves a way through entangled bushes and swamps, following the left bank of the Barima. With the exception of two rivulets, the tributaries which the river received were of inconsiderable size. Its bed is frequently traversed by granitic dikes, over which the water precipitated itself impetuously; and its current was so rapid, that it would have been difficult to make any way in ascending, even in a small corial. I admired the number of noble forest trees, among which I observed the Bullet tree, the Locust tree, the Crab-wood, Curahara, Hupu, Cuyama, Yarurá, and its allied species Paruacussana; the Suari or Impa, and Makaraballi; but the most remarkable appeared to me the Tunkara, which measured in circumference from twenty-eight to thirty feet. Its trunk rose free from branches, smooth and round, to about seventy or eighty feet; and I was told by some of my Indian Guides that the Warraus used the tree for making canoes. It is soft and white, and the Colonists prepare staves from it. The Warraus make their bark or shell-canoes from the bark of the Bullet tree and Makaraballi.

[1] Tukwari

[2] Whanama-paru River.

[3] The Whana.

[4] In the published version (p. 184), these directions have been changed to 'south-west' and 'west' respectively. They are undoubtedly the general directions from which these rivers flow, but may not be at the actual point of junction.

Incessant rains rendered our travelling through the woods, and over swampy ground, by no means comfortable. We continued our march next morning (the first of July), and passed a large rivulet, which, – the Indians of the inhabitable part of the Barima, below its falls, never having ascended so far – and in the absence of any other name, we called Rocky river,[1] from the numerous blocks by which its course was obstructed.

Our stock of provisions being now completely exhausted, we had to return towards our camp. I halted, therefore, at eleven o'clock; and having marked a tree with Her Majesty's initials, we returned to the spot where we had encamped the previous night, under severe rain and thunder. The river was about thirty feet wide when we had left to follow its bank; its course frequently obstructed by rapids and falls, and upwards, west-north-west. The land adjoining on either side was fertile, consisting of clay mixed with sand and vegetable earth. The forest scenery was luxuriant, and hillocks of inconsiderable height, perhaps not more than fifty to sixty feet above the Barima, appeared particularly adapted for the cultivation of coffee and cacao.

We reached, on the following day, the camp at the junction of the two rivers; where Mr Glascott, during our absence had only succeeded in taking meteorological observations;*

*

	Barometer in inches	Thermometer		
		Attached	Detached	Wet
Highest	30.088	80°.5	80°	78°.5
Lowest	29.942	71.4	71	70
Mean of 37 observations from 6 am to 6 pm	30.007	75.56	75.30	74.02

the unfavourable weather having prevented him from determining its geographical position astronomically.

Having once more reached the corials, we floated down the river, and our return was rapid. While it had taken us six days to ascend from Manari to the Mohokawaina, we accomplished our return in two days and-a-half.

An Indian messenger awaited us here from the lower Barima, with the news that a party of Venezuelans, headed by the Commandant of the Orinoco had proceeded to the mouth of the Barima and the Amacura, and cut down the boundary posts which, in the execution of the service confided to me, I had planted there.

How far this information was founded in truth, I cannot assert. However, the appearance of these boats, which were said to be armed, has created a panic among the Indians; and those of the rivers Aruka and Amacura were fled into the woods.

Our departure from Manari was delayed in consequence of the indisposition of the first Coxswain Peterson; and Mr Echlin, attached as Artist to the Expedition, but to whom, from his study of medicine and his knowledge of the diseases of the Colony, the medical treatment of our sick had been entrusted, reported that, in consequence of serious indisposition, Peterson would be unable to journey with us overland. From the information which I had procured, the road promised to be of the most fatiguing description; and as I was anxious that the Chronometers, of which two had hitherto preserved a fair rate, should reach safely the coast regions, in order to prove by remeasurement of Georgetown how far the observations taken by their means were to be trusted, I desired Mr Glascott,

[1] As it is still so called on the reference map.

the Assistant Surveyor, to proceed with the Coxswain by water to the coast; while Mr Echlin, and the men best fitted for such an undertaking, were to accompany me overland to the river Cuyuni. I had another object in view in sending Mr Glascott by the route alluded to; as, should the weather have proved favourable, he might be enabled to determine, by astronomical observations, some of the more important points on the coast.

According to our observations, the settlement Manari is situated in 7°35′34″ north latitude, and 60°00′36″ west longitude, or one hundred and nine miles west of Georgetown.

The extract of our meteorological observations gave us the following result:–

	Barometer in inches	Thermometer Attached	Detached	Wet
Heighest	30.176	89°	88°.8	83.4
Lowest	30.018	72.2	71.5	68
Mean of 37 observations, taken hourly	30.092	78.5	78	75.2

We were joined by a number of Warraus and several Waikas from Manari, whose services we had engaged to assist in carrying our luggage from Manari to the Barama, which flows into the Waini. We were told that we would have to ascend the Barama in boats for four days before we should reach the path that leads to the Cuyuni.

An Indian carries scarcely more than twenty-four pounds' weight on journeys overland. While the Negro carries invariably his burthen on his head, experience has taught the Indian that by doing so he would not be able to make much progress through the thick woods; and his load is, therefore, slung on his back. For that purpose they have baskets, which are made of the stems of a Calathea,[1] or of some species of Palm.

Our preparations had been completed; the loads distributed according to the appearance of strength possessed by our carriers; and, after Mr Glascott had left with his party, in a boat which was hired for the purpose, we commenced on the eighth of July our march overland.

The forest through which we now began our march, appeared to have less underwood; and I noted numerous specimens of that valuable timber tree the Siruaballia, which affords one of the best timbers for the planking of vessels and the construction of gigs, boats, &c. I saw trees of this description of which the trunks might have measured seventy feet before they branched off. Through the whole of our day's journey, Cedar and other forest trees, many of them of the most gigantic dimensions, were abundant; besides numerous Hya-hya trees. The latter is the remarkable tree which yields, by incision, a milky fluid that forms a good substitute for cow's milk. The Indian, to whom it is inexplicable how man can make use of milk after having been weaned from the maternal breast, does not attach any value in that regard, to this fluid; but the younger community prepare from it balls of caoutchouc: and it has now become of such vast importance, as to be considered almost a necessary of life, the vegetable Milk tree adds another to the number which furnish this valuable substance.[2]

Our path led over hillocks from fifty to sixty feet high, extending in longitudinal ridges; their intermediate valleys formed, generally, swamps; on crossing which we frequently sunk to our girths in mud and water. After four hours' march we crossed the

[1] A genus of the family Marantaceae, which contains rhizomatous perennial herbs, among which is Arrowroot.

[2] *Sapium* spp., known in Creole as the Gumtree or Milkwood.

Caruwavu, here merely a rivulet; and arrived in the afternoon at a small settlement consisting of two houses inhabited by Waikas. I had in the morning the mortification to find that the Mountain Barometer, which I took with me, had materially suffered from the land journey, and was for the present unfit for use. After having continued our march for two hours, we halted at Paripu, a settlement likewise inhabited by Waikas. We found the inhabitants in great tribulation: a messenger had arrived from the Cuyuni, informing them that some Spaniards had come across from Angostura, and were building corials at the banks of the Cuyuni, for the purpose of surprising the Indians of that river: that they intended to kill the adults, and lead the younger portion away into captivity. This Messenger was sent to urge the Waikas of these regions to assist them in making war against the Spaniards. Not only here, but likewise in the sequel, where I found that this alarm had spread, I showed them the inutility of such a violent measure, as neither in number nor in the means could they cope with their assailants. But I advised them, provided the report were true, to be on the alert, and, on the approach of their oppressors, to retire into the woods.

The Cassada grounds around the settlement were extensive, and the magnitude of the plantains, and of the Indian corn or maize, struck me with astonishment. Some of the ears of the latter were twelve to thirteen inches in length; those which are produced on the coast regions do not reach more than five inches. The soil is here a rich, black mould, mixed with white sand, and would produce anything. Considering the extent of this productive soil, and the importance to British Guiana that she should avail herself of her natural resources, and thereby produce food for her inhabitants without relying on importations from foreign countries, it is a subject of astonishment to me that the Maize of the interior should not have been cultivated on a larger scale than merely what the Indian uses for the supply of his individual wants. To prove the importance of the Maize I would only observe, by the way, that the importation of this cereal grain from the Slave-states of North America into the British West Indian colonies amounted, in the year 1836, to one hundred and twenty six thousand, six hundred and eighty bushels; of corn meal, to thirty-six thousand, one hundred and sixty-eight barrels; valued, together, at sixty-one thousand, three hundred and forty-one pounds sterling.

The neatness and order in which we found the provision fields around the settlement, showed that there presided over them an Indian who distinguished himself from the generality of his brethren. Paths led through the fields; the yams were trailed against poles; lime and orange trees, so seldom to be met with amongst the Indians, increased the favourable idea I had of its inhabitants, and induced me to suppose that they were some of the scattered remnants of those fugitives from the missions, who, during the revolutionary war, were obliged to fly to save their lives. We found only an Indian and some females at home; the rest, with their chieftain, were gone to work for a period at a woodcutting establishment on the river Pomeroon, with a view to earning sufficient money to procure themselves such articles as have become almost necessaries of life with them; namely, clothes and other apparel, implements for working at their fields, guns, powder, and shot.

Leaving Paripu, we continued our march; and, in the afternoon of the same day, arrived at another large settlement, judging, at least, from the number of the huts. Here, also, the male inhabitants were absent, having gone to work at the Pomeroon. The fear of "La Patrias", as the Venezuelans are invariably styled by the Indians bordering on the republic, and who still with shuddering think on the massacre to which their bretheren were exposed when the lawless hordes entered the mission, and spread devastation under the cry "Por la patrias"! – prevailed likewise here; and the raised voices and violent gesticulations of the females, when they told our Guides of the Reports which had come from the Cuyuni, were a speaking proof of the wrongs which had been committed upon these once happy people. Unprotected as they were, they intended, they said, to leave their settlement, and to seek their way to the Pomeroon, where their husbands were working. Our interpreter stated to them the object of my mission, that I was then on the way to the Cuyuni; and that, if I should meet any Venezuelans there, I promised every exertion to prevent them from crossing over to this part of the Colony.

While passing through the village I noticed at the farther end a house which was abandoned. Two heaps of ground, thrown up near the middle of the house, and one covered with a large earthen vessel, attracted more particularly my attention. I made enquiries, and learned that they were the graves of a father and his child, both of whom had been killed by the mal-practices of a Piaiman or Conjuror. – When is the period to come, at which the Christian religion shall inlighten these poor benighted beings, and prevent the recurrence of such dreadful scenes, with the effusion of blood in their train? The accusation that the victim has died through the agency of a Piaiman, is sufficient to awake a revenger of the deed among his relations.

After we had passed the village, we had to wade to our necks for upwards of a mile through water. The rivulet Parapaimai[1] had inundated its shores; and as the rain descended in torrents, we were glad when, towards evening, we arrived at the Caribisi settlement Cariacu,[2] situated on the banks of the Barama, which is here about sixty yards wide. The Barama flows, about forty miles farther below, into the Waini; and is the largest tributary of that river. It is inhabited by Waikas, Caribisi, Warraus, and a few Arawaaks, whose aggregate number I estimate at five hundred individuals.

The men at Cariacu, like those at the two settlements previously passed, were absent at a wood-cutting establishment on the Pomeroon, and we found only a few of the female community, who, with a Carib, had been left in charge of the place.

Several of my Indian carriers and guides declined going any further. The reports which they had heard in the course of the day, were repeated at Cariacu; and as they consisted mostly of Warraus, the most timid of all the Indians tribes, such reports could not fail of having their effect. I had to replace their number from among the Caribisi and Waikas of the vicinity, which occasioned a delay of two days. The weather during this period was so unfavourable, that I could not procure astronomical observations.

We had to ascend the Barama to a distance of four days' journey hence, before we should meet the path which leads to the Cuyuni; and as there was only one boat to be had which afforded place to four individuals, we had to resort, for a conveyance, to shell- or bark-canoes, called by the Colonists of Demerara *Wood-skins*, and by the Spaniards *Conchas*. They are made merely of the bark of divers species of trees; that portion being stripped off, and manufactured into the boat. They are generally from twenty-five to thirty feet long, and, when laden, seldom draw more than three inches of water. Light, and the most simple of construction, they can be easily carried on the head, over rocks and other impediments which might obstruct the navigation. Indeed, they are the only craft with which the Indian navigates the upper parts of rivers, but

[1] Parapimoi.
[2] Kariakau.

require proper management, as they are dangerous; and a false movement, when sitting in one of them, may cause it to sink. However, we could not procure any other conveyance, and we confided our persons and luggage to these frail vessels.

We departed from Cariacu on the 11th. of July. The Barama resembles much the upper Barima: its banks are clothed with similar vegetation, and it is equally serpentine in its course. I noticed a good deal of Potter's clay, used by the Caribisi for the manufacture of pottery, which, for its durability, is highly appreciated by the Colonists. This clay has greyish colour, and is mixed with the loose materials of decomposing granite.

The rivulet Nakuwai was the largest tributary which we passed in the course of our first day's ascent; it joined the Barama at its left bank. We noticed the first rocks which were lying in the river's bed, above the rivulet Abocotte. About a mile and a half above this, the Erawanta and Mazuwini[1] join, close to each other, the Barama. During the rainy season, when the bed of the river is full, it forms numerous off-flows, which adopt a more direct course than the river itself, and join it again at some distance on. The Indians who are acquainted with these branches, navigate them, and thus shorten the ascent materially.

We passed, on the afternoon of the thirteenth of July, some hillocks, and, soon after, the first rapid, formed by dykes of granite; and reached a settlement of Waikas, called Cadiu, which we were told was the last inhabited place below the great fall. We were here struck with an air of plenty; the cassava grounds were extensive; yams, sweet potatoes, plantains and bananas, were abundant; also the Paripi palm, and Papayas, of which the fruit resembled a large melon, some of them measuring 28 inches in circumference. Sugar cane, cashew, and cotton trees grew around the huts. A number of wild fowls was observed; marodies, powis, parrots of all plumage; several sun-birds[2] – all tame, and associating amicably with one another.

I succeeded in procuring a set of circum-meridian altitudes; according to which the settlement was in 7°19′ North latitude. We heard quite distinctly during night, the roaring of the great fall Dowocaima, which is about two miles distant, and bears S.58°W.

Having engaged three more Indians to accompany us from Cadiu to the Cuyuni, we started next morning at an early hour; and, after passing some rapids, approached the great fall. We had to unload near the island Wayaruima, and carry the crafts and luggage for the distance of two miles overland. These cataracts surpass in grandeur the great falls of the river Demerara; to which in their structure they bear some resemblance. The whole fall on the Barima[3] amounts, in the given distance of two miles, to about 120 feet; but, from the sinuosities of the channel, there is no one point which affords a *coup d'oeil*. The grandest scene is offered by the three upper falls, where the river, narrowing in to about eighty feet, rushes turbulently down the precipice in three jets, and forms in the distance of about hundred yards, a fall of thirty-five to forty feet perpendicular. This part is called Dowocaima; and as we saw it at the height of the rainy season, when the river was full to overflowing, the scene was sublime indeed. The banks were bordered by a primitive forest, and foliage of every hue; among which the bright red of the young Mora leaves formed a striking object. Lianes reaching from boughs sixty feet high, down to the water's edge; a thousand creepers, so closely enveloping whole rows of trees, as to give them a

[1] Although there is a River Mazawini, this is probably the Mawukani of the reference map, and the Massiwindui, referred to below, the Mazawini.

[2] Possibly the Sunbittern (*Eurypyga helias*).

[3] Clearly a mistake for Barama. It has been corrected in the published version (p. 189).

fanciful resemblance to old massy columns crowned with ivy; white festoons, and clusters of purple and yellow salver-shaped flowers trailing from tree to tree; all combined to form a vivid picture of tropical vegetation. The uproar of the masses of water which rush over the ledges of rock, and envelop in foam the surrounding scenery, added the characteristic feature of the landscape.

The ledges of rock which form these striking scenes of nature, are composed of gneiss; their stratification being S.33°W. They form an impediment to all further navigation; and one which, if a denser population should render such a step necessary, could only be overcome by canals or railroads. In the absence of these, our Indians took their light bark-canoes on their heads, and carried them to that part of the river where there were no serious obstacles to its further navigation.

We passed, next day, the rapid Massiwinidui,[1] and several others of less consequence, and encamped in the evening at the foot of the fall Aunama, from whence the path leads to the Cuyuni. The river Aunama[2] joins the Barama just below the fall. The latitude I determined to be 7°14′N.

At a day's journey above this fall, there is a Caribisi settlement; further up, the Barama is uninhabited. It is said to have its source in the same parallel of latitude with the Barima and Amacura; namely, in the extensive savannahs north of the Ikruyéku mountains.[3]

We commenced our over-land journey on the morning of the sixteenth of July, and traversing occasionally hillocks from one hundred to one hundred and fifty feet high, followed the valley through which the small river Aunama flows towards the Barama. We reached at noon an Indian settlement. The provision-grounds around it were in good order, but the houses were tenantless. Our tortuous path continued in a west-south-west direction, still following the Aunama.

After a march of six hours, we arrived in the afternoon at a settlement of Caribisi Indians; called, from the rivulet on which it is situated, Aunuma; and, according to circum-meridian observations of the *α Gruis, in 7°9′ north latitude. On the ridges which we were this day crossing, and which generally stretched north by west and south by east, I observed several tracts of granitic blocks, the direction of which was north-west by west. The trees we met with on our journey were lofty, and there was less underwood than along the banks of the rivers. The Mora, which had been so abundant, became scarcer the farther we receded from the Barama, and was replaced by a greater variety of timber trees; as Kakaralli, Determa, Siruaballi, Cedar, Yaruri, Souari, &c. We did not observe any Greenheart – a wood much and deservedly esteemed by Ship-wrights and House-builders.

Our course on the seventeenth of July continued West-south-west. We crossed, at ten o'clock in the morning, the Aunama for the last time; and having passed a ridge of small hills which stretched south-by-west, we stood soon after on the western branch of the rivulet Acarabisi.[4] We had now reached the most elevated spot between the Cuyuni and Barama, and entered another system of rivers, which, instead of flowing northwards, towards the Waini and Barama, tend in an opposite direction – to the south – and, uniting with the Cuyuni, find ultimately an outlet into the Atlantic by the Essequibo. From this

[1] The reference map shows the Mazawini River joining the Barama at this point.

[2] The same name on the reference map.

[3] The Barama and Barima rivers rise on the eastern flanks of the Imataka Mountains, but the sources of the Amakura are not so far west.

[4] Akarabisi River.

ridge of hills, the natural configuration of the ground is sloping towards the banks of the Cuyuni, southward; and I estimated the highest ridge, which separates the two systems, at 520 feet above the level of the sea. Heights which really deserve the name of mountains commence twenty miles further westward: nevertheless, these ridges of hillocks are of importance in the determination of the boundary on the principle of natural divisions. I claimed them, accordingly, to form the limit from the source of the river Amacura, passing south-eastward the sources of the rivers Barima and Barama, and continuing in that direction until the ridge meets the river Acarabisi. From the Amacura, consequently, the northern slope of these hillocks belongs to the British Colony of Guiana; the southern slope, to the westward of the river Acarabisi, and along which the rivulets flow to the Cuyuni, would belong to Venezuelan Guiana.[1]

The Aunama and Acarabisi are only divided from each other by those hillocks which rise not more than sixty to a hundred feet above their level: both rivers, if properly cleared of trees which have fallen across, would afford a navigation to canoes and punts; and as the portage is not more than two miles, these rivers present, at the very frontier, the means of connecting the Pomeroon and Morocco coast with the Upper Cuyuni, where that river is comparatively free of obstacles.

Having claimed the right bank of the Acarabisi, as forming part of the western limit of British Guiana, I had several trees, which stood along its course, marked with Her Majesty's initials. Towards evening we reached a Caribisi settlement, the latitude of which I found to be 7°4′N. It consisted of six houses, and seventy inhabitants. Its height above the level of the sea was ascertained by Wollaston's[2] barometric thermometer to be five-hundred and ten feet.

We followed the valley of the Acarabisi – by no means a comfortable path, as at this season of the year it formed an almost continued swamp, and we fell sometimes to our girths in the mire. A rich retentive soil qualifies these regions peculiarly for the cultivation of rice. It rained almost incessantly, and we were truly rejoiced when we arrived on the morning of the nineteenth of July at the Caribisi settlement Haiowa, about two miles distant from the left bank of the Cuyuni. The general feature of the country between the Barama and Cuyuni, is that of a series of narrow valleys, situated between hillocks of no great altitude. The principal valleys are those which follow the course of the rivers Anauma and Acarabisi. The general direction of the others is at an oblique angle to these, and they vary considerably in extent: sometimes they are merely defiles, and the greater number of them do not expand more than about a quarter of a mile. I am fully persuaded that there can be no soil better qualified for the cultivation of coffee than that of this part. The zones of granite, sometimes in spherical blocks, and the vitrified and ferruginous masses of clay which I observed frequently to traverse the mountains, are favourable to the cultivation of that plant.

The productiveness of the soil nearer to the banks of the Cuyuni is evident from the specimens of Sugar-cane, Cotton, and Plantains, which were brought to me while at Haiowa. I saw a cane measuring fifteen feet long and seven inches and-a-half in circumference. The cotton, too, was of excellent quality and staple; and the few tobacco plants

[1] This is today the current, if disputed, line between Guyana and Venezuela. This boundary follows the Amakura, then skirts round the headwaters of the Barima and Barama rivers, until it reaches the Akarabisi which river it follows before turning westward along the Cuyuni.

[2] William Hyde Wollaston (1766–1828), British physicist who designed a mountain barometer.

which the Indians raised for their own use, were remarkable for their large leaves, and, as I was assured, for their fine flavour.

Haiowa consists of four houses and thirty-five inhabitants of the Caribisi nation. The Caribisi, like the Waikas, are a superior race. They are fairer in colour than the Warraus and Arawaaks, and their average height is five-feet five-inches. The female sex vie in symmetry of form with the men; their features are more regular than those of other Indian tribes; and a profusion of hair, the tresses of which nearly touch the ground, contribute to their good appearance. Both sexes are great smokers: children, indeed, commence at an early age to indulge in that bad custom.[1]

We now learned that the rumour as to the Spaniards having come to the Cuyuni, had had some foundation: two individuals had arrived from the neighbourhood of Angostura at the Cuyuni, with the object, as I afterwards understood, of finding whether cattle could be driven from the savannahs near the river Caroni to the British Colony. – Whether their appearance had caused the alarm; or, whether, by unguarded words, they had given reason for mistrust, I know not; but the Indians here were under the same apprehensions as those of the Barama.

We met with several Indians who spoke, still, with feelings of the greatest respect for the Missionaries who were formerly settled at the Caroni; and when relating the relentless cruelties of the self-styled patriots towards these innocent victims of the civil war, it became evident that even to this hour, after the lapse of tens of years, their prosecutors are held in the utmost abhorrence. I was assured by an old Waika that nine Missionaries who intended to escape by the Cuyuni to Demerara, were taken by the patriots and shot in cold blood; that the missions were destroyed; and the Indians hunted down and sacrificed by a relentless soldiery.

We expected to meet here the party which was to have been sent with a supply of provisions up the Cuyuni: for our stores had long since given out, and we were reduced to Cassava bread and what game chance brought into our hands. We were, however, disappointed in our expectations, and, in the absence of any craft, I had to send my Coxswain a journey of two days higher up the Cuyuni, where I was told there was a corial large enough for our use. On his arrival at the settlement, its inhabitants considered him to be one of the Spanish party returning to execute their threat, and the men rushed out, armed with guns and cutlasses. However, they were soon assured that our party came as friends to the Indians; and having bargained for the corial, the Coxswain returned with some additional guides, on the morning of the twenty-second July; and after embarking our baggage, we commenced, a few hours after, the descent of the Cuyuni.

While at Haiowa, I proceeded to the mouth of the Acarabisi, which bears from the settlement N.75°W., – distant about a mile and a half; and took, formally, possession of it in the name of Her Britannic Majesty, as the point where the western line of limits meets the Cuyuni. The line stretches from thence across the Cuyuni to its right or southern bank (where another tree was marked with Her Majesty's initials) and continues upwards to its source.[2]

[1] The next two paragraphs were omitted from the copy sent to the Royal Geographical Society.

[2] This was later revised so that the boundary turned south, along the right-bank tributary of the Cuyuni, the Wenamu. In 1843 Schomburgk proposed that Great Britain should cede its right to the whole upper basin of the Cuyuni in exchange for a foothold at the mouth of the Orinoco (see below p. 111).

Several meridian altitudes of stars gave me as mean result for the latitude of Haiowa 6°56′N. The boiling point of the barometric thermometer was .47th of a degree higher than at Acarabisi, and .50th of a degree lower than at the mouth of the Barima, which would give the approximate height of 260 feet above the level of the sea. I attempted to repair the barometer, and made a number of observations while at Haiowa: but until this instrument shall have been tested in Georgetown, it cannot be relied on. The mountains west of the Acarabisi rise to a considerable height, and the summits of the Ekreku[1] are estimated at 2,000 feet above the level of the Cuyuni. Catiya, or Curumu,[2] where in the Royalist time, a Spanish military post (Destacamento de Cuyuni) was established, is about twenty miles to the westward of the Acarabisi, but the Spaniards penetrated during the revolutionary war, as far east as the river Airékuni,[3] only eight miles above the Acarabisi. All the old inhabitants, both Waikas and Carabisi, concurred in the assertion that the Spaniards, up to that time, had never penetrated farther eastward than the Airékuni river: whilst Father Caulin, in his "Historia Coragrafica de la Nueva Andaluçia y vertientes, del Rio Orinoco" (1779)[4] has observed that, at the period he visited the Orinoco, while attached to the Expedition of Limits, the Dutch had already carried on an extensive trade, by means of the Cuyuni, with the Indian tribes at the Caroni and Parawa; and it was at that period (1750–1760) that the Dutch possessions extended to the foot of that series of falls, of which *Kanaima*[5] is the most considerable.

The Cuyuni presented, where we embarked, a magnificent sheet of water. I estimated its width at from four to five hundred yards. Its current was rapid – perhaps three miles and a half in an hour; and its bed full to overflowing. A small chain of hills, called Macapa,[6] bore nearly west: they are distant about a mile. Our progress was rapid, and in the afternoon we had safely passed the dangerous fall of Kanaima, and rested at an abandoned settlement on the river's right bank. There were some other settlements in the neighbourhood, the inhabitants of which came to visit us. We did not observe any Mora trees along the banks; they were replaced by another equally majestic tree, which the Indians called *Tá-au*. The islands with which the river was interspersed, were almost covered with bushes of the *Quassia amara* or bitter ash.[7] The stream itself continued as if cut up by a multitude of large channels, which are not seen from each other, thickly-wooded islands intervening; and no accurate idea can be formed of their total breadth. Sometimes a little range of densely-wooded hillocks approach the river's edge.

We generally found that in the morning, with sunrise, a strong breeze set in against the current, and that it changed by degrees to east-south-east, or east-by-south. Descending at the rate of five miles an hour, we passed numerous rapids; where the river was free of

[1] The Ekreku Mountains lie south of the Cuyuni, some 40 miles west in a straight line from the Akarabisi.

[2] On the Guyanese map the Curumu River appears as a left-bank tributary of the Cuyuni and one of its left-bank tributaries is the Botonamo River. On the Venezuelan map, no Curumu appears and the river is called the Río Botanamo.

[3] Perhaps the Erakri.

[4] Antonio Caulin, *Historia coro-graphica natural y evangelista de la Nueva Andalucia, provincias de Cumaná, Guyana, y vertíentes de Río Orínoco*, Madrid, Dada á luz de Orden, y á Expen^s de S.M., 1779, Bk III, Chap. XXXI.

[5] The Kanaima Falls are about eight miles down stream from the mouth of the Akarabisi.

[6] Makapa Hills.

[7] *Quassia amara* is Bitter bush in Creole. The Bitter ash is *Quassia simarouba* (*Simarouba amara*).

impediments, it was about 600 yards wide. We passed the Otomong hills,[1] and avoided, by narrow passages between islands numerous large cataracts, which, in our small canoes, it would have been dangerous to attempt to descend. At the cataract of Poinka-marka, or *Wommipong* of the Caribisi we had to unload and draw the crafts over a portage of about three-hundred yards extent. The perpendicular fall of this cataract is not less than thirty feet, and it is generally called the *Canoe-Wrecker*, in consequence of many fatal accidents which have occurred here. We halted in the evening at six-o-clock at a single hut inhabited by a Waika, his wife, child, and a dog. He shared his hut with us, although we were rather a numerous party for a single house. At a short distance hence, the rivulet Aracuna enters the Cuyuni. It is inhabited by a few Waikas, and a path leads from it to the river Puruni,[2] which flows into the Mazaruni. The latitude of the hut is 6°46′N.

The rapids and falls now became less frequent, and still water commenced. The tract of granite & gneiss which causes these impediments, extends, therefore, from the Aracuna hills, uninterrupted, to the small range of hillocks called Macapa. It is about fifty to sixty miles in length, and constitutes the second large series of falls. About eight miles below Arakuna, and opposite some small hills which rise on the river's right bank, is the island Tokoro (Tokoro-patti),[3] where, towards the close of the last century, the furthest out-post of the Dutch was situated. Although generations have elapsed, the circumstance that a Dutch Post-holder once resided here, has remained traditionary; and our Guide, an old Waika, assured me that his father had frequently mentioned it to him, and that the Post-holder's name was "Palmsteen". The post was afterwards destroyed by the Spaniards, and the Post-holder withdrawn nearer towards the cultivated part of the Colony. A little below Tokoro-Patti, on the left bank of the Cuyuni, is the rivulet Iroma. The rivers Rupa and Appa[4] join the Cuyuni from the north: they were the largest tributaries we had met in the course of our descent. We reached in the afternoon the Toroparu, a rivulet from whence a much frequented path leads, in a day, to the Purumi.[5] We had anxiously looked forward to meet the party which we expected with supplies of provisions. We heard of them to-day, at a settlement opposite the Toro hills; but only to have the disappointment of learning, that, on ascending the previous day the dangerous fall Wakupang, they had lost everything, and saved only their lives and the corial. Among the luggage lost was one of the instruments – Massey's log[6] – and a new tarpauling. Thus disappointed in our hopes of meeting comfort, we had for some days longer to continue our scanty fare. We paid off our Guides, who had accompanied us from Haiowa, as, with the men who had come up from the Essequibo, our crew was sufficiently strong to reach that river.

The dangerous fall Wakupang, where our stores were lost on the preceding day, was passed without accident. This is the commencement of the second series of falls or rapids. The river is studded with islands. Green-heart and Purple-heart, both

[1] Not marked on the reference map, but the Otomung River is a right-bank tributary of the Cuyuni.

[2] The Aracuna is not marked, but the Puruni flows for most of its course between the Cuyuni and Mazaruni before turning sharply south to join the latter.

[3] Tokor-opati Island.

[4] The Iroma River is a much larger and longer river than the Rupa. The Appa does not appear on the reference map.

[5] The headwaters of the Toroparu are close to the Kartuni River, a tributary of the Puruni.

[6] Edward Massey (1768–1852), instrument maker of London who patented the first mechanical log in 1802.

most valuable forest trees, become abundant along its banks; but the impediments which the numerous rapids throw in the way, will for some time render these treasures unavailable to the Colony. We passed in the afternoon the Cutuau hills, along which a river of the same name has its course.[1] The Cutuau offers a communication with the river Waini, and is much frequented by the Indians of both rivers; eight miles further eastward is the rivulet Wayarimpa,[2] whence another path leads to the Puruni. The river is here free of impediments; its breadth from 600 to 700 yards; its water clear, and of a brownish colour. The circum-meridian altitudes of three northern stars gave me 6°43′N., as the latitude of this creek; a few miles from it is the cataract Tonomo, where the Post-holder resided, after his station had been withdrawn from Tokoro Island.

We had commenced this morning (July twenty-sixth) the descent of the third series of falls, caused by a small range of mountains, through which the river has broken itself a passage. It rained almost incessantly; and as the wind was against us, it endangered our descent of numerous rapids; and the Coxswain could scarcely look forward. We had to unload at the cataract Aruaka-matubba, and to haul our corials over land. We passed, soon after, the Woku or Powis mountain[3] which rises on the river's right bank to a height of 500 to 600 feet: this ridge extends west-north-west, and east-south-east, and it can be seen from the junction of the Cuyuni and Mazaruni. Little islets, consisting of heaped-up masses of rock, divide the river into numerous channels. We had to pass the fall Camaria;[4] and, as it did not afford any portage, we attempted to descend it in our crafts. It nearly proved our destruction. As it was, the craft filled with water; and it was only the presence of mind of some of our crew, to which, under the Almighty, we were indebted for our safety. We reached, on that evening, Ematubba,[5] generally called "the great fall", where we had to unload, and to haul our corials overland; and encamped at the foot of the small island, whither the corials had been drawn. Continued rains precluded the possibility of any observations, and we started on the morning of July twenty-seventh, under the same unfavourable weather. An hour and a half after, we were at the foot of the last fall, called Akaya; and saw before us the junction of the three rivers, Essequibo, Mazaruni and Cuyuni.[6]

We were received with demonstrations of hearty joy by the remainder of our party, who had awaited us at the Protestant Mission at Bartika Grove, near the junction of the Mazaruni and Essequibo; and the Missionary, the Reverend J. H. Bernau,[7] joined in their welcome. The mission over which this worthy man presides, was founded by the Church

[1] Kutuau River.

[2] Waiarimpo River.

[3] Oko Mountains. In some Carib languages, 'Oko' is the word for powis or bush turkey.

[4] Kamaria Falls.

[5] Matope Falls.

[6] The whole of the next passage, of over five manuscript pages (to the end of the second paragraph on p. 38 here), relating to the education and evangelization of the Amerindian was omitted from the text sent to the Royal Geographical Society, but a copy was sent to the Church Missionary Society.

[7] John Bernau, a German, was the missionary at Bartica Grove from 1837 to 1848, and from 1850 to 1853. He left a full account of his first ten years there in *Missionary Labours in British Guiana* (London, Edward Stanford, 1847). For an account of the early years of the Bartica Mission, see Peter Rivière, *Absent-minded Imperialism. Britain and the Expansion of Empire in Nineteenth-century Brazil*, London, Tauris Academic Studies, 1995, pp. 10–17.

Missionary Society about the year 1831, and Mr Bernau commenced his labours in 1837; since which time improvements have followed successively. The mission now consists of about one hundred and twenty inhabitants, who are all located in neat cottages, surrounded by gardens and provision grounds. I am too well acquainted with the Indian character, to expect a perfect reform in the adult Indians; and the Missionary has no doubt reaped a similar experience. His chief object is, therefore, to inculcate in the minds of the *youth* entrusted to his care, religious precepts, the benefits of industry, and civilization; and in this his labours are successful. The school is attended by from forty to fifty children, mostly Indians: some of the latter (I think twenty-four) are maintained by the mission, and instructed in the normal school, that they may hereafter return among their tribes and relations, and assist in working out the great objects of conversion and civilization. Their progress is, indeed, encouraging. Some of them, in the short space of four months, have learned to read and to write, and the copy-books which I saw, would not have dishonoured an European school of even higher pretensions. One of the boys, an Arawaak, had advanced to the Rule of Three in Arithmetic.[1] Their psalmody is sweet; and when, on the evening of our arrival, we attended prayers, we were much pleased with the attention they paid to the exhortations of their religious teacher.

Mr Bernau's great plan is, to induce the Indian parents who live at any distance from the mission, to send to him their children, when only four or five years of age; thus to alienate them from Indian life and manners, and, from their earliest youth, to point out to them the beneficial examples afforded by the Christian religion and Civilization. An asylum for female Indian orphans has just been erected, and is under the guidance of a respectable female teacher: the house is spacious, and has all the comforts required for such an institution. The orphans boys are under the guidance of a teacher sent from England. It is not, however, religious principles alone, which are sought to be implanted in their breasts. The Missionary shows them, by example, that the destiny of man is not indolence; the children are encouraged to manual labour; they assist in keeping in order the surrounding garden; and I was astonished when Mr Bernau told me, that the wharf near the mission had been constructed solely by him and his disciples during leisure hours.

The religious service is at present held in a house; but a church, to which the Colony has contributed, is in the course of erection; and a hospital for the sick has been added to the house where the boys are located.

This is a cheering picture of the good work which has been commenced among the remnants of the aboriginal tribes; and, if I am entitled to an opinion, it is the only means by which the great objects of the Christian religion and Civilization can prosper. During the late expedition, I had frequent opportunities of meeting Indians who had been brought up at the former Spanish missions; and although so many years had elapsed, it was nevertheless evident that they belonged to an improved race. Philanthropy must mourn that civil strife should have felled with one stroke the effects which the religious and industrious endeavours of the good fathers had produced during numerous years amongst these children of the forest: effects, of which the fruits were yet visible, although more than a score of years had elapsed;

[1] The Rule of Three is the method for finding a fourth number from three given numbers, of which the first is in the same proportion to the second as the third is to the unknown fourth number.

and those who show that superiority in their manners and labourious habits, must have been mere children when they lived in the missions of the Capuchins of the Caroni.

The continuance of the principle which was implanted in them at their youth, though it may be, still, only a mere spark, gives me every confidence that the mission at Bartika grove will produce the happiest results; and I have only to express my sincere hope, that similar Institutions may yet be founded in other parts of British Guiana, and that aid tendered to the remnants of the aboriginal tribes within the British Territory, which will place their rising generation much above their present religious destitution. This is the only recompense which England can tender to them for the loss of their lands, and for the miseries which Europeans have inflicted upon them. There are about two thousand four hundred Indians now inhabiting the rivers on our western boundary, who walk in perfect darkness, but amongst whom an institution like that at Bartika Grove would produce the most beneficial results. From my intercourse with these tribes, and from conversations I have had with the more intelligent members of them, they appear willing to receive instruction, and to send their children to a normal school, if such an Institution were to be founded in that part of British Guiana.

Our party left Bartika Grove on the twenty-eighth of July, in two corials, and we arrived safely in Georgetown on the second day ensuing, after an absence of three months and a half during which period we had made upwards of seven hundred miles: and although that period presented but a continuation of the most unfavourable weather, we nevertheless determined *twenty-one* points astronomically, and acquired a true knowledge of the course of the rivers Waini, Barima, Amacura, Barama and Cuyuni, all of which had never been visited before by any person competent to delineate them on a map. No wonder, therefore, that their actual course should be almost opposite to what it is represented on extant maps.

The fertility of the tract we have explored has been pointed out in various places in this, as well as in my former Report. The lands adjacent to the rivers Amacura, Barima and Barama, and beyond the reach of the tides, are superior in quality to those of any other district hitherto visited; and this refers equally to the Cuyuni, where I met sugar-canes of the finest description, and native cottons of superior staple and quality. But the difficulties which the Cuyuni presents to its navigation, and those tremendous falls which impede the river in the first day's ascent, will, I fear, prove a great obstacle to making the fertility of its banks available to the Colony. The Amacura, Barima and Waini are, for a great distance, free of such impediments; and a thicker population only is wanted, to render this part of British Guiana one of the most productive throughout its whole extent; towards which, the numerous natural canals and connections between its chief rivers would materially contribute.

These tracts are at present inhabited by the following tribes:–

Warraus along the coast, from Pomeroon to the Amacura.

Arawaaks, intermixed with the former, chiefly at the rivers Waini, Barima and Amacura.

Waikas and Chaymas, sister tribes of the Wacaawais, at the upper course of these rivers, and the regions between the Barama and Cuyuni.

I estimate their whole number, as already stated, at two-thousand five-hundred. Many of them assist in felling timber, or in working on the estates; and if the system which only of late years has been followed, namely, that of treating the Indian as a rational being, and giving him a fair remuneration for his work, shall be generally adopted, the aborigines, there is no doubt, would prove most useful labourers to the Colony.

No person has had more opportunities, than myself, of becoming acquainted with the tyrannical conduct which has been practised within the last ten years towards these poor beings; and it is not too much to say, that the treatment of them, which we now reprobate in the adjacent territories, many of the former Colonists even of the British Territory have been guilty of. What wonder, therefore, if these children of the forest, who still recollect the wrongs which in times past were inflicted upon them, should pause before they trust to the fair promises now made to them. They already recognise the protection of the Superintendant, under the existing regulations; and it is my full persuasion that, if the attention and paternal provisions which the aborigines of Guiana have of late years enjoyed at the hands of Her Majesty's Government be continued, and means adopted to afford them religions instruction, the relict of the once numerous Indian population may be yet rescued.

Robert H. Schomburgk
Her Majesty's Commissioner for
surveying and marking out the
boundaries of British Guiana.

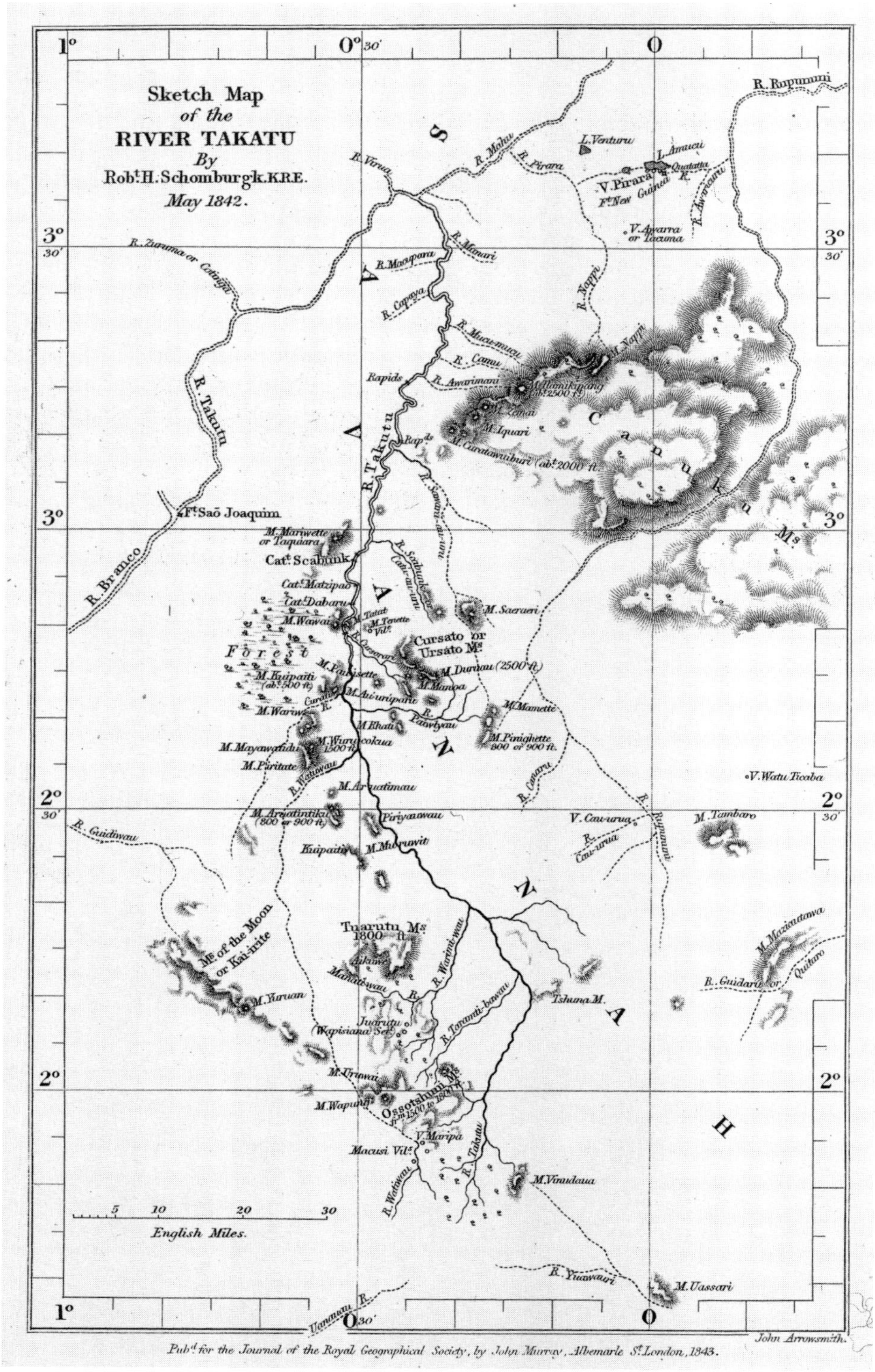

Map 2: Sketch map of the River Takatu. This copy is taken from a proof; in the original published version 'Takatu' was corrected to 'Takutu'. Reproduced by permission of the Royal Geographical Society.

CHAPTER 5

Brazil and the Western Frontier: 1842–1843

Following the return to Georgetown of the first expedition of the Boundary Commission at the end of July, Schomburgk presented to Light a plan for the next stage, the survey of the frontier with Brazil, that envisaged it getting under way in September. There were, however, the inevitable delays. These were partly due to continuing doubts concerning the funding of the survey, and once again intransigence on the part of the Combined Court nearly brought the whole thing to a halt, only to be rescued by the governor's contingency fund.

A second reason for delay resulted from the resignation of Glascott, Hancock and Echlin from the commission and the need to find replacements for them.[1] A young man, Edward Goodall,[2] was recruited in London to replace Echlin as artist. In the place of Glascott was William John Fryer, who had recently migrated to British Guiana. He had served with Colonel Sir George de Lacy Evans's volunteer corps in the Spanish Carlist wars, both as a doctor and subsequently as an officer. He was taken on as Schomburgk's secretary and assistant, and was expected to provide the medical care. The Treasury in London, irritated by the colony's continued refusal to contribute to the cost of fixing the boundaries, declined to pay Fryer's salary and Schomburgk had to do so out of his own pocket until after the whole expedition was over when it relented and

[1] Richard Schomburgk wrote (*Travels*, I, p. 213): 'Within a few days after our arrival [in Georgetown after the first boundary survey] our hitherto fellow travellers, Lieut. Glascott and Secretary Hancock, tendered their resignations to my brother and the Governor. Both had been none too pleased with the perils and hardships of an expedition like ours, and as the most dangerous of the journeys were still to be performed they thought it wiser to withdraw before they started. Glascott intended settling in Georgetown as a Land Surveyor, especially as Emancipation had brought about considerable changes in the relations of property and opened a profitable field for his activities: Hancock on the other hand wanted to renew at the Colonial Hospital the medical course that he had thrown up a long time before, and later on to practise as a full-fledged physician. Unfortunately he fell into bad company, left the Hospital a few days later, and indulged his propensity for spirituous liquor unchecked: within three weeks of leaving us he was down with fever and beyond hope of recovery.' Echlin was only a temporary member of the commission, who volunteered to replace the artist, Walton, at short notice, when the latter resigned soon after his arrival in Georgetown.

[2] Edward Angelo Goodall (1819–1908). For a brief biographical sketch of Goodall and the reproduction of a number of the paintings he made in British Guiana, see M. N. Menezes, 'Introduction' to Goodall, *Sketches of Amerindian Tribes 1841–1843*, pp. 11–30. There is a collection of Goodall's work in the British Library (Add. MSS 16936–16939). Goodall kept a diary from his arrival in Georgetown on 28 July 1841 until it abruptly ends on 13 June 1842 (see Edward Goodall, 'The Diary of Edward Goodall Esq. during his sojourn in Georgetown from 28th July to 11th December 1841', *Journai of the British Guiana Museum and Zoo*, 35, 1962, pp. 39–53; and 'Edward Goodall's Diary 23rd December 1841–13th June, 1842 (on the Essequibo & Rupununi Rivers)', *Journal of the British Guiana Museum and Zoo*, 36, 1962, pp. 47–64. Hereafter Goodall, 'Diary' and 'Goodall's Diary' respectively).

reimbursed him. For the same reason Schomburgk remained on half-salary, as did Goodall, although both of them finally had the deficit made good by the Treasury.

The delay was also occasioned by the fact that the boundary commission was to operate in conjunction with the military expedition to retake Pirara from the Brazilians. Schomburgk had been against a military operation, and of the view that the whole thing could be much better achieved and managed with a police sergeant and two or three men – a force which he judged to be quite adequate to hold Pirara. Light had overturned this view and the Boundary Commission was forced to wait for the military force to assemble.[1]

The military detachment, drawn from the 1st West India Regiment of Foot, which had white officers and black men, left Georgetown on 11 January 1842. In total this force consisted of forty-five people; it was under the command of Lieutenant Edmund Bingham, with two other offices, an assistant-surgeon and two civilians as commissariats. There were two sergeants, a drummer and thirty-six other ranks. They were transported in a fleet of ten canoes paddled by eighty Amerindians who were under the command of William Crichton, the Inspector General of Police and William McClintock, the Postholder and Superintendent of Rivers and Creeks of the Pomeroon District.[2]

The Boundary Commission had set out over two weeks before the military expedition, on 23 December 1841, and having made a more leisurely journey arrived at Pirara Landing on 9 February, only three days before the soldiers. The officers of the boundary commission were Schomburgk, Fryer and Goodall, and there were nine other permanent members, four Germans (Nicholas Reuter, Gottlob Stöckle, Henry Peterz, and Tiedge[3]), three coloured men including the coxswain Hendrik or Henry Chessman, the black cook Hamlet Clenan, and Sororeng, the Amerindian interpreter who had been to England with Schomburgk in 1839. Richard Schomburgk accompanied the expedition at the request of the Prussian government. The one other person, who travelled partly with the boundary commission and part of the way with the military expedition, was Thomas Youd who was to be reinstated in his mission at Pirara. This proved relatively easy as the Brazilians had virtually abandoned Pirara and the British re-occupied it without resistance.

The two British parties settled down at Pirara to await the arrival of the Brazilian authorities from Fort São Joaquim who had been summoned to a meeting. This did not take place until the end of February and it was not until early March that the Brazilians agreed to retire, under protest, from Pirara. Schomburgk had ascertained that there was no objection to his survey on condition that any markers he set up would be recognized as merely being for scientific purposes and not as boundary markers. Further delays meant that the expedition, a survey of the Takutu river, one of the presumed frontiers, did not get under way until near the end of March. It returned two months later, and in the meantime the military detachment sat in Pirara with very little to do.

[1] A full description of the military operation is to be found in Rivière, *Absent-minded Imperialism*, pp. 92–138, so that only an outline account is provided here.

[2] William C. F. McClintock had over forty years as Postholder, Superintendent of Rivers and Creeks, and finally as Special Magistrate. He never ceased campaigning on behalf of the Amerindians and he seems to have been an exemplary incumbent of the various posts he held (see Menezes, *Amerindians in Guyana*, *passim*).

[3] We know next to nothing about Tiedge. Robert Schomburgk never mentioned him by name, and Richard Schomburgk makes only occasional reference to him (*Travels*, II, pp. 238–9, 270).

Figure 3: Pirara, a Macusi village. The picture also features the buildings of the military detachment and the survey team. Picture taken from the copy of Richard Schomburgk, *Travels in British Guiana*, I, p. 304, in the editor's possession.

Schomburgk's report on the exploration of the Takutu was sent to Light and duly transmitted to the Colonial Office. For four weeks from 10 November 1842 there was an exchange of letters between the Colonial Office and the Royal Geographical Society as the former declined to give permission to publish it without Schomburgk's express approval. Finally, however, the Colonial Office handed the original manuscript over to the Royal Geographical Society, as a note in the index to volume CO 111:195 records. Oddly, the Colonial Office does not appear to have kept a copy of it, or, at least, there is not one amongst its papers. The Royal Geographical Society received the manuscript on 6 January 1843 and its shelfmark is JMS 6/37. It consists of 199 pages, and the page numbering is straightforward until page 189 at which point pages 190–99 are numbered pages 80–89. It is not in Schomburgk's handwriting. It was published, in an abbreviated and heavily edited form, in Volume 13 of the *Journal of the Royal Geographical Society* (1843) as 'Visit to the Sources of the Takutu, in British Guiana, in the year 1842'. Although the deletions, mainly relating to boundary matters, are marked on the manuscript, most of the editorial changes are not. This suggests that a further copy was made for the printers. The published version is full of misreadings of the manuscript; for example, von Martin rather than von

Martius in the first paragraph and throughout. Also some rather curious mistakes have crept in; for example, on p. 23, 'W[m] Swainson' has had a gender change and becomes 'Mrs. Swainson'.

JOURNAL OF AN EXPEDITION TO THE SOURCES OF THE TAKUTU UNDER HER MAJESTY'S COMMISSION, FOR SURVEYING AND MARKING OUT OF THE BOUNDARIES OF BRITISH GUIANA, BY ROBERT H. SCHOMBURGK, K.R.E.

Various circumstances which need no explanation in these pages, delayed the departure of the Guiana Boundary expedition from George Town, to a period of the year which was least calculated to ascend the rivers of the interior; combined with this delay arose another on our arrival in Pirara where it became of importance to sound the opinion of the Brazilian authorities at the frontier what their sentiments were towards us, and at the time a satisfactory result had been obtained in this regard, the month of March had approached, or, in other words, the dry season had fully set in.

It was impossible at this advanced period to pursue my first plan, of ascending the river Cotinga, in order to reach the culminating point of the system of mountains which connected with the Parima Mountains of the Orinoco, divides the basin of the Orinoco, the Esssequibo, and the Amazon; which chain has been named Pacaraima by recent Geographers.*

* v. Humboldt's Personal Narrative Vol.V p. 797, 841, 857, vol VI p. 514.[1]

I selected therefore the river Takutu for our exploration which river according to the instructions of Her Majesty's Government,+ was to form part of the provisional boundary between British

\+ Parliamentary papers relative to British Guiana ordered to be printed May 11th. 1840.[2]

Guiana and the northern part of Brazil.

According to the information which I had collected from Von Martius'[3] and Humboldt's works, it appeared no easy task.

The attempts of the Portuguese, as Von Martius understood from Don José Joaquim Victoria da Costa, the Governor of Rio Negro, to ascend the Takutu in Canoes, were frustrated by innumerable rapids and cataracts; and the dry state of the small rivers

[1] Schomburgk used the seven-volume English translation by Helen Maria Williams. Alexander von Humboldt and Aimé Bonpland, *Personal Narrative of Travels to the Equinoctial Regions of the New Continent during the Years 1799–1804*, trans. Helen Maria Williams, 7 vols, London, Longman, Hursts, Rees, Orme & Brown; J. Murray; and H. Colburn, 1814–29.

[2] See Parliamentary Papers, 'British Guiana', *Parliamentary Accounts and Papers*, Vol. XXXIV, Ordered, by the House of Commons, to be printed 11 May 1840, pp. 315–31. This, of course, was the boundary as recommended by Schomburgk in the first place.

[3] Karl Friedrich Philipp von Martius (1794–1868), the Bavarian botanist and explorer.

Awaricuru, Quatatta, and Pirara, rendered it impossible to convey our canoes across the Portage of Pirara.[1]

Fray José,[2] the Brazilian Missionary of the Rio Branco, promised however to furnish me with a small canoe from the Rio Branco, which at the earliest opportunity he promised to send to the mouth of the Pirara, and which I intended to use for the transport of the most indispensable instruments.

In the expectation of receiving information that the Canoe had arrived at the mouth of the Pirara, several weeks elapsed, and, if I wished to execute my plan of tracing the Takutu to its source before the setting in of the tropical winter or the great rainy season, I was aware that no time was to be lost.

I engaged therefore a number of Macusi Indians to accompany us as carriers, and guides, and commenced our march towards the mouth of the Pirara on the 26th of March.

The place of assistant surveyor which in consequence of the multiplied occupations which fall upon the commander of a similar Expedition in the wilds of the interior, is of great importance, had not been filled, I had individually engaged Mr Fryer, formerly an Engineer Officer in the Spanish Legion, and who possessed some medical knowledge, to accompany me, under the hopes that he might in some regard fill the office of assistant surveyor and his knowledge of disease might prove of additional consequence.

Mr Goodall, the draughtsman of the Expedition, Mr Richard Schomburgk, who on application of the Prussian Government to Her Majesty's Government, had received permission to accompany the Expedition in order to make collections for the Royal Prussian Museum and the Botanical Garden at Berlin; and nine canoe-men, partly Europeans, who were permanently engaged, composed the persons of our Expedition; Sororeng, who accompanied me in 1839 to London, acted as on former occasions, as interpreter.

There is something enlivening in a departure like ours from the village which had been our head quarters, for almost two months; all is bustle – the Indians fix their burthens, which in no instance ought to be heavier than thirty pounds; women and children take leave; and the canine race are far too important to be overlooked on such an occasion; they show by their loud barking their delight of being of the party; but it has been differently decreed by their owners; and above all resounds the shrill voice of the squaw which recalls the recreant dogs, which in spite of cudgel and throwing of stones, try to outflank the opposition, to join the numerous party.

Indian like, we followed the small path one by one, and stood soon at the ford where we had to cross the river Pirara after it issues from the lake; this was at this time connected with no difficulty, as the water reached scarcely to the knees.

We had selected a different path from the one I had followed on a former occasion,

[1] Johann Baptist von Spix and Karl Friedrich Phil. von Martius, *Reise in Brasilien auf Befehl Sr. Majestät Maximilian Joseph I, Königs von Baiern in den Jahren 1817 bis 1820*, 3 vols, Munich, Lindauar, 1823–31, III, p. 1300. The governor's name was José Joaquim Vittorio da Costa.

[2] José dos Santos Innocentes, a discalced Carmelite friar. He is a central character in the whole Pirara incident (see Rivière, *Absent-minded Imperialism*, pp. 58–60 and *passim*).

and passed the small lake "Venturu" now almost dry, but not the less visited by numerous vicissi (Dendrocygna viduiata)[1] which disturbed by our approach, rose from their feeding place, and encircled the small basin of water in rapid flight.

If it were it not for the two mountain chains "Canuku, and Pacaraima", which at the distance of twenty miles bound the view to the south and north, the march over the savannahs would be monotonous to the extreme.

During the dry season no bird, scarcely an insect, enlivens the scene; a few stunted shrubs, almost leafless, and some grass cover the savannahs, the reddish soil of which, in consequence of the strong east wind, and want of moisture, is rent in numerous crevices; clouds of dust raised by the strong breeze ascend the air in columns and cover the traveller momentarily, and almost blind those who do not guard themselves against the small particles of dust entering their eyes. During noon-tide the heat rises on these savannahs to 120° and 125° Fr. in the sun, and the *mirage* plays upon their surface, and in the distant horizon the trees and other object appear to hover in the air.

We were glad when we reached (after three o'clock in the afternoon) the river Pirara for the first time, where we could quench our thirst; we had been marching eleven or twelve miles without finding any palatable water.

The best of our pedestrians reached the mouth of the river at five o'clock in the afternoon, some followed us only the next morning.

It is a distance of seventeen English miles from Pirara; by no means an easy march under the Tropics.

We had encamped near the junction of the Pirara with the Mahu; two small corials which we secured in the vicinity would have been quite useless for our purpose; if they had not been previously thoroughly repaired, and besides this, a delay of at least three days was required to observe what effects the journey over the savannahs had produced upon the Chronometer.*

*Chronometer – Frodsham, 389 – Arnold, 6062 – were carried by one of the most careful of our canoe-men in a small tin canister, which he slung by means of a strap across his shoulder.[2]

Two of the Canoe-men (Europeans) went out shooting for their own amusement, and without having first requested permission to leave the camp; on their return the gun of "Petry"[3] – one of the men – got entangled in some withes, and impatient to free it from this obstruction to his progress, he dragged it forcibly after him, by which the cock of the percussion lock was raised, and the gun went off; the whole charge lodging in his back below the shoulder blade.

We heard the report, and his cry in our camp, and on my reaching the spot where the accident had occurred, I found the man weltering in his blood, and having inspected his wound, I must confess that I had little hopes of his recovery; I gave him under Mr Fryer's Charge, and the unfortunate accident detained us until Saturday the 2nd of April at our camp, when Mr Fryer thought "Petry" might be carried upon a stretch to Pirara.

[1] The White-faced Tree-duck.

[2] Charles Frodsham, an instrument maker of London, and John Roger Arnold of Arnold & Dent, instrument makers of Strand, London.

[3] Peterz. The other hunter was Stöckle.

The longitude of our camp was, according to the two Chronometers 15′3″ (in arc) west of the village of Pirara.*

* The longitude of the village of Pirara being not determined as yet to any satisfaction; I have adopted it as the first meridian of my survey. The Chronometers which I had during my last Expedition under the direction of the Royal Geographical Society gave 14 miles west as result; I have more confidence in the present result;

Arnold 6062	giving	$59^{s}.50$ in time West of Pirara.
Frodsham 389	"	1^{m} 00. 88 in time.
		$01^{s}.38$ difference

In order to employ the time which the accident caused us to remain at the mouth of the Pirara, in the most useful manner, I ascertained the distance, of some of the most remarkable mountains in the Canuku, and the Pacaraima chain which were visible from here; while the men were occupied repairing the Corials, and to render them stronger by additional knees.

A tree of the natural order of Dilleniaceae, the Curatella americana, furnished us natural knees for this purpose; it is almost the only tree which is to be found upon the savannahs; and so stunted of growth, and its branches so crooked that besides knees for boat-building, it might be used for military saddles; however, this is not the only use; its leaves are rough or scabrous, and the Indians use them for purposes where our tradesmen employ sand paper.[1]

The Indian smooths his blow pipe (Cura) to such a degree with this leaf, that it appears as if polished; from this circumstance the tree has received the name curatakie;*

* Whether Aublet has derived the generic name from "Curatus" worked in consequence of their use, or from the Indian name, is to be questioned; I believe the latter.

he uses the leaves likewise for polishing his bow, war-club, &c.

Among other botanical curiosities near our camp, that splendid tree the Martia excelsa (Benth)[2] – one of my former discoveries – was covered with blossom.

The Indian is a great naturalist, anything that strikes his eye, or attracts otherwise his curiosity in Flora's empire receives its specific name; it is seldom that he generalises.

The Martia is called Ayawiye, by the Macusi Indians; Tshiwipa by the Caribs.

Along the margin of the Pirara, a Eugenia attracted my attention, with fruits as large as a Damson; doubtless this species of that extensive genus produces the largest fruit; I observed two varieties, one with purple fruit the other yellowish with a red spot; it has an agreeable acid, and the Macusi Indians who mix it with their drink, call it Casami.

The heat was extreme while we were encamped near the mouth of the Pirara. Although the thermometer was placed under the tent which was surrounded by shadowy trees, it rose between two and three o'clock in the afternoon, from 97° to 100° F.; its lowest stand at 6 o'clock in the morning was 73°; our trigonometrical

[1] It is known in Guyana as the Sandpaper Tree.

[2] *Martiodendron excelsum.*

operations on the savannahs were therefore not without their fatigue; fortunately they had in some regard their recompense, some spots of the savannahs were covered with a small plant (Psidium turbiniflorum Mart.)[1] the berries of which had the aromatic taste of Strawberries; we did not spare them, whenever we fell in with their fruit, which the Macusi Indians call Piriko; another species of a larger size, and approaching in its appearance the Guava (Psidium pomiferum)[2] only that the shrub is scarcely above 3 or 4 feet in height, is called Canung; the fruit is however equally tasteful as the former.

I was rather astonished to hear the cry of the larger shrike, so common near the coast, that whoever has visited George town, will seldom forget their almost perpetual cry, which has been resembled to the sound "que-e'-que-dit, and from which the bird has received its trivial name.

It is very likely the Psaris Cayanus, (Cuv.)[3] and our Macusis called it from its cry "Setté-qui."*

* The mean of my thermometrical observations while at the camp near the confluence of the Pirara, consisted

Period 1842	6 a.m.	9 a.m.	12 a.m.	3 p.m.	6 p.m.	Max:	Minim:
March 27 to April 2nd	75°.72	83°.13	91°.82	94°.92	86°.07	98°	73°

The thermometer had been compared with an excellent standard thermometer of Troughton and Simms. A strong easterly wind is generally blowing from sun rise to 8 or 10 in the evening.

April 2nd Mr. Fryer having reported that the wounded man might be carried to Pirara, I gave him under his charge and desired him to remain in Pirara until he considered Petry out of danger.

Although I was thus deprived of the only assistant in my manifold occupations, I was too well aware that Petry without some surgical assistance would have little chance of recovery.

During our detention at the mouth of the Pirara, we had repaired three small corials, which though they afforded only room for two or three men each, were calculated to convey my instruments with more safety, than if they were carried on the backs of the Indians.

I had restricted myself to the most useful instruments, namely;

2) An artificial Horizon
1) A sextant by Jones[4]
A Troughton and Simms New Telescope
prismatic compass
An horizon with levelling screws

[1] *Psidium striatulum = Eugenia punicifolia.*

[2] *Psidium guajava.*

[3] *Tityra cayana* is the present scientific name for this species. The 'qu-e'-que-dit' or ''Qu'est-ce qu'il dit', for which the 'trivial' English name is the kiskadee, is *Pitangus sulphuratus.*

[4] Thomas Jones and Sons, instrument makers of Charing Cross.

Hansteen's[1] apparatus with two horizontal needles, which Colonel Sabine[2] kindly lent to me.*

*They are the needles L (a) and L (b) mentioned in Table LII. of the Magnetic Survey of Great Britain.

A chain of 100 feet, divided in decimals, and compared with a standard chain by Troughton & Simms.

Three thermometers, by Troughton and Simms, compared with the standard

Two pocket chronometers; Frodsham N°. 389, & Arnold N°. 6062.

Knowing from experience how easily during an overland journey the Mountain Barometer is put out of order, and desirous to be enabled to carry this instrument on my next expedition to Roraima, I had left it in Pirara – how short sighted we are in forming plans – although every precaution had been taken in putting it up; Mr. Fryer on entering one day the house, which during my absence was not occupied, found the instrument upset and broken.*

*Fortunately a barometer by Bunton[3] N°. 430 compared with the Barometer of the Royal Observatory at Paris and that of the Royal Society in London, remains in good order.

These instruments were embarked in the Canoes, which were entrusted to the guidance of the Coxswain, while we forded the Pirara, and continued our march to the junction of the Mahu with the Takutu over Savannahs, as monotonous as those I described previously.

The Takutu like other rivers which flow through the Savannahs, is fringed with trees which show a more luxuriant vegetation than the generality of the vegetable productions on these plains.

The wood, through which we had to pass before we reached the river, was nearly a mile in breadth, interspersed with high trees, and an arborescent shrub, which, like the mangrove (Rhizophora) of the sea-shore, had taken root in the alluvial soil, and formed dense thickets almost down to the verge of the river; Its branches, similar to the celebrated Banyan tree, throws down roots into the soil, and formed vegetable arches, which scarcely admit the Sun's light, under which we passed, as if under the dubious light of an early misty morning.

The large trees belonged mostly to the genera Cordiaceae, Malpighieae, and Mimoseae; the first was the Cordia tetraphylla of Aublet,[4] the table-tree of our colonists, and the second a Malpighia, a tree of considerable height and extent; its berries of a deep orange, were then ripe, and a large number being on the ground, our Indians eagerly collected them, and eat them apparently with much relish; although to our palates they appeared austere.

[1] Christopher Hansteen (1784–1873), Norwegian physicist who invented the method of measuring the earth's magnetic field using two needles.

[2] Edward Sabine (1788–1883), later Major-General, Fellow of the Royal Society and a recipient of honorary degrees from both Oxford and Cambridge. He was an expert on magnetism.

[3] Bunten, an instrument maker of Paris during the first part of the nineteenth century. There are very few personal details known about him but he invented in 1824 an important modification for the barometer; an air trap which prevented air getting into the vacuum in the mercury tube.

[4] *Buchenavia tetraphylla* of the family Combretaceae. *Cordia* species belong to the family Boraginaceae.

The handsomest tree was however a Mimosa, its trunk of a light ash grey, its branches spread to a great extent, and its fine pinneted foliage of a vivid green added to its remarkable appearance.*

* I am not acquainted with the botanical name of this tree of which there are no specimens among my former collections; it had neither blossom nor fruit when we visited the Takutu.

A Loranthus[1] with bright scarlet flowers, covered a great many of the trees near the mouth of the river, and formed a strong contrast with the white flowering risps of Desmanthus.[2] Our camp was selected on the left bank of the Mahu, where it enters the Takutu.

I have some where else observed*

*Journal of the Royal Geographical Society, Vol. X.[3]

that the Takutu appears more like a tributary of the Mahu, than the recipient of the latter; and it is in reality less in breadth.

A trigonometrical measurement in 1838 gave me for the width of the Takutu 192 yards before it receives the Mahu, and the latter was found to be 263 yards wide at its mouth.

As I had not previously determined the latitude of the junction of these two rivers, and as it was to form the first boundary mark, it became necessary to determine this point with some exactness by astronomical observations, we remained therefore several days at the Takutu.

The sky was much overcast at night; hour angles in the morning gave me however as difference of longitude between Pirara and the confluence of the two rivers $1^{m}.36^{s.}11$ W. in time. The latitude is 3° 35′ 8″ N.+

+This is merely approximate; time has not permitted me as yet to calculate the means of upwards of one hundred circummeridian altitudes of north and south stars. The long[t] is the mean as given by the two Chronometers.

The Southern Cross, that beautiful constellation, sung by one of the immortal poets of European's southern sky,[4] afforded me the best opportunity for circum-meredian altitudes, as it was less clouded at that period.

The Indians themselves have not passed this constellation without notice, and the Macusi Indian dedicates it to the Spirit of the Savannah Powis (Powis, Ourax erythorymhus Svv.);*

* Animals in Menageries, W[m] Swainson. London, 1838; p. 187.[5]

[1] The tropical species of the Family Loranthaceae have now been assigned to genera other than *Loranthus*.

[2] *Desmanthus* is a genus of the Family Mimosaceae.

[3] 'Journey from Fort San Joaquim, on the Rio Branco to Roraima, and thence by the River Parima and Merewari to Esmeralda, on the Orinoco, in 1838–9', *Journal of the Royal Geographical Society*, 10, 1841, p. 192.

[4] It is not clear whom Schomburgk has in mind here, but it could be Dante, who in *Purgatorio*, Canto I, ll. 23–27, is thought by some commentators to be referring to the Southern Cross.

[5] The book in question is by William Swainson, *Animals in Menageries*, London, Longman, Orme, Brown, Green & Longmans, 1838. Swainson and Schomburgk appear to have been in contact as several entries in Part III of the volume, entitled 'Two centenaries and a quarter of birds, either new, or hitherto imperfectly described', are attributed to information from 'Mr. Schomberg'. Swainson also published *On the Natural History and Classification of Birds*, London, Longman, Rees, Orme, Brown, Green & Longman, 1836–7, which work seems to have influenced Schomburgk. For example, Swainson is one of the very few ornithologists to classify the Powis as *Ourax erythrorynchus*, rather than the more usual *Crax elector*. In fact, Swainson's classificatory system was not accepted by the scientific world.

the Pauituima of the Macusis. He asserts that when the cross stands erect that bird commences its low moan; the first slow, the two next in quick succession; on another occasion I had ascertained the truth of this assertion, when the Southern Cross was on the meridian at 4 o'clock in the morning; but as this is nearly the hour when that bird commences its booming (to use the phrase to which that moan comes nearest) I did not at that put much value upon it.

The banks of the Takutu are famed for the number of Savannah Powis, and I had here a second opportunity of ascertaining that there was some truth in the Indian's assertion.

On the fourth of April 1842, while I was taking circum-meridian altitudes of α Crucis, the solemn moan of the Powis commenced, and lasted about a quarter of an hour; that star was on the meridian that day at twenty five minutes past eleven o'clock at night, by no means a usual hour for that bird to moan; the coincidence therefore was the more remarkable. The Draughtsman of the expedition, who noted the time of my altitudes, and to whom I had some days previously related the information I had received from the Indians, was equally astonished at the strange coincidence.[1]

The Takutu was dwindled to a small rivulet near its mouth; its waters, of a light blueish cast, appeared on many spots almost stagnant quite a different aspect afforded the Mahu, which although lower than generally, pushed its mass of dark brown water far beyond the junction with the Takutu.

Our Indians amused themselves in wading through the water to the opposite shore; however cautious to guard against the ravenous Pirai (Serrasalmo niger*)

*Naturalist's Library, Natural History of the Fishes of Guiana, part i. p. 225, plate 18.

and its numerous congeners, among which the S. piranha is perhaps next to the former the most dangerous; we ourselves enjoyed our bath only under apprehension, as the Pirais which our Indians caught with hook and line, proved that they were numerous.

Another fish which was equally plentiful at the mouth of the Takutu, was the Pimelodus insignis,+

+Ibid. 180, plate 6.

so remarkable for its second dorsal fin in length, which occupies nearly the whole space between the first dorsal and the tail.

We had selected one of these beautiful trees of the genus Mimosa, of which I spoke previously, to engrave the mark of the survey upon it, and the Artist's pencil and brush was set in requisition before we parted, to take a sketch of the confluence of the two rivers.[2]

[1] That Native Lowland South Americans are extremely knowledgeable about natural and astronomical phenomena is accepted as a matter of fact today.

[2] Goodall's painting of this scene is in the British Library (Add. MS 16939, pl. 6). It is reproduced in Rivière, *Absent-minded Imperialism*, opposite p. 115.

The sky had been mostly clouded during our sojourn, and a fresh easterly breeze lessened in a great measure the heat which we would have otherwise felt, the more since an extensive sand bank, then 2600 feet long, bordered on our camp.*

*Mean of Thermometrical Observations at the mouth of the Mahu.

Period 1842	Forenoon 6 hour.	Forenoon 9 hour	Noon	Afternoon 3 hour	Afternoon 6 hour	Remarks
April 3rd to 5th	72°.50	80°.12	89°.12	89°.43	82°.50	Except some slight drizzly rain, there was no rain.

April 6th. The instruments and other packages which found conveniently place in the small boats having been embarked, we commenced to ascend the Takutu.

I preferred the first day to walk in company with the Indians along its shore; Mr Goodall, the Artist, and Mr Richard Schomburgk, went in two of the small canoes; the Corials ascended slowly, and we pedestrians soon out walked them.

At the ford which the Brazilians of the Rio Branco generally make use off in crossing the Takutu, we waded to the left bank of the river, fringed with underbrush, Bamboo, called Rappu in the Macusi language, and a few trees of the same description as those I have already mentioned.

The Sawarai palm (Astrocaryon Iauari, Mart.) occupied whole spots among these thickets, and we had to guard ourselves against its large black prickles.

We kept along the outskirts of these thickets; by no means an easy matter to push our way forwards through grass of a man's height, and swamps which although dry at present, offered by their rank vegetation every obstacle to our progress. These spots are overgrown with numerous Mauritia Palms, and our Indians profited by this opportunity in providing themselves with new sandals.

The Mauritia palm, or Ita of the Colonists (Mauritia flexuosa) is one of the most useful among the interesting order which Linnaeus has so appropriately called "the princes of the vegetable kingdom."[1] There is scarcely any part of that majestic tree which is not made use of for domestic purposes; therefore we ought not wonder that Father Gumilla, the illustrator of the Orinoco,[2] calls this beautiful palm "Arbol de la Vida"; Its leaves folded like a fan, furnishes thatch; the fibres, thread to weave hammocks, and make cordage; the trunk furnishes a kind of sago, which the Warrau Indians make use of in time of scarcity, and at certain seasons the Indians of the Savannahs, and the Warraus of the Orinoco, draw from it a liquor of vineous taste, which fermented is intoxicating; the fruit which is tessellated, is of a dark brown outside, and when arrived at maturity, yellow within.

The Indians immerse this fruit in water until the yellow within becomes mellow, when it is considered fit to eat; and although it is not generally liked by Europeans at

[1] Where Linnaeus so described palm trees has not been ascertained, but Stephen Endlicher did so (see *Genera plantarum secundum ordines naturales disposita*, 1 Vol and 4 supplements, Vindobonae [Vienna], Apud Fr. Beck Universitatis Bibliopolam, 1836–40, p. 245) and introduced Principes as the formal name for the order of plants that includes palms.

[2] Padre Joseph Gumilla, *Historia natural, civil y geografica de las naciones situadas en las riveras del Rio Orinoco*, Barcelona, Carlos Gibert y Tutó, 1791. The words 'árbol de la vida' appear on p. 151, and pp. 145–53 contain a discussion of the multifarious uses of the Mauritia palm.

the commencement, it soon becomes an acquired taste; for my own part, I became soon partial to these fruit.

The deer of the Savannah, the Peccary[,] Monkeys, Aguris, Parrots; all appear equally fond of it, and the heavy Tapir on its way to this feeding place, leaves numerous paths across the rank vegetation of such a swamp where an Ita palm has ripe fruit which, one unacquainted with the fact, would consider to have been made by human feet.

The savannahs between the Rupununi and Rio Branco, are covered with numerous angular quartz rocks, and it would prove painful to walk with unprotected feet over such places; the Indians use therefore sandals (salza in Macusi) made of the half sheathing bases of the Ita leaves, which are more calculated for this purpose, than those of any other palm; they are afterwards fixed in a manner not much different from the way of the ancients, and such a pair of sandals last at least a few days' journey over these rocky plains.

Even when decaying, the Ita proves of some use; when its trunk lies prostrate on the ground, a large beetle, the "Calandra palmarum", deposits its larvae in it, which are considered a great delicacy not only by the Indians, but likewise by many Europeans, and chiefly by the French in Cayenne, Martinique, and Guadeloupe.

The Indians frequently fell these Palms, and having cut several holes in the trunk, he is sure to find numerous larvae of the Calandra in its proper time; nor must I forget to mention, that the Indians prepare a kind of salt from the ashes of the burnt leaves.

The Mauritia grows only in moist soil or swamps, and whenever we failed to procure water by digging at the foot of their trunks, we knew that our search would prove hopeless any where else in the neighbourhood.

We continued to follow the small fringe of wood which bordered the Takutu; sometimes the Savannahs stretched to the river's edge, and formed steep banks, which now, where the river appeared nigh on its lowest level, rose from forty to fifty feet above the water.

Such natural sections, as a Geologist would call them, show the structure of the adjacent Savannah; white and ochreous clays, much mixed with rounded pebbles formed the substratum, scarcely covered a few inches with mould or fertile soil; indeed I considered it hopeless to extract any fruits from these Savannahs, which only may prove useful as grazing grounds.*

*The Macusi of Pirara have planted a piece of Savannah soil, in the immediate vicinity of the village, with Cotton and Indigo, and when we arrived at Pirara, the crop was over. During the dry season the shrubs had the appearance of dry sticks; but since the rain has set in, both plants have sprouted out most vigorously, and promise another crop. The Indigo is used for colouring the Cotton blue for hammocks.

I took a series of angles on one of these raised banks, which the Indians called Iperaghiri.

We started several deer which were grazing in groups of from four to ten: these parts being but seldom disturbed by the presence of man, they were more numerous, than I have seen them generally; our guns were put in request and a fine buck was soon procured, and many others might have followed, if we had had the means to transport them.

The river, on ascending makes a considerable bend to the east, and is joined on the right bank by the small river Manari;[1] its course upwards turns afterwards towards the south & west.

We reached, towards four o'clock in the afternoon, the mouth of the small river Macupara, which falls into the Takutu from the left bank; it has received its name from a tree which I have no doubt is Aublet's Macoucoua (Ilex macoucoua Pers.);[2] the fruit is of the size of a small apple and has a velvety appearance from the number of soft short hair with which it is covered.

The boats reached this place, where there had been formerly an Indian settlement, only after six o'clock in the evening. The Indians had set the Savannahs on fire on the opposite bank of the river, and the large columns of red flames, which advanced against the wind, afforded after sun set a beautiful aspect.

April 7th. On starting from Macupara I prefered taking a seat in one of the Corials, as it appeared to me that in this way I could trace more effectively the course of the river; we had to pass a small rapid, which no doubt does not exist when the river is high. Whenever the river turned towards the mountains, a fine prospect opened to us; the higher elevations visible of the Canuku chain were: Iquari, Zemai-epping, Ilamikipang, with its precipitous rock; but above all towered towards the east the wall-like mountain Nappi (Batata Mountain).[3] The latter bore in the morning east by south, our course being nearly south; at the end of this reach the river makes a sudden bend to the east, the Savannah approaches on the left to the very verge of the river, and the western angle of the Canuku mountains become visible.

The river Capaya (Papaw *

* Carica papaya.

River)[4] fully as large as the Pirara, joins at the bend from the left bank; near its mouth is a peculiar formation of rock, (specs. 1 Takutu journey) it consists of caverneous shelves of sandstone mixed with clay, exhibiting on its horizontal planes a number of marks, which may have arisen from one or the other species of the order Isopoda;[5] the most remarkable part however, was the upper structure which more or less exhibited clusters of cells, some square, some rounded, and from six to eight inches in diameter. About two miles further up the Takutu makes another eastern bend, and adopts now, on ascending a more decided southern course deviating merely a point or two to the west. We halted in the evening near the mouth of the small river or creek Mucu-mucu,[6] which has its source on Guariwaka (mountains of the mist) one of the large mountains of the Canuku Chain. I resolved to remain here for some days, as Mr Fryer had been directed to join us here, if the state of Petry's wound permitted it.

While we were yet occupied to erect our tents, and make the other necessary arrangements for our Camp, we heard some of the Indian females cry out that a large

[1] The River Manari rises in the Kanuku Mountains and is the first of numerous streams that flow from those mountains into the Takutu.

[2] *Ilex guianensis*. Pers is Christiaan Hendrik Persoon (1761–1836), Dutch-German botanist, born in South Africa and died in Paris.

[3] The sweet potato.

[4] Probably the Rio Arraia.

[5] An order of Crustaceans with seven pairs of equally sized legs.

[6] Moco-moco River.

snake was lieing coiled up where they had intended to sling their hammocks; we found it was a large rattle snake, by no means a pleasant guest in the vicinity of a Camp.

During my stay, I took, as frequently as opportunity would permit it, astronomical observations, and ascertained by trigonometrical measurement the height of Ilamikipang; according to a rough calculation, its height is 2500 feet above the Savannahs, which is about the height of the Cradle Mountain in Brecknockshire.[1]

Our party was successful in fishing, and hunting; and as we had sent to the neighbouring settlements, we procured likewise an additional supply of Cassada Bread, and ground provisions. Among the Pirais (Serrasalmo niger) which our fishermen brought in, were several of a uniform black; of equal colour was the eye; the iris surrounded with a golden ring; one measured 1 foot 5 inches in length, and 8 inches in breadth; I considered it merely a variety of the common Pirai.

The Colite (Platystoma tigrinum),*[2]

*Fishes of Guiana in Naturalist's Library, part i., p. 185, Plate 8.

one of the most beautiful of its congeners, was very abundant; one weighed 16½ lbs and was 2 feet 11 inches in length, and the girth of the body amounted to 1 foot 8 inches.

I marked one of the trees near our Camp at the bank of the river with a boundary mark; for this purpose we selected a tree of that beautiful and curious genus which through some misunderstanding, has been called 'Etaballia', by Mr.Bentham;*

*V. The Journal of Botany by Sir W.J.Hooker. London 1840 p. 99.

on transmitting specimens of that genus which I had first discovered at the Essequibo, I was told on enquiry by my Coxswain, who was a native of the river, its name was Etaballia – later I discovered that I had been misinformed; and that the Etaballia of the Arawaak Indians is the Vochy Guianensis of Aublet.[3]

The heart of the tree which is misnamed Etaballia, is uncommonly hard, and would prove as ornamental as the Tiger wood (Machaerium Schomburgkii Benth.)[4] it has nearly the same colour, and the intermediate woody mass is equally soft; the tree exudes between the wood and bark a thick red gummy substance which upon paper turns into a deep orange; this tree was in its full blossom, and the bright yellow of its flowers gave some variety to the banks of the river.

I consider April the commencement of the flowering season in the interior of Guiana, whether the rain has set in or not. It has been usually asserted that it requires first several days' severe rain before vegetation is awakened out of its lethargy.

The month of March and part of April had elapsed without a drop of rain falling in Pirara,*

*The rainy season set in uncommonly late that year, and it was only in the first week in June, that it commenced earnestly to rain.

[1] Pen Cadr Fawr, the highest peak of the Black Mountains.

[2] Kulet or Black-striped Cat-fish or Tiger Fish.

[3] The tree to which Bentham gave the name *Etaballia guinensis* (*E. dubia*) belongs to the family Fabaceae whereas the trees, which in Arawak are called Iteballi, all belong to the genus *Vochysia*.

[4] *Mimosa annularis*.

Figure 4: Mount Ilamickipang. Schomburgk surveying near Mount Ilamikipang. Watercolour by Edward Goodall, reproduced by permission of the British Library (Add. MS 16936/60).

nevertheless the banks of the Takutu were covered with trees, and shrubs in blossom. Numerous were those shrubs which like our Hawthorn push forth, first their blossoms, and afterwards their leaves.

A splendid Tecoma,[1] with its large yellow flowers, was seen now as an arborescent shrub, now as a tree; and although its blossoms almost covered the tree, there were only a few among the many which had any leaves.

The splendid white flower tinged with rose of the Gustavia, the cyan-blue of the Jacaranda, and the pure white of some Ingas, gave the whole picture an appearance which the Tropics can only offer to the eyes of the traveller.

The palms were represented by the graceful Cucurit (Maximiliana regia), the Ita, and the Sawarai; the latter certainly not a pleasing sign to the agricultarist, who considers the soil sterile where it grows.

Over the adjacent Savannahs spreads the Malpighia verbascifolia[2] with its silvery leaves, and appeared to usurp whole spots exclusively; two arborescent shrubs of the same genus, one of them with yellow flowers, the other of the colour of our peach blossom, composed the whole Flora, and did not assist to remove the idea of the sterility of those arid plains, which have no further resemblance to the green meadows of the temperate zone, than that they are flat and covered with grass and sedges.

The mean of my observations gave for latitude 3°21′37″N., and the Chronometers showed a difference of 20.58 miles west of Pirara.*

*Mean of Thermometrical Observations at our Camp near Ilamikipang.

Period 1842	Forenoon 6 hour.	Forenoon 9 hour	Noon	Afternoon 3 hour	Afternoon 6 hour	Remarks
April 8th to 10th	74°.57	79°.67	86°.17	91°.17	84°.83	it was generally clouded & a strong breeze from N.N.E.

The wind blew here with great force; towards sunset it generally lulled, but commenced to blow hurricane like from the northern quarter towards eight o'clock. I observed a black cloud rise about that time in the southern hemisphere, which, when it reached the altitude of about 25° degrees expended and the strong breeze set in.

April 11th I had understood that the next place we would come to on our ascent, was a settlement of Wapisianas Indians, near the Cursato Mountains,[3] which we might reach by water in four days, while it took only three days' march across the Savannahs; I resolved therefore to send the greater part of the canoe-men and the Indian crew by land, and to pursue the course of the Takutu upwards in the Corials.

We continued our journey on the 11th of April, without having been able to procure any information from Mr Fryer, or Petry, and although through Mr Fryer's absence the whole direction of the expedition devolved upon me, I did not wish to defer the further ascent of the Takutu, as that river continued to fall daily; Mr Goodall, the Artist, had been meanwhile instructed how to note the time during my astronomical observations, which he willingly and satisfactorily undertook; and the rest of the additional business

[1] The genus *Tecoma* has been renamed *Tabebuia*.
[2] *Byrsonima verbascifolia*.
[3] Kusad Mountains.

for which I had engaged Mr Fryer's assistance, was overcome by a little exertion from my part.

Soon after our departure from the Camp, we passed a great many blocks of a ferruginous

[*margin*: v. Geological Spec^ns^ T N° 2]

conglomerate, which had almost a vitrified appearance, and might been resembled to slacks coming out of a furnace; such heaped up blocks, sometimes of considerable extent, are very common along the rivers Rupununi, Rewa, and Guidaro. We passed the small rivers Camu[1] and Awarimani; the former has its source on Mount Ilamikipang.

The Takutu winds considerably where it approaches the Canuku Mountains. This chain in which we recognise the mountain of Cumucumu in the map of Pontes,[2] the Cerro del Dorado, or Cerro Ucucuamo of the journal of Santos, and the Acucuamo of Caulin,*[3]

*v. Humbolds Personal Narrative vol vi, p. 517

divides the Rupununi from the Takutu. The former river has forced its way through the Chain; The Takutu however on arriving opposite its western angle, is turned slightly south-east and then westward; and having received its tributary the Mahu, makes a sharp angle, and turns South-West towards the Rio Branco. A single glance on the map, proves that the Mahu ought to be considered the recipient of the Takutu; its continued south western course after it issues from the Pacaraima Mountains to its junction with the Branco, and the larger mass of water which empties itself into the Takutu, entitles it in Geographical respect to be considered the continuation from the junction with the Takutu to the Rio Branco, and which portion ought to bear the name of the Mahu (Ireng of the Macusis) in lieu of the Takutu; however custom has sustained an opinion diametrically opposite, and it would prove useless to change now a name which has existed for centuries. Nicholas Hortsman, who no doubt was the first European who visited these regions, names that branch which contains the united streams, "Mahu", and considered the Takutu a tributary of the former.*

*v. Humbolds Personal Narrative vol v. page 800.

The Curatawuiburi, a mountain of about 2000 feet in height, forms the western point of the Canuku Chain; from here they stretched on their northern side N.E. (more accurately N.43° E.) and on their southern side S.E. by E. towards the Rupununi; their loftiest summits, Nappi, and Curassawaka are placed on their northern side.

In its advance towards the South east, the southern range stretches from the right bank of the Rupununi eastward towards the natural pyramid of Ataripu, and the right bank of

[1] Kuma River.

[2] This is a reference to the 1781 map of Silva Pontes and Ricardo Franco d'Almeida Serra. A copy of it is to be found on pp. 44–5 of *Questions des limites du Brésil et de la Guyane Anglaise a l'arbitrage de S.M. le Roi d'Italie. Atlas accompagnant le premier Mémoire du Brésil*, Paris, Ducourtioux et Huillard, 1903–4.

[3] Caulin, *Historia Coro-graphica*. The volume contains Luis de Surville's 1778 map of the region on which Lake Amucu and Lake Parime are depicted as two different bodies of water, the former to the northwest of the latter which is far more extensive. The Acuquámo Mountains, known as the Cerro Dorado by the Spanish and Portuguese (see p. 176n.), are located to the north-east of Lake Parime. There is no particular reason to identify this range as the Kanuku Mountains.

the river Guidaru or Quitaro, a tributary of the Rewa; the greatest breadth of the range from Curatawuiburi to the banks of the Guidaru is sixty miles. The mountains which stretch southwards, are an irregular grouping of mountains, seperated from each other by Savannahs, and of which I shall have occasion to speak hereafter.

On approaching Curatawuiburi, we found the Takutu so shallow, that we had repeatedly to unload the Corials, and push the crafts by force over the sand.

Large banks, consisting of heaped up river sand, changed with beds of rounded pebbles, so smooth and accurately rounded, as if their form had been effected by instruments; they consisted chiefly of quartz pebbles of a yellow and white colour; but amongst them I found some very fine

[*margin*: v. Geolog.[l] Spec.[s] T. N°.3]

Agates, principally of that kind with interior Zig-Zag parallel lines, which are known to mineralogists under the name of fortification Agates, from the resemblance these lines bear to the plan of a modern fortification; others were of globular structure, and transversed with veins of jasper similar to the Egyptian pebbles.[1]

Below these beds of pebbles or sometimes in their vicinity, I noticed a black sand with minute quantities of gold of a light yellow; but whether it will be worth to wash for it must be left to closer investigations, than I could dedicate to it.

We were obliged to leave the corials in order to make them lighter, and to follow the bed of the river by wading through the shallows, or march over the sand banks. Between two and three o'clock in the afternoon the heat of the sun rose generally from 126° to 128° F.; this enormous heat, the glare of the white sand, which the rays of the sun heated to such a degree, that it became painful to walk over it, but above all, legions of sand-flies, against the bite of which we could not protect ourselves, combined to render our progress the more fatiguing.

Although inured to the heat under the Tropics, my face and hands were blistered, and the white sand and the dazzling mirage which hovered over the banks and the Savannahs, caused inflammation of the eyes; the others fared equally bad, and we were glad when with the setting sun, the freshness of the evening cooled our burning faces, and drove those blood thirsty insects – the sandflies – (different species of Simulia) from us.

April 12th. We met this morning the first rapids, they were formed by a bar of clay slate which crossed the river in the direction of S.6½°W. the dip appeared N.60°E. at an angle of about 12°;

[*margin*: v. Geol.[l] Spec.[s] T N° 4. 5. 6.]

Several other rapids followed and our progress became so slow, that if we had not fortunately purchased a good stock of provisions in Ilamikipang, we should have been in want ere we reached the Settlement.

It became evident that it would be impossible to reach the Cursato mountains at the time we expected on leaving our camp at Ilamikipang, moreover the river was well stocked with fish and frequented by numerous ducks and widgeons. The ducks which we found in numbers of ten to fifteen, belonged to the well known species of our poultry yards; the Musk Duck which through an error of some late Ornithologists, who considered it introduced from Russia, is more generally known under the name of

[1] Brown mottled jasper from Egypt.

Muscovy Duck. This species appear to be spread over the greater part of South America, as we observe from the work of Azara,[1] that it is likewise to be found in Paraguay. It feeds upon algae and other fresh water productions; but I never heard, as Azara observes, on the roots of the manioc or cassada, which is well known proves fatal to animal life. Nor do I think that it feeds upon young Indian corn or maize; but I have frequently seen them nibbling on the vegetable productions along the margins of the river.

The male birds are of a glossy black, with a few white spots on the smaller wing coverts, and some white feathers near the scapulars. The warty skin commences at the base of the bill and surrounds the orbits; it makes its nest along the banks of rivers in a hollow tree, frequently in the jungle which Bambusae and rushes form near rivers and stagnant waters. I have seen young ones in May, and in September, and I am inclined to think it breeds in its wild state twice a year; I have accounted from eight to ten young ones. The female watches over the ducklings with great care; and at the approach of any danger, the young ones disperse, and hide themselves among the jungle and bushes which border the river. As soon as the danger is over the anxious mother collects her brood by a peculiar note.

During the breeding season bloody fights occur between the Aspirants of a favourite female bird, which on more than one occasion I have witnessed in the distance, until I settled the dispute by firing my gun amongst them. The river is covered for some extent with feathers, where these fights have occurred, at which the female bird remains passive, and no doubt bestows her affection upon the conqueror. They roost at night in trees, to which they likewise resort when disturbed on the water. Their flight is heavy, and accompanied by a loud noise, caused by the wings. The young birds of a years age are delicious and highly flavoured; the meat of the old ones is however not always tender.

It is asserted to have received its name Musk Duck (Anas moschata L.) from the musky smell which it is said to possess; however I have never been able to perceive that smell.

The second species, the pretty vicissi (Dendrocygna viduata)[2] is mostly to be found in large flocks; When disturbed they fly up and encircle the feeding place, making a whistling kind of noise, not unlike the sound "vicissi," from which it has received its name. As frequent as they are on the rivers and ponds, chiefly in the Savannahs, I have never been able to discover one of their nests. They are easily tamed, and frequently raised by the Indians, who sell them to the Colonists; I never heard that they bred in the tame state, although the Muscovy Duck, reclaimed from its wild state, is known to do it.

There are two other species which frequently feed with the vicissi, the Dendrocygna lugubris (*Swains*)[3] and the D. Autumnalis;[4] but if it should happen that they are disturbed while feeding together, they divide in their flight.

The Roppong of the Macusi (Chauliadis pallida *Swains*)[5] belongs more properly to ponds, and is seldom found on rivers.

[1] Felix de Azara (1746–1821), *Apuntamientos para la Historia Natural de los Páxaros de Paragüay y Rio de la Plata*, 3 vols, Madrid, Ibarra, 1802–5, III, p. 410.

[2] The White-faced Tree-duck.

[3] Probably the Fulvus Tree-duck (*Dendrocygna bicolor*).

[4] The Black-bellied Tree-duck.

[5] The Brazilian Duck (*Amazonetta brasiliensis*).

The trees of the river were frequently ornamented with numerous blue macaws (Macrocercus ararauna)[1] which by their cries interrupted the silence that otherwise prevailed around us. As they make an excellent soup, we did not spare them; they are found sometimes in couples, and at other times in flocks of from ten to twenty. When it occurred that the male or female of a couple had been shot, the survivor would fly round the tree with piteous cries, and looking in vain for its mate, fly off to the opposite shore; still returning to the spot, where it last saw it alive. So much affection makes the hunter repent the good luck of his gun; Man the most selfish being, as resulting from it is likewise the most cruel of God's creation.

According to the observations I took this night the latitude of our camp was 3°12′53″N.*

* Stars observed l & a of Ursa Maj. l & h of Argus.

The mean of the two Chronometers gave 26.6 miles (1^{m}. 58^{s}.38s. in time) west of Pirara.

April 13th We met the first blocks of granite in the Takutu this morning. I noted here a remarkable geological feature; a large block of granite-gneiss(?) was surrounded as it were, by

[*margin*: v. Geologl. Specs. T. N°. 7.8.]

contorted masses of gneiss which rose about two feet six inches above the water. The granite-gneiss is stratified apparently at an angle of 45°, pierced by veins of quartz, and large blocks of quartz lie upon it; The river upwards takes a sudden bend, and here large masses of granite, some twenty to thirty feet in height, are heaped up; Curatawuiburi bears N.73°E which appears to be the direction of the strata.

We had now to pass a succession of falls, and rapids; legions of sand flies plagued us; and the Thermometer showed at noon 108° F.*

*That is, not in shade, and the bulb freely exposed, but not directed towards the sun.

– no breath of wind was stirring, as anxiously as we wished that more than a common breeze might set in, when the sand flies vanish as if by magic. I had observed that the wind dies regularly away about 9 o'clock in the morning, and only sets in again towards evening.

These cataracts are only caused by ridges or bars of rocks which do not form real mountains, but traverse the Savannahs in more or less an easterly direction.

The cataract of the Rupununi the "Corona", is formed in a similar way in the flat Savannahs; and it is worthy of remark that where this river has forced its course for nearly sixty miles through the Canuku mountains, it runs comparatively smooth, and forms only a Cataract where it issues from the mountain pass, near the former mission Urua or Curua.[2]

We passed, soon after noon the mouth of the small river Sawara-au-uru*

†Sawara is the name of the prickly palm which I have already mentioned (Astrocaryon Iauari) Au-uru the name for a river in the Wapisiana language.

[1] Possibly the Blue and Yellow Macaw (*Ara ararauna*).

[2] Also known as Urwa, the site to which the missionary Thomas Youd had withdrawn in 1839 when forced by the Brazilians to abandon Pirara (see Rivière, *Absent-minded Imperialism*, chap. 4).

(Sarauri or Sarauru in extant maps)[1] by means of which and a portage, the Rupununi is reached in three days.

This was the portage traversed in 1739 by the surgeon Nicholas Hortsman,[2] who left Demerara in search of Gold and diamonds, which he supposed to find in the interior of Guiana; and in 1793 by Francisco José Rodriguez Barata, Lieutenant Colonel of the first regiment of the line at Para, who, when ensign, was sent twice with dispatches from the Amazon to Surinam.*

*v. Humbold's Personal Narrative, vol. v., p. 480; vol. vi., p. 515.

This portage is still frequented by the Indians of the Takutu, and the Brazilians of the Rio Branco. The river itself is about the size of the Pirara, and has its sources at the Pinighette mountains.[3] The direction of the strata which formed the numerous rapids we had to cross, varied generally from S.10°E. to S.10°W. I considered it gneiss which sometimes was incrusted with indurated clay and whole masses of decomposing quartz rested on it.

The Kaimen, the crocodile of these rivers (alligator Sclerops) are by no means uncommon in the Takutu. While our Corials were advancing, we saw a great commotion about a hundred yards before us in the water; we increased our force in paddling, and soon came up with a large Kaiman which had just secured one of its own kindred, the smaller Kaikutshi. It had seized it by the middle of the body, so that the head and tail protruded out on each side of its immense mouth; it still defended itself there; but without effect. At our nearer approach the Kaiman dived, and as it cannot devour its prey in the water, we saw it soon after climbing up the banks of the river to enjoy the flesh of its own kind.

April 14th. The Thermometer stood at eleven o'clock freely exposed 90° F. it was cloudy and the wind E.b.S.

We passed a high bank of the river, perhaps fifty to sixty feet high, formed of conglomerate of rounded quartz pebbles of different sises, the cement consisting of ferruginous clay. Numerous

[*margin*: v. Geolog^l^ Spec^s^ T. N° 13.[4]]

blocks, which through the influence of the weather were losened, were lieing heaped up in great confusion on the side of the hilly bank.

I could trace these rounded pebbles which formed the conglomerate, not agglutinated or cemented, for some distance in the savannahs, though there are no marks, or any likelihood, that the river reached this height in our time. Our[5] path was almost obstructed to day by six otters (Lutra enudris? F.Cuv.),[6] which appeared to contend our further progress; they raised themselves partly out of the water, and making a peculiar snore, showed their

[1] Sauriwau River.

[2] In fact it was not. Horstmann went via the Pirara portage.

[3] Pinniyet Mountain.

[4] There are no notes referring to geological specimens 9, 10, 11, and 12. This also applies to many other geological specimens that go unmentioned.

[5] In the margin at this point is written 'This seems to refer to the Crocodiles* must have been transcribed in a wrong place//*otters'. Beneath this is a further note in a different hand which reads 'NB. This note was on the paper when received from the Colonial Office. J.R.J.' The initials are those of Julian R. Jackson, Secretary of the Royal Geographical Society, 1841–6.

[6] This is almost certain the Giant Otter (*Pteronura brasiliensis*). The smaller otter, referred to in the accompanying note, is probably the River Otter (*Lutra longicaudis*).

formidable teeth. They so little cared for our presence, and came so near our Canoes, that the Indians attempted to strike them with their paddles. I have little doubt but that in this instance, they had their young ones in the neighbourhood; although, generally not shy, it is but seldom they show so much courage as on this occasion*.

*There is some uncertainty among naturalists to which species the two kind of otters belong, which frequent the large rivers of Guiana. The second species is much smaller than the one which attacked us in such a determined manner.

While the corials had to pass some rapids, I walked to an isolated hillock which rose to the height of about 130 feet at the distance of a mile from the river's right bank. The view which I enjoyed was lovely: the continued chain of the Canuku mountains to the N.N.E.; to the E.S.E. the three capped Saeraeri, that strange object in the landscape, which forms its principal feature from whatever point it may be looked at, in a circle, extending thirty to forty miles in circumference. A number of isolated hillocks interrupted by Savannahs extend from Saeraeri in a S.S.E. direction for thirty or forty miles.

The highest summit of the Cursato mountains rose a point eastward of South, and far to the South-west bounded the mountains of the Moon (Kai-irite of the Wapisianas)[1] the horizon: a singular dome shaped mountain towards the S.S.W. which I recognised as the Taquari, of my former expedition, but which the Wapisianas call Mariwette,[2] appeared to be only five or six miles from my station, four isolated hills extended from there in a South western direction, and as the Takutu appeared to flow along these hills, it promised us hard work for the following day.

We encamped on the left bank of the river, near an inlet, which, when the water is high, unites with the river and forms a large island.

Numerous shrubs of that species of Eugenia, of which the Macusi Indians call the fruit "Casami", bordered the inlet, the branches bending nearly to the ground under their load of fruit. I observed several trees of Outea accaciaefolia*[3]

*Described since in Mr Benthams Enumerations of my Guiana Plants, Journal of Botany vol ii and sequel.

and a new species of Genipa; some trees of the latter were so umbrageous that I erected my tent under one.

The inlet or Kirahagh was stocked with that delicious fish, the Arowana,+[4]

+The Naturalists History of the Fishes of Guiana i part page 205, plate 12.

and we saw whole shoals going in and out. This fish appears to abound in the Takutu; we hemmed them frequently in among the falls, and the Indians rushing armed with cutlasses, and bludgeons among them, did excellent execution to the advantage of our daily meals; indeed our success in fishing had been astonishing; besides the Arowana, we secured numerous Pirapoco or Monowai (Xiphostoma ocellatum),**[5]

** Ibid part I page 245, plate 23.

[1] Probably Serra Caiirid, part of the Serra da Lua.

[2] Serra Taquara or Mariwete.

[3] *Macrolobium acaciifolium*. Bentham's description is to be found in *Journal of Botany*, 2, 1840, pp. 94–5.

[4] *Osteoglossum bicirrhosum*.

[5] *Boulengerella lucia*.

Patha (Hydrocyan? Armatus);++[1]

++ Ibid page 249, plate 25. This plate is not well executed the second dorsal fin has been omitted and the appearance of the fish has suffered in the reduction. The head is flatter than represented.

The first is generally met with on sandy shallows, and we had here many a hunt, armed with cutlasses, to prevent them from escaping through the only narrow channel, where some water was flowing.

The Patha, or Baiarra frequents Cataracts and rocky places, where the swiftness with which they swim, frequently pierced by an arrow, five to six feet long, has astonished us. Their strength is wonderful and two of the front teeth are so long in the lower jaws, as to protrude through the upper into sockets, prepared for their reception.

Of great interest proved to me a fish which we caught to-day among the falls; it belonged to the genus Hypostoma and to that section where the interoperculum is very moveable and is furnished with tufts of rigid spines. The fins and the tail are margined with beautiful bright orange Colours, and spotted with black; its first dorsal fin is uncommonly large, and its integuments spotted with black. The strong contrast of its general colour, a dark brown, approaching to black, with the edging of bright orange, and the formidable aspect of this marked fish, when it raised its dorsal fin, rendered it so interesting, that I was quite delighted with my discovery; and I resolved, if it should prove new – as I have little doubt – to dedicate it to my kind friend Sir Wm Jardine B^{t}, whose ichthyological researches and scientific acquirements are so well known among the friends of Natural History, that he is justly entitled to this public proof of my respect and friendship towards him.[2]

Our camp was in Lat. 3° 01′ N.*

* Stars observed γ Argus and α Ursae Majoris.

and 29.6 miles west of Pirara.

April 15th Our progress was to day interrupted by the fall "Scabunk" the largest we had passed hitherto; it has received its name from the small river Scabunk-oute[3] which falls at the foot of the Cataract from the east into the Takutu; it is called Catu-aw-ru by the Wapisianas, which signified like Scabunk-oute, Sandy river, from the large quantities of sand which form its banks.

We had transported the first two of our Corials over this obstacle, and had considerably progressed with the third, so that we flattered ourselves to overcome this difficulty before noon; when an accident occurred which for the present stopped our further progress.

The Takutu abounds in stingrays (kaja), which bury themselves partly in the sand, and prove dangerous to those who wade through the river. Although the low state of the water had obliged us to wade at least one half the way, we had hitherto made, and to dragg the corials by force over the shallow places, we had escaped being wounded. However near Scabunk they were so numerous, that one of the Arecuna Indians was wounded twice above the instep; he appeared to suffer excruciating pains. We were yet

[1] The Patha or Baiara is *Hydrolycus scomberiodes* (*DCEU*).
[2] Unidentified. There appears to be no species of *Hypostoma* named after Jardine.
[3] Skabunk River.

occupied in administering to his alleviation, when another Indian, – a young Macusi of about thirteen years of age – was likewise wounded; not possessing so much resolution in bearing pain as the former; he gave completely away under it; he threw himself upon the ground under piercing cries, burrowed his nails into the sand, and in his paroxysms went so far, that he commenced to bite the sand, and buried his face in it. He was wounded in the sole of the foot; but he suffered the greatest pain in the groins, the region of the heart, and under the arms. In both instances I had a ligature applied above the wound, and not only pressed it as much as possible, but even had it sucked, in the instance of the younger Indian. I applied poultices of Cassada Bread; and towards evening the pains were much alleviated. It is but seldom that fatal instances occur, where wounds have been inflicted by the formidable weapon nature has given to the Sting ray for its defence. The serrated or jagged nature of this instrument causes a dangerous wound; I doubt however that there is any injection of a deleterious liquid connected with the infliction of the wound. It must be confessed the pains and symptoms resemble those of snake poison; and so late as last year (1841) a valuable labourer on the Plantation Zelandia at Wakenaam, died in consequence of the wound which a Stingray had inflicted upon him.

I noted at Scabunk, where the water had scooped out the banks, a white sandy grit, about twenty feet below the surface. The formation of the rocks are in every respect remarkable near this Cataract, which in consequence of the low state of the water was entirely exposed; The strata consisted of Specimens N°14 their direction being S.20°E. they were traversed by Specimens N° 15 forming dykes about two feet in breadth of such a regular appearance, that one might have supposed, not nature, but man had laid them in.

The direction of the dykes is N.60°E. the strata is sometimes traversed by veins of quartz which had a direction of E.15°S.

April 16th. Our patients were better and although they could not walk, they were well able to sit up in the Corials, and as our stock of Cassada bread was at the lowest ebb, we broke up our Camp at an early hour.

We continued to toil against the rapids, and Cataracts; the river spread considerably and huge blocks of granite were lieing about in great confusion; some were dome shaped, others cylindrical, but all were more or less rounded at the edges; the larger amongst them were 40 to 50 feet in diameter.

We passed the fall Matsipao;[1] While we were toiling here to get our Corials over the only place where some water flowed over the rocks, we heard human voices, and soon after saw to our joy Sororeng and another Indian making their appearance. He had been walking with the other Indians from Ilamikipang direct to the Cursato Mountains, where they had arrived the third day after they had left us. Alarmed at our non appearance, he came to look for us. We learned from him that three hours good walking over the Savannahs would bring us to the Wapisiana village; as much as we were rejoiced at this information, as unexpected was it to us to learn that the provision grounds of the village were so badly provided that the Inhabitants satisfied themselves with eating once a day a little Cassada Bread, and lived mostly upon wild fruits, and palm seeds.

[1] Possibly the Machid-pawa Falls.

I had ere this given up every idea of prosecuting our way in Canoes; the fatigues we had undergone, and the little progress we had made during the last seven days, showed how impracticable it was to follow the Takutu, at this advanced dry season; it was therefore merely my wish to reach the landing place of the village, from whence, it is about an hour's walk eastward.

I noticed at the fall "Dabaru"[1] an immense plateau of granite with large flakes of quartz: veins of

[*margin*: v. Geolog[l] Spec[s] T. N°.17]

transparent quartz traversed it parallel with dykes of a similar composition, as the one I noted at Scabunk.

We halted half a mile beyond Dabaru, and sending Sororeng forward to the village, ordered him to bring our Indian Crew, and as many of the inhabitants as he could procure next morning, in order to carry our baggage overland, it being my intention to abandon the Corials.

April 17 [th]. Our Indians having arrived, the Corials were unloaded, and we proceeded towards an isolated hillock which the Wapisianas call Tenette; at its foot on the south eastern side is the village. We had first to walk through a grove which extended partly up the hill, from which the most delicious odour was wafted towards us.

"Sabean odours from the spicy shore
Of Araby the blest."[2]

I might have compared it to the fragrant smell of Hyacinths. We soon discovered the origin; numerous trees of a large size, almost covered with white blossom, and dark green leaves of a lucid appearance, dispersed this delicious odour; Its seed vessels resembled Aublet's Apalatoa, and I would not hesitate to pronounce it to be Tomhiroa aromatica,[3] if there were not some anomalies in De Candelle's description;[4] It may prove a new species of that interesting plant, as the inflorescence is terminal, and the leaves alternate.

We soon after entered the village which consisted of six round houses. The captain, and a great many of the inhabitants were absent; it appeared scarcity had forced them to pay long visits to their friends who were better off in their provision grounds than they. I found here messengers from Pirara, who had arrived this morning, and who brought information that "Petry" the wounded man, was by no means in such a state, that M[r] Fryer could have safely left him.

A great acquisition was a supply of rice which was sent by him, and as I succeeded in purchasing three baskets of Farinha, I was enabled to execute my design of connecting by trigonometrical operations the Canuku Mountains with the Cursato, and the isolated groups, where, we were told, our path would lead us to. The localities appeared to be well suited for this purpose. A short distance North-West from the village rises the hill Tenette 124 feet above the Savannahs; It is partly clothed with large trees, but its

[1] Perhaps near the Dabarwau River.

[2] 'Sabaean odours from the spicy shore / Of Araby the blest'. John Milton, *Paradise Lost*, Bk 4, ll. 162–3.

[3] Almost certainly a copyist's mistake for *Touchiroa*. *Touchiroa aromatica* = *Crudia aromatica*. Apalatoa is probably *C. spicata*.

[4] Augustin Pyramus de Candolle (1778–1841), Swiss botanist and author of *Prodromus systematis naturalis regni vegetabilis*, 18 vols, Paris, Treuttel & Würtz, 1824–73.

top is almost bare; there being only a few Curatella trees. Before the summit is reached, one has to cross two granitic platforms upon which a few Cacti (Cereus and Melo Cactus) were growing. Near the platform among the fragments of rock are numerous Myrtaceae, and a species of Bursera (gummifera?). The prospect is very lovely, and the Saeraeri mountains form again the chief object. I found here the tree in blossom, which gives the wood known under the name of Poa da Rainha by the Brazilians; it is of a dark orange colour approaching almost to red, and is used for many domestic purposes and likewise for building small Canoes. As it takes a fine polish, it would prove highly useful to the cabinet makers; there are numerous trees round Pirara, but I had not seen it before in blossom. The most remarkable part is the large winged fruit vessel, prickly near the base.*

*If I remember right, I have seen it pictured either by Jacquin[1] or Aublet.

The flower is papilionaceous, and of a pale orange colour.[2]

Near the path which leads through the copse at the foot of Tenette is a remarkable silk cotton tree which astonished us by its immense size, and the extent of its ligneous buttresses or excrescences; its height was only 102 feet, but it extended its branches to129 feet; the circumference of its trunk about one foot above the ground was 57 feet and the breadth of one of the tabular excrescences amounted to 8¾ feet. The Macusi call that kind of Bombax "copae-ye." It is the Bombax globoso of Aublet,[3] of which Bombax ellipticum (H.B. & K.)[4] appears to be only a marked variety.

I had soon selected a spot which appeared to be qualified for measuring a base line, for the determination of the height of the Cursato mountains, and the situation of the surrounding isolated groups of hills and mountains. A larger base line was from here determined, which resting its end on the hills Tenette and Manoa, extended in a north-western direction about 7500 yards.*

*It must be understood that the angles were merely ascertained by the Sextant; but as they were repeated on different days, and at different hours, and the mean of these observations used only for the calculation, I have tried to equalise the error which such an incomplete instrument for such operations might otherwise produce.

The small mountain Chain of Cursato or as some of the Indian tribes call it Ursato is of no great extent. Its whole length from north to south is scarcely five miles, and its highest summit (in 2°47′N. lat.) does not extend 3000 feet above the river Takutu. The mountains are densely wooded, and the granitic plateause or walls which we observe in the Canuku Chain are almost entirely wanting. South east of Cursato is the mountain Duruau, and in front of it Manoa. The former is very rugged, and forms a sharp angle; Manoe has received its name from resembling in its outlines an ant-bear,[5] it is necessary to be endowed with a strong imagination to trace even the smallest resemblance in that unshapen mass to the animal whose name it bears. A number of isolated groups the

[1] Nikolaus Joseph Jacquin (1727–1817), Austrian botanist.

[2] Richard Schomburgk (*Travels*, II, p. 33) identifies it as an *Ormosia*. The Brazilian name is 'pau da rainha', literally 'wood of the queen'.

[3] Wild Silk Cotton Tree (*Bombax globosum*).

[4] Humboldt, Bonpland and Kunth.

[5] Ant-bear is the Guyanese name for ant-eater, although it is usually only applied to the giant ant-eater (*Myrmecophaga jubata*) (*DCEU*).

largest of which are about a mile and half or two miles in length, tend it may be said to join the more extensive groups of Tuarutu[1] and Kai-irite, either by intermediate groups or by ridges, while the latter send counter slopes towards the Pacaraima Chain to the north west and towards the mountains of the Essequibo to the South east.

The heat was excessive during the time I carried on these trigonometrical operations; and in consequence of the great extent of the sides of angles, and having no assistant, it proved very fatiguing, I was therefore always glad to meet one of those copses of wood on the Savannahs, which occur more frequently on the left bank of the Takutu than on the right; they afforded shade, and there were at that time numerous trees loaded with fruit, which in consequence of its vinous taste proved highly refreshing; it was the Melicocca bijuga,[2] a tree of considerable size, which the Macusi Indians call Macu.*

* There are from one to two seeds in each berry; ("Bacca una sperma") in the systematical description must be therefore modified. I have seen as many as three seeds in one berry; that side of the seeds where they touch each other is flat.

This tree is cultivated in the West India Islands, but seldom to be met with in Demerara. Indeed I know only of one tree in Georgetown. Among other fruit which were then in season was the Duroia eriopila (Genipa merianae Rich.[3]) and Genipa edulis: the first is known in the Colony under the name of the marmalade box, and is delicious. The Macusi Indians call it "umpa." The fruit of the Genipa edulis (Rich.) resembles a mellow quince, but cannot vie with the former, according to my taste, although they belong to the same genus.[4]

In our peregrinations in the Savannahs we met frequently the nests of wild bee, they belonged to a species, which the Macusi Indians call Wampang; the Wapisianas, Camuiba. The hives or nests are fixed generally to branches of trees, and are from two to three feet in length; the materials with which they build, are bits of wood mixed with glue to a paste-like form, which by means of their feet they tread until it becomes of the consistence of paper, and of firm texture. Their cells are hexagonal, and contain only what is vulgarly called worms, and honey. The latter is uncommonly sweet, but it is remarkable that in none of their nests, as many as I have seen opened, any vestage of wax is to be found.

The bee is small, being not more than four to five tenths of an inch in length, the body is dark brown, hairy from which it receives a velvet-like appearance; the wings of the abdomen are black, margined with a rust coloured almost yellowish band, the superior wings transparent, at their outer edges brown, and marked with eight or nine cells. They sting severely; and in order to secure nests, the Indians make fire under it, when they abandon their fabrics "en masse". I have seen however an Indian, who was the conjuror or Piaiman of his tribe, merely approach the nest, and knocking with his fingers against it, drive all the bees out of it, without a single one injuring him; I saw him first leading his fingers under the pits of his arms before he knocked against the hive.

[1] Serra Tuarutu.

[2] *Melicoccus bijugatus*.

[3] Louis Claude Marie Richard (1754–1821), French botanist.

[4] *Genipa edulis* is not reported from Guiana and is probably being employed as a synonym for *Genipa americana*. The genera *Duroia* and *Genipa* both belong to the family Rubiaceae, and the name 'marmalade box' appears to be used to refer to both species.

A second species of honey bee is without a sting, and prepares not only honey, but likewise wax; the honey is slightly acidulous and they build in hollow trees. The Macusis call this kind "mapa."

I was five days occupied from morning to night before I finished my operations in the Savannahs; much to the astonishment of the Wapisiana Indians, who at the commencement could not conceive for what purpose I underwent so much fatigue; but when I had given to understand through our interpreter "Sororeng", that it would lead to secure them against the atrocious slave-hunts of the Brazilians, they accompanied me on my excursions, and most willingly carried the instruments and fixed the signal poles. One rather advanced in age, who appeared to have travelled a good deal in the neighbourhood, made me acquainted with the names of the numerous groups of mountains which I could see from my chief stations, and proved thus of such great service, that I gladly engaged him to accompany me to the sources of the Takutu.

The Wapisianas are taller than the Macusi Indians, and their heads are small in proportion to the body, and their necks short; the molar or cheek bones are more prominent than with the Macusis, and their nose straighter. Their language differs materially from that of the Macusis, but it resembles the Pauixana,[1] who border on their territory to the south west; they are less industrious than the Macusis, and negligent in their houses which are generally filthy. These houses are built in the form of domes or Cupolas and covered with leaves of Palms, chiefly with those of the Mauritia or Ita, of which there are many on the Savannahs.

The circular hut is about twenty five to thirty feet in diameter, and is inhabited by several families. The entrance or door is the only aperture in this fabric, which by night is shut by a door made of palm leaves; there are no partitions to divide the properties of different families; a few stones as the hearth, is the only token of the family's right to that peculiar spot.

Three beams fastened to uprights by means of bushropes or Lianas, traverse at a height of five or six feet the hut, and serve to sling their hammocks upon, or to keep their bows and arrows, and that singular instrument, the "Blowpipe" out of harm's way, which in the general bustle of domestic Indian life, might get injured. As every family considers it the peculiar right or duty of the other, to clean their common habitation, I need not observe that none do it, and filth accumulates to a disgusting degree. The smoke from four or five fires, not being able to find an outlet but through the narrow crevices, which the thatch may have left, circles in mazy columns through the hut, and irresistibly brings tears in the eyes of those, who are uninitiated to Indian life; Fowls which are found in almost every Indian settlement, Parrots innumerable, and other domestic animals, contend for the partial possession of the hut; but the large number of half starved dogs, always ready to make acquaintance with the stranger's calves, constitute the greatest nuisance. Numbers of Fleas, and that insidious insect. "the Tshigo", form the frame to the comfortable picture which awaits the strangers sojourn in such a hive. I made it therefore a rule where no hut could be given up to me for our sole occupation, (which naturally had first to undergo a thorough purification which likely it had not received since it was built), I decidedly prefered to sleep under the tent cloth, or rather in the open air.

[1] As previously mentioned, both the Wapishiana and the Pauixana are Arawak-speaking, whereas the Macushi are Carib-speaking.

Das Innere einer Wapisiana Hütte.

Figure 5: Inside a Wapisiana house. Picture taken from the copy of Richard Schomburgk, *Travels in British Guiana*, II, p. 16, in the editor's possession.

The Wapisianas wear their hair short; I have never seen a Wapisiana with his hair hanging down to his shoulders, as this is often the case with the Macusis. Those who can maintain several women, practise polygamy: it is not so frequent however as among the Warraus and Accawais. In their domestic manners, they do not differ from the Macusis or Indians of Guiana in general.

In consequence of the survey, I had been detained until the 23rd of April at Tenette, when scarcity of provisions forced us to continue our journey; the weather had been fair, but the mornings were generally so much clouded, that observations became precarious; the mean of several Meridian Altitudes of North and South Stars, gave me as latitude 2°49′40″N* ;

* I possess 45 circum-meredian altitudes of North and South Stars which want of time has not permitted me to calculate as yet.

and the difference of longitude between Pirara amounted to 29.13 miles west; I rely with the more certainty upon the correctness of this difference of longitude, as by measurement of the difference between Tenette and Pirara, I received 29.23 miles as the result; the mean would be 29.18 miles, or $1^{m}\ 56^{s}.76$ in time.

During our stay we had an instance how easily the Takutu swells from merely a thunder-storm; the rain it is certainly true, fell in the night of the 18th to the 19th of April in torrents, but I was astonished to find the rocks of the neighbouring Cataract Cocoya[1] covered, and the water rushing over it with great force; Two days previously I walked to the opposite shore, the water reaching scarcely above the ankles.

I vibrated here for the first time since we left Pirara, the magnetic needles L (a) & (b), and found that the magnetic force exhibited itself, by 100 oscillations in $2^{m}51^{s}.25$ at 87° F. per needle L (a), and the same number in $3^{m}56^{s}.21$ at 86° F. per needle L (b); the first result is the mean of two sets, each of 360 vibrations, the second only of one set, as a thunder storm interrupted the experiment.*

* Mean of Thermometrical Observations at Tenette.

Period 1842	Forenoon 6 hour.	Forenoon 9 hour	Noon	Afternoon 3 hour	Afternoon 6 hour	Remarks
April 18th to 23rd	75°.87	79°.83	87°.63	90°.10	81°.33	These observations were made in an hut, and the Thermometer guarded against any reflection of the suns rays, it was clouded during this period

April 23rd. We were obliged to have our baggage carried upon the backs of the Indians, as we could not make any progress by water; I had therefore to increase our number considerably, in which I readily succeeded.

We left on the morning of the 23rd of April and reached, after an hour's walk over the Savannahs, the small river Cursarari where it enters the Takutu; we found here a Canoe,

[1] Kukuipaw Falls.

by which means we crossed to the left bank of the Takutu. Numerous trees of the Elizabetha coccinea (Mihi)+[1]

+So named in honour of her Majesty the Queen of Prussia. V. Journal of Botany by Sir W. Hooker, vol. II. page 92.

with its bright scarlet flowers, skirted the river; and I found here, for the first time, ripe seeds of this interesting tree, which I hope may germinate in our Stove houses in Europe. The tree bore buds, flowers, and seeds; the pods of a scarlet velvety appearance add to the beauty of the tree.

Our course was directed towards the singular peaked mountain Au-uruparu. Interested by some plants, I loitered behind the marching column, when an approaching thunder storm warned me to hasten myself to come up in time with the Indian, who carried my cloak; unfortunately I missed the way and did not get up with the others until I was perfectly drenched with the rain. In order to avoid some swampy ground, the guides had led across a pathless tract. It was one of those tropical torrents, when the quantity of rain amounts to a couple of inches in the course of two hours; and although we had reached a copse of palms and wild plantains*

*Phenakasperum Guianensis (Endlicher).[2]

(parime of the Macusis) near the brook Totowau, it could afford us only a trifling protection. When we issued from this copse, I was much pleased with the aspect of a number of hillocks, which clothed with a vivid green, and encircled by woods, contrasted so strongly with the general appearance of the Savannahs. It appeared the woody tracts became more prevailing towards the West, where we traced a large fringe of thick forest, no doubt extending for a considerable [distance] to the westward. We entered soon after a wood, exhibiting the exuberance of tropical vegetation; numerous palms; the broad leaf of the wild plantain, so gigantic among the Endogens, that our European vegetation has to offer nothing similar. Passion flowers with bright scarlet flowers and blueish rays; Psychotricaceae with orange, & Cephalideae[3] with rose coloured and blue bracts, stamped this forest with a tropical aspect.

The vegetable mould upon the ochreous clay appeared to possess every fertility; I was therefore not astonished to find in the middle of the wood, a spot which the Indians had cleared, and planted with provisions; Sororeng pointed out to me cuttings of a plant, very probable a Euphorbiaceae which had been planted among the Cassada, and which he told me were used to intoxicate fish with; it may be a Phyllanthus, but as only cuttings without leaves were to be seen, I cannot decide whether it was so or not.[4]

We halted near a place where blackened posts, and heaps of ashes a number of broken pots, and other Indian utensils, proved that some accident or atrocity had been the cause of destruction; we learnt unfortunately that human hands raised against their fellow beings had committed the destruction of this Settlement. This was one of the

[1] There are two species: *Elizabetha coccinea* M. R. Schomb. ex Benth. var. *coccinea*; and *Elizabetha coccinea* M. R. Schomb. ex Benth. var. *oxyphylla* (Harms) R. S. Cowan.

[2] *Phenacospermum guianensis.* Usually known today as *Phenakospermum guyannense.* Endlicher. Stephen Ladislaus Endlicher (1805–49) was Professor of Botany and Director of the Botanic Gardens, Vienna.

[3] The genera *Psychotria* and *Cephaelis* both belong to the family Rubiaceae, but the latter genus has now been subsumed by the former.

[4] At least two species of *Phyllanthus* are used as fish poison: *P. acuminatus* and *P. brasiliensis*.

villages which was surprised by the Brazilians during the slave-hunt in 1838, and those of the inhabitants which could be secured, were led away, after their huts had been burnt to the ground. A small hut, scarcely affording room for two or three men, rose on the side of the spot, where the circular mark of black ashes showed that there was formerly a house of some extent, according to the comparative ideas of those children of the wilderness; whether the small hut was built by one of those, who fortunately escaped the fetters of the Brazilian Slavery, and finding himself still attached to the spot of his former happiness, satisfied himself with so small a hovel; or whether erected by a relation of those who lived, and acted here, remained undecided; as its inhabitants were absent. At present it was temporarily occupied by a Macusi family from the Rio Branco, in whom to my great astonishment and joy, I recognised two Indians, who accompanied me on my journey across the mountains to the Orinoco, and one of whom we were under the necessity to leave behind in a village on the banks of the Kundanama, being in such a weak a state from low fever, that we had been obliged to carry him in his hammock;*[1]

* V. Journal of the Royal Geographical Society, vol. X.

on his recovery he traced his way back to his village, although it was between four and five hundred miles, and looked now hearty and well. After I had given them, and their families some presents, we continued our journey towards the naked hill "Kuipaiti", having understood we should find a settlement there. It consisted of a single hut built near the foot of the hill Au'uruparu. The inhabitants were mostly old, and the Chief man was one of those, who upon my application through Pedro Ayres to the district authorities of Rio Negro, was realeased from his slavery in 1839.

I observed an old woman with snow white hair, the second whom I have seen among the Indians, with that indication of great age. Her skin in wrinkles, and her body attenuated almost to a skeleton, afforded an appaling picture; her curiosity induced her in spite of her weakness to leave her hammock to view the first Paranaghiri, as they call the light haired Europeans, who ever visited their abode. Old age among the Indians appear in more frightful forms than with the Europeans of the temperate Zone; and as if to make in this instance, the aspect the more repulsive or commisserating; she was led out of the hut by an idiotic boy, who with a vacant look, his mouth open, and his tongue protruding, stared alternately at us and her. Idiotcy is considered sacred among the Indians; they look upon these, who are possessed with it, as marked beings and their doings and sayings are considered oraculous. In this case the spirit did not move him, and oppressed by the picture, I will not say disgusted, I turned away.[2]

A young female, indeed the only one, who by appearance and age contrasted strongly with those around her, brought me some Mapa, or wild honey which she had collected in a bottle gourd, and which she tendered with a smile; a few strings of mock coral beads which she received in return as a present, seemed to make her happy.

Our expectation of procuring a large supply of Cassada bread. or other provisions, seemed to suffer disappointment; only a few cakes were promised us, but as if to make

[1] See Vol. I, p. 353.

[2] There is little support in modern South American ethnography for the claim that idiots are individuals possessed by spirits, although in the past shamanic possession was often interpreted as the result of psychological conditions.

up for this defeat of our hopes, we were told that large numbers of Deer were in the neighbourhood: I resolved therefore to halt next day, and despatched our best huntsman in search of game; while the naked hill Kuipaiti, one of the points determined from Tenette, would afford me opportunity to verify, and extend the angles of my survey.

April 24th. The Wapisiana Indians call all those hills by the general name of Kuipaiti, which consist of solid rock, and are only sparingly covered with vegetation. The hill to which we directed our path this morning, does not strictly belong to this class; the base consists of granite, or granite- gneiss, forming (about two hundred and fifty feet above the Savannah) an immense wall from fifty

[*margin*: Geolog$^{l.}$ Spec$^{s.}$ T. N° 18]

to sixty feet high, extending South West for several hundred yards; the rock consists near the summit or about fifty feet highest up (vide Specimen N° 19) of [blank.]

The seams of the stratification, where the rains have washed off the soil, run E.N.E. being

[*margin*: v. Geol. Spec$^{s.}$ T N° 21]

apparently traversed by others in a N.E. direction; from which it receives the appearance of Lozenges; as the rock has a basaltic appearance I have little doubt that it has pierced the granite. Large veins of Quartz traverse this rock; At the foot of the hill lie numerous loose rocks, chiefly

[*margin*: v. Geol. Spec$^{s.}$ T N°. 22]

near a Rill, which although dry at this period, may be a mountain torrent during the rainy season, when every hill sends its contribution of water to the plain below.

On our way to Kuipaiti we observed many hundreds of a beautiful white lilly, which looked like stars among a sea of verdure; they are generally found in Moist Savannahs; and their stem, upwards of thirty inches long, bears from one to three lillies, (generally two), ten and a half inches long, and six inches wide; they have a faint smell during the morning and evening; the Macusis call them Manasero; the Wapisianas Guatuppu; it is a Hippeastrum belonging to the order Amaryllidaceae or Narcissus tribe.[1]

Our prospect from this hill was more extensive than from Tenette or Manoa, and as we were favoured with an almost clear sky, my desire to procure angles for the survey, was fully realized; I think the summit of Kuipaiti is about five hundred feet; and our view extended southward to the far distant Tuarutu, while the Mountains of the Moon, the Kai-irite of the Wapisianas, bounded our prospect to the South; towards the north we observed the Canuku, having the appearance of a blueish line here and there interrupted by towering clouds which rested upon the ridge of mountains; at our feet the Savannah through which the Takutu meandred in numerous windings, fringed by a line of bush, and high trees; thick forests extended from Mount Mariwette along Kuipaiti towards the Takutu and Fort San Joaquim. The Watuwau,[2] a tributary of some size joined the Takutu towards the South West.

[1] According to Richard Schomburgk (*Travels*, II, p.46), it was *Hippeastrum solandriflorum* (*H. elegans*).

[2] On the Guyanese map the upper course of this river is marked Watuwau, the lower Urubu. On one Brazilian map (1983) it is called the Urubu and on the other (1963) the Tiquirre. Watuwau is Wapishiana for the vulture which is called urubu in Portuguese.

I received here a proof how fallacious it is to trust implicitly to Indian information, chiefly if double interpretation is necessary to come to an understanding; when I visited in August 1838 the Caruma Mountains an Indian whom I understood to have much travelled, accompanied us to the summit, and from him I learnt that by means of the river Guidiwau[1] the Indians of the Rio Branco reached the Takutu. I further understood him, or at least so it was interpreted to me, that the Guidawau approached the Caruma within a few miles, (and indeed he pointed out the water course to me) and afterwards made a circular sweep to its junctions with the Takutu, which he said was about a day's journey from the Ursato or Cursato mountains. I mentioned this information in my report to the Royal Geographical Society and inserted the course according to this information in dotted lines into my Map.*

*V. Journal of the Royal Geographical Society, Vol X.[2]

To my great astonishment I learned here, and received undeniable proofs, that the Guidiwau is no tributary of the Takutu but falls into the Rio Branco, but that it approaches the small river Curati,[3] which falls a little south of Au'uruparu into the Takutu, so near, that a much frequented portage exists between the Guidiwau and the Curati. The name of the river Watuwau which falls into the Takutu, and the portage between Guidiwau and the Takutu, or rather one of the tributaries, no doubt has given rise to this mistake, and proves that my informant had never been there.

A traveller cannot guard himself sufficiently against false information; partly the wish of the Indian to be considered well acquainted with his country, or much travelled, as they express it; partly the wrong interpretation of a language which leaves much to conjecture from the form of speech or the emphasis with which it is pronounced, gives frequently rise to great mistakes in Geography and natural history; chiefly if the traveller is under the necessity to use double interpreters.[4]

When we arrived in our Camp we found that our hunts-men had been most successful; seven fine deer had been shot in less than three hours; we were thus provided with sustenance for several days, and I resolved at once to continue our journey next morning, as we had to march four days before we were to meet another habitation.

The Deer of the Savannahs, the "Beyu" of the Colonists, is allied to, or perhaps only a variety of the Cervus Mexicanus of Pennant;[5] it is found in couples, but often in numbers of five or six upon the Savannahs; it produces a Young one in March or April; I know not whether it is peculiar to the Savannahs of the Rupununi, or whether it is the same species which grazes upon the Savannahs near the coast; so much is certain, it does not inhabit woods; its wide branching antlers would little qualify it for such an abode.

[1] Rio Quitauau.

[2] 'Report of the Third Expedition into the Interior of Guayana, comprising the Journey to the Sources of the Essequibo, to the Carumá Mountains, and to Fort San Joaquim, on the Rio Branco, in 1837–8', *Journal of the Royal Geographical Society*, 10, 1841, p. 187.

[3] Rio Kurati. The Quitauau does not approach the Kurati but one of its main tributaries, Rio Capivara, does.

[4] Whereas it is clear that there are innumerable ways in which mistakes and misunderstandings arise under such circumstances, it is curious that Schomburgk did not pick up the error as he must have passed the mouth of the Quitauau in April 1839 on his return journey from Esmeralda.

[5] Thomas Pennant (1726–98), Welsh zoologist. The deer is a species of either *Odocoileus* or *Mazama*.

April 25th. We started at half past six; our guides led us first over pathless Savannahs, and directed their course afterwards towards Mount Wurucokua. We passed the stream Curati, which as I have already observed, offers a portage to the Guidiwau, and Rio Branco. A tree which had been felled in a such a way that its base remained on the left bank, while its branches rested on the opposite side, served us as a bridge; those who feared giddiness, and did not wish an involuntary bath, preferred to wade at once through it. We passed the small hills Wariwea, they are well wooded, and extend for about a mile and half N.N.E. & S.S.W. We halted near a place, where marks of fire gave evidence that huts had likewise here been burnt down; I learnt however that it was done by the Indians themselves, who, when they learnt of the misdoings of the Brazilians in their vicinity, rather set fire to their own huts, and fled further eastward than allow themselves to fall into the hands of their persecutors. The situation for a Settlement had been well selected; the small chain of hills, covered with forest trees, abounded in rich vegetable mould for provision grounds, and the eminence upon which their huts had been erected, afforded an extensive view. A pity that remorseless man, manstealers, as well they may be styled, should destroy the peaceful domestic life of the red man, and carry the torch and destruction of happiness to those, who satisfied with so little, do no harm to their oppressors and usurpers of their country.

A number of detached hills, among which Wurucokua is the highest, extend in an E.S.E. direction for several miles; they are only sparely wooded, covered with fragments of rock, and their sides furrowed by Tropical torrents, exhibit a dark red clay, quite in keeping with the yellow appearance of the few grasses and sedges, which are the only representatives of vegetation upon the hill. The small brook "Urucokua" has its source among these hills, and falls into the Watuwau. A wall consisting of fragments of rocks, extends for nearly a mile S.60°E. from Wurucokua over the Savannahs; it reminded me of the granitic wall of the Caquire, near Esmeralda; but where was the majestic Duida with its cloud-topped summit and colossal walls of sandstone? that beacon which guides the traveller upon the Orinoco for hundreds of miles! Wurucokua could not vie with it in height,*

*The Duida rises 7147 feet above the Savannahs, I estimated Wurucokua 1500 feet above the Savannahs.

or historical interest! No Humboldt had botanised in its vicinity or executed geodesical measurements; no Spanish legend told of treasures of diamonds, and Emeralds, burried in its bowels!

A walk of half an hour further brought us to a formation of indurated red clay, mixed with angular quartz pebbles, upon which blocks of decomposing granite were lieing in great confusion.

The hill Piritate forms the most southern knot of the irregular groups of Wurucokua, Wayawatiku, and Wakuroite.[1] We had in no instance traversed as yet a hill of any consequence but as no regular path led to the Tuarutu Mountains, which were for the present our determination, we followed along the eastern side of mountain groups, which stretched southwards towards Tuarutu and Ossotshuni; and South westward towards Kaiiri. Similar groups among which, Pauisette, Rhati, Duruau, Pinighette,

[1] None of these hills is marked on the reference map but they lie between the Takutu and Watawau Rivers.

&c.,[1] were the most considerable, stretched from North to South along the right bank of the Takutu; that river flowing at a distance of from five to ten miles from these groups.

We entered about noon low Savannahs; small copses of shrubs, perhaps not more than fifty steps in circumference, rose here and there out of the Savannahs; the soil was very moist, and these green spots had a peculiar refreshing appearance, where every thing else looked yellow from the continued drought. I was not a little astonished to discover among several Epidendreae, the curious Epistephium parviflorum (Lindley) which six years ago, I discovered upon the Savannahs of the Tapacoma-lake, a distance of upwards of three hundred miles from its present spot of growth; in the vicinity I saw likewise the pretty Bachia (Mihi – or Cleistes rosea Lindley) which is found in the same situtation in the Savannahs of the Tapacoma and Capooy;[2] but so much as I was pleased with these remembrances of former days, and of a delightful ride in company with one of the most scientific planters of Demera,[3] I would not have been a botanist, if my joy had not been surpassed by the discovery of a new Orchideae, among the moist places of the little copses. At present, where I have explored British Guiana from North to South, and from east to west, a new species added to the number of known Orchideae has its rewards; how much more pleasing is it to find a new genus, which distinguished itself by habit, beauty, and fragrance!*[4]

*Drawings and dried specimens, with a more detailed description, have been sent to Dr Lindley, to whose excellent monograph,[5] Botanists and the Amateurs of this interesting family of plants must be ever thankful, and I have no doubt, that if new, a more detailed account will be published in another place.

We reached at noon the river Watuwau,†

†It has received its name from the Common Vulture (Cathartes aura).[6]

about 150 yards wide a foaming torrent; although its bed was scarcely filled with water, and reached us only on two or three spots beyond the waist, we found it uncommonly difficult to ford it, its current being so strong, and its whole bed a continuation of loose rocks.

After having halted for an hour, we continued our march, and passed Savannahs covered with fragments of angular Quartz rocks; The highest mounts of Kai-irite bore S.54°W; the bright rock upon which the sun reflected its rays, shone with a dazzling whiteness; The outlines of this rock being crescent-like, when looked at in the distance; the Wapisianas have resembled it to the moon (Kaira in their language) and call it in consequence Kai-irita or Mountains of the Moon. The rock appears to be white quartz, and like others of a similar description in the Canuku Chain, the Duida &c being colder

[1] Rhati is the Raad Mountain and Pinighette, Pinniyet. The others are not marked.

[2] Tapacuma and Capoey Lakes lie between the Essequibo and Pomeroon Rivers.

[3] It is not clear who this was or when exactly the ride took place although, if it occurred six years earlier, thus in 1836, the most likely date was between Schomburgk's first and second expeditions.

[4] It has not been possible to identify this orchid, and it is noteworthy that Richard Schomburgk fails to make any mention of its discovery.

[5] John Lindley, *The Genera and Species of Orchidaceous Plants*, 6 pts, London, Ridgways, 1830–40.

[6] Also known as the Turkey Vulture.

than the atmosphere is constantly moist in consequence of the condensation of vapours, which the thickly vegetation produces. As soon as the sun reaches therefore a certain height, and throws his rays under a certain angle upon the moist rock, it shines with a dazzling white, and may be seen at the distance of fifty or sixty miles. Such rocks shine frequently only periodically, according as the sun has North or South declination. A similar rock lies on the side of one of the mountains of Pacaraima, bearing N.29°W. from Pirara; it shines only from May to August; earlier or later than the stated period; the sun throws his rays in an angle which does not fall upon the rock to reflect its moist surface. These rocks may be called natural Helioscope, and like that instrument, have served me in my geodaesical operations to determine their exact situation and the mountains upon which they lie.

The Canuku Mountains have similar rocks at Mount Curassawaka and guwariwaka; Humboldt tells of others at the Duida, and other mountains of the Orinoco, which the Indians have called the sun and moon. The rock of Kai-irite, however is the largest of this description, I have seen in my journey. At the distance of about 25 miles, its height makes still an angle of 18' from which I suppose it cannot be less than 700 feet high; The mountain chain of which Kai-irite is the highest, is called at present by the Brazilians "Serra da Luna"; it is perhaps the "Serra Yauina" of older maps, at least it corresponds with the situation given in Arrowsmith's map to that serra,[1] and although it is not far distant from the Rio Branco, I have been astonished that, as well among the Brazilians as among the Indians, it is a perfect "terra incognita"; In all my travels, I have not found one individual, who had visited it, and the Indian banishes all evil spirits in[2] this region, while the Brazilian considers it the abode of wild Indians, who massacre any person who dares to come within their precints.

The chain draws from North towards South-east, it is not directly connected with the Mountains of the upper Essequibo; isolated groups and Savannahs, where forests dispute their supremacy with grass covered flats, interrupt the Chain. However Kai-irite forms the highest pinnacles of any chain in the neighbourhood; the angle of that mountain amounts at the distance of 23 miles to 1°16′55″ and as I was then on rising ground, its height cannot be less than 3100 feet but this measurement was made near Aruatintiku (Tiger hill) a hill about 8 or 900 feet high; it is well wooded, and has a singular peak, being no doubt an aiguille of Granite. It gave the hill such a remarkable appearance, that it was easily recognised at a great distance, and became therefore of great service in my survey. Although it has received its name from the number of Jaguars or Tigers, which are said to prowl about it, we selected a narrow woody valley which it formed with the next hill, as our night's quarters. In the neighbourhood of Aruatintiku is a remarkable rock poised on the top of another; it resembles the celebrated rocking stone in Cornwall;[3] Another singular rock, columnar in its shape, lies on the side of the other hill, nearest to Aruatintiku.

[1] This is the Serra da Lua, which is marked as Yauina on A. Arrowsmith's 1811 map and subsequent editions, as well as on J. Arrowsmith's 1834 map.

[2] An editorial hand has written here 'to or from?'. Clearly the context requires 'to'.

[3] There is a number of rocking or logan-stones in Cornwall. The most famous, and the one to which Schomburgk is probably referring is that on Treryn Dinas promontory, near Land's End. This stone, which weighs some sixty-five tons, was toppled over in 1824 and, although soon replaced, apparently no longer rocks as easily as it did. I have found no record of Schomburgk's having ever visited Cornwall.

With the exception of a blueish thick water which we found in little pools, we were not able to discover any other in the vicinity of our Camp, and necessity forced us to quench our thirst with it, as bad as it looked. The Indians found numerous trees of the Melicocoa,[1] the fruit of which assisted in quenching our thirst. At night a severe wind blew from the South West, and the trees on the hill bent under its force, it lulled about midnight; however once awoke, I found it difficult to drop asleep again; I was amused with the regular notes of a night-hawk which went through the scale of a musical octave; omitting only one note in the gamut.

April 26th. We broke up our Camp at an earlier hour than generally, anxious to fall in with flowing water. We passed the isolated mountain groups Tabaitighu, the latter forms almost a half circle, and appears to be connected with the western part of Aruat-intiku; Mount Wayawa-tiku, which we passed yesterday rose above the former. The little chain Muruwit rises perfectly isolated out of the Savannahs, it consists of several eminences which in the distance have the appearance of Towers; the most western is a solid mass of granite, from four to five hundred feet high, and like all those hills of solid stone, is called Kuipaiti "the rock par excellence".

The wish to procure some angles for my survey, induced me to mount it; and I must confess there was something like a foreboding, that I should make a new botanical discovery which urged me not to mind the steep and perilous ascent. I had ascended for about two hundred feet, when I found myself divided from the summit by a deep hollow; the place where I stood, exhibited still the features of the Savannah, some spots of grass and a few stunted Curatellas.

The almost naked mass before me was only covered here and there with a species of Clusia, some Epidendreae and a Monochanthus;[2] from the ravine or hollow rose some large trees, and by dint of climbing and descending I reached the ravine; a small rill of water meandered through it, forming a succession of Cascades; it was a romantic spot, but what enhanced it, was the finest odour of numerous wild pineapples which grew in clusters along the margin of the stream;

"Her luscious fruit ananas rears,
Amidst a coronet of spears."[3]

However the spiny leaves rather opposed our progress, and the delicious odour, the citron yellow of the fruit, raised the gastric nerves of our tongue; The wild pine is small

[1] Presumably *Melicoccus*, a genus of the family Sapindaceae.

[2] *Monachanthus* have been reclassified as *Catasetum*, a genus of Epidendroid orchid.

[3] This is almost certainly a misquotation of Mary Russell Mitford's *Christina, the Maid of the South Seas* (1811), Canto one, verse XII:

Those gardens every beauty yield
Of Indian isle, or Europe's field,
Prone on the ground the melon lies,
Of different clime, of varied dyes;
That, of the hint of Hymen proud,
This, rosy as a summer cloud:
There, her rich fruit Anana rears
With coronet of verdant spears:

I am grateful to Nigel Rees and a visitor to the website of the BBC Radio Four programme, 'Quote ... Unquote' who proposed this source for the quotation which, for four years, had resisted any attempt at identification by me and many other people.

and of a brighter yellow than the cultivated fruit; its taste, when perfectly ripe is highly aromatic; the only draw back is the great number of seeds which it contains, and which are so seldom met with in the cultivated pine. When we reached the granitic platform, a fine prospect opened towards Kai-izite and Tuarutu; and the blueish outlines of the Canuku indicated the direction of the path, we had so far successfully accomplished. A Eugenia, some Cacti, and Melo-cacti, Cyrtopodium Andersonii, which so appropriately has been styled the Sugar Cane orchidea by the Indians, from the resemblance of its leaves, and general habit to the root stocks of the Sugar Cane, diversified the otherwise not numerous Flora of Kuipaiti. I found on the summit a Gesneria, not in flower, and was already on the point of returning having procured the desired angles, and rather disappointed in my botanical discoveries, when some brigh scarlet flowers attracted my attention; the side view which I had merely of the first flowers, and the yellow disk on the petals, made me almost believe, I saw an Abutilon before me; but a nearer examination persuaded me that it was the most splendid of the family of Helictereae;[1] Its large flowers, their bright scarlet and yellow disks, as well as their peculiar form, will render it a great acquisition to our collection of stove plants; it is a shrub from 12 to 15 feet in height, the branches straggling of a greyish colour; the leaves are cordate and serrated; above of a dark green, but in consequence of the soft hairs, playing according as the light falls upon them, they have a velvety appearance; the lower part is downy; the calix almost yellowish green; the petals of a bright scarlet, irregularly formed and the upper two marked with a yellow disk; the shrub was thickly covered with flowers; it appears however, they are not of long duration, for which a succession of buds recompenses. It is known to Botanists that the fruit is spirally twisted from which circumstance it has been resembled to a screw. At the foot of the hill, I observed numerous specimens of Epidendrum pachyanthum (Lindley)[2] with large green flowers, which I have seen on the banks of the lower Essequibo, and at Roraima; indeed those copses of underbrush on the Savannahs among which it grew, remind me much of the Savannahs of that remarkable region, "the sandstone mountains of Roraima"; however their pride, the Sobralia Elizabetheae (Mihi)[3] was not to be seen among the copses near Kuipaiti.

A large column of smoke pointed out the direction, where the others were halting; some had set the Savannahs on fire, and fanned by a smart easterly wind, the flames advanced with great swiftness, until it seized the fringe of wood which bordered a small rivulet, which we passed afterwards. The naked trees with their blackened trunks, the ground covered with whitish ashes, and a number of Carrion Crows, and hawks, disturbed by the heat of the flames, flew in circles around their former roosting place,[4] increased the desolate appearance which a Savannah offers during the height of the dry season.

The barking of several dogs told us that Indians were near us; we soon after reached an eminence consisting of solid granite, and discovered an old Indian who with a certain nonchalance in his features calmly expected our arrival; He was sitting on a large rock,

[1] Richard Schomburgk (*Travels*, II, p. 49) identifies it as *Helicteres glaber*, although this plant is listed as *nomen nudum*, i.e., without a description.

[2] *Encyclia pachyantha*.

[3] *Sobralia liliastrum* Lindl.

[4] While this may have been the reason, such birds frequently congregate round savannah fires in order to prey on the ground animals caught by or fleeing from the flames.

and having thrown a glance on our number, turned his face. At a temporary hut, we observed fire, and marks that several people had been here; on enquiry the old Indian remained mute, until one of the Wapsianas came up, when his strange behaviour was soon explained. We had been taken for Brazilians, but our approach had only been observed, when the dogs commenced to bark, and betrayed their presence; The old man hurried therefore his son in law, with his wife, and a Child, into the neighbouring wood, while he calmly expected our arrival. When we asked him, why he remained since he considered us Brazilians? he gave us to understand, that he did not expect they would have led him into slavery, as he was too old, and not able to work.

He hallowed into the wood, and his daughter came soon after with a fine boy of about seven years of age; her husband followed some time after; they had only arrived that afternoon from the Settlement of Tuarutu, for the purpose of hunting Deer, of which there are none in the mountains; it had taken them nearly three days to reach the spot where we found them.

This granitic hill was covered with numerous Agava (A. Vivipara) some in blossom, surrounded by a large number of young ones, which having fallen on the ground from the mother stem, had taken root. The Agava vivipara is by no means so handsome as the common Agava (A Americana) the bright orange of the latter, and its number of flowers, surpass in beauty the greenish blossom and the straggling shape of the viviparous Agava.

Torrents of rain accompanied by Thunder kept us during the latter part of the night awake; although we were partly protected by huge trees, our tent curtains were not strong enough to withstand such a severe fall of rain.

April 27th. The Indians whom we had met the day previous, had left their sleeping place long before daylight in order to arrive at their settlement before us, partly I presume, burning with a desire to communicate the news of the arrival of Paranaghiris, and partly to remove the fear which otherwise our appearance would have caused.

The severe rain of last night had set the Savannahs under water, and we had to wade for several hours; I observed near our path an Oncidium with yellow flowers, which appeared new to me; at least I collected it for the first time in Guiana; the prevalence of yellow amongst this genus is remarkable, and it will be observed that the leaves of the yellow flowered species correspond in form, while the rose-coloured Oncidium Lanceanum, and the spotted O papilio, differ in their form of leaves from all the others; indeed these two species and their kindred stand singularly alone among the Orchideae.

On approaching the Tuarutu mountains we commenced to enter a wood; and passing a defile, were among hills of about four hundred feet high, and where Savannahs appeared to struggle with forests, which of the two was to prevail in the landscape. These hills were encircled by others of much greater height; scarcely any traces of a path were visible, and I gave the necessary orders that our people from the Coast, should keep close together to prevent their losing the way; we thus passed from wood to Savannah, and vice versa, now mounting hillocks or winding our path among large blocks of granite. The scenery bore the stamp of great wildness. On the left we observed a remarkable rock, rising solitary to a great height much in the form of the natural pyramid of Ataraipu its summit was covered with vegetation; and here and there found a shrub a place in some of the clefts which the granite formed; this strange object bore N.85°E. from our path, and was about two miles and a half distant. The Wapisianias called it Aikuwé, but I could not learn whether any traditions were connected with it.

We had again to climb a hill covered with enormous blocks of granite between which we had to force our path; some of these rocks were covered with Cyrtopodium Andersonii, Schomburgkia Marginata, Vanilla, Brassevola, and a number of the beautiful Catleya Superba (Mihi)[1] of which I had not seen any Specimens since I left the Rupununi. This force of vegetation upon a naked rock, which already from its peculiar form would attract the attention, is surprising to the stranger who is accustomed to the slow progress of northern vegetation, where the blocks of granite afford seldom nutriment to flowering plants and only some lichens spread over its surface.

At the verge of the wood stood a Colossal Fig-tree, spreading its branches to a great distance, and like the celebrated Banyan tree, its horizontal branches throwing down roots into the ground. The numerous blocks of granite which were lieing below it, might be resembled to natural seats, while the wide spreading branches formed an arbour that afforded most delightful shade, and invited us to rest our fatigued limbs. We had to traverse an immense solid pile of granite, perhaps a mile and a half in circumference, and two or three hundred feet high, numerous cavities, resembling the Kettle and pans in Cornwall,[2] proved the influence the weather had exercised upon this solid mass.

Cereastri,[3] and the strange Melo-cacti, and a few specimens of the Viviferous[4] Agava, covered the highest summit; shrubs of Desmanthus, Clusia, and a small leafed Cassia formed the under shrub, about fifty feet below the summit. The Cereus is generally a token of sterility; and its upright six sided columns, leafless as they are, do not add to give diversity to the surrounding scenery; nevertheless its long whitish blossom [is] of an elegant form. The inside of a yellowish colour, is formed like the rays of a Star, and diffuses during night a delightful odour; and as only at the midnight hour they appear in their greatest splendour, there is a mystic interest attached to these plants which their peculiar form increases still more.

"Queen of the dark whose tender glories fade
In the gay radiance of the noon-tide hours."[5]

Long before the sun has reached the meridian, they are faded, and hang down quite decayed. The large purple fruit follows which is eaten by the Indians, and the [pulp] of which has a sweetish taste.

We descended and entered another basin, enclosed with mountains. Our path led now over Savannahs, now through wood, and as we had not met with any water for the last four or five hours, our thirst was considerable; we halted at noon; the

[1] *Cattleya violacea.*

[2] I am grateful to Jens Andersen (personal communications, 6 and 7 April 2005) of the Camborne School of Mines who first informed me that circular or semi-circular rock basins, some 20–50 cm deep and one or two metres across, occur on top of some of the granite tors in Cornwall. He then drew my attention to the following reference from Peter Stanier, *Cornwall's Geological Heritage*, Truro, Twelveheads Press, 1990, p. 42: 'Granite tors on the summit exhibit rock basins, such as the Giant's Crocks and Kettles'. This is presumably what Schomburgk had in mind, although, as mentioned above (p. 78, n. 3), there is no record of Schomburgk's visiting Cornwall.

[3] Unidentified, but presumably *Cereus.*

[4] Perhaps 'viviparous' was intended.

[5] The source of this couplet has not been located.

greater number of our men had not come up as yet, and we sent some of our guides forward to look for water; they returned without success; they brought however numerous fruits of the Cucurit Palm (Maximiliana regia) which have a vinous taste, to which a European soon accustoms himself, although few like them in the commencement.

After having walked three miles through wood, we met the river Manatiwau, glad to quench our thirst. It has dark water, and falls into the Takutu. We halted for nearly an hour to allow the stragglers to come up with us, and knowing that some Wapisianas were still behind, whom we suspected would guide those who had not come up at that time, we continued our march, and reached, after three o'clock, the Wapisiana Settlement, which is named after the mountain chain near which it lies. "Tuarutu".

The Settlement consists merely of two huts and its inhabitants were mostly old people. The young man whom we had met the day previously, was a Macusi, who having married a Wapisiana, had settled among them. They had three handsome boys; the eldest perhaps seven years, and resembling each other so strikingly, that if the difference of their age had not contradicted such an opinion, one might have thought they were born at one birth.

We soon ascertained that there was not much bread to be procured at this settlement; but as there were several others in the vicinity, where we were told they had plenty of Cassada in the fields, I resolved to rest for several days, and erected our camp in the vicinity of the settlement, near a small copse. Late in the evening we received information that all our men had come up, except the cook; a black man from the Coast.[1] Sororeng had remained behind near the river Manatiwau, in case he should come up in the course of the evening.

April 28th. We had received no tidings of the lost man, who had to ascribe his misfortune to his own imprudence, as he was warned repeatedly not to lose the Indian guide out of his sight. He had started before the hindmost party, who had rested themselves, and they had seen nothing of him since. I despatched a strong party in search of him, and burned large fires on hills in the neighbourhood during the night.

A number of Wapisianas arrived this morning to greet the first white men, who ever visited these regions. Some certainly afforded a most grotesque appearance; one, who as we were told, was a Chieftain, had a piece of Calico wrapped round his loins; and his hair being turned backwards at the front of the head, that part was covered with masses of roucou,[2] ornamented here and there with the white downy feathers of the Powis. One of his followers carried a low stool, on which he took his seat, as soon as he had made his Congé, waving his hand twice or three times across my face.

I have already observed, that the Wapisianas are taller men than the Macusis; this was likewise here confirmed. They looked with great curiosity at the different objects which we had brought with us, while their Chieftain remained placidly sitting at the entrance of my tent, and received the reports of what the others saw remarkable among my baggage. It appeared to be below his dignity to go and look himself, and as there was no doubt some difficulties, to describe to him such things as they saw for the first time in

[1] Hamlet Clenan.

[2] Urucu, the red paint from *Bixa orellana*.

their life, a most animated conversation was carried on, which drowned every attempt to put in a word from our side.

Forks are for those Children of the Savannahs, who see them for the first time, the most miraculous things; one passed from hand to hand, and when I showed them the use, they broke out into a boisterous laugh, making very likely their own commentaries upon the use of an utensil far inferior to the one, nature has given to them. The reserve of the Chieftain melted away, when he left the tent and went to the temporary hut which we had erected for our kitchen. He inspected now in his own person the numerous utensils, which his eyes likely saw here for the first time, and the wonder of the party had no end, when Sororeng, our interpreter, explained to them what use we made of them.

Among the numbers who had come to visit us, was a Capoucre; an offspring between a Negro and an Indian Woman. That race is generally stronger and more muscular than the Indian, and their hair is woolly like that of the Negro; there are however very few in British Guiana, but large numbers are to be met with in Surinam; where it appears a great many run-away slaves have intermarried with the Caribs.

I had seen this individual some years ago at Pirara; he was now on a visit at the Tuarutu. His father was a Slave at Santa Maria on the Rio Branco, where his mother a Wapisiana, was likewise kept in Slavery; and becoming acquainted with her, he married, and had several Children by her. Tired of the oppression of their master, the whole family escaped through the forest, and over Savannahs, and safely reached the tribe of the woman's family; where he has settled, and remained there since. I presented the Chieftain some trifles, and he promised to bring us a supply of bread the next day.

There are several Settlements in the neighbourhood, consisting merely of one or two huts. The soil is uncommonly fertile in the Copses, which intersperse the Savannahs; and a wish of self independence may have given rise to settle seperately. Among the presents which were brought to us, were large goobies[1] filled with dried fish.

The Indian is well acquainted with the effectual method to preserve the fish, by subjecting it to the smoke of a wood fire, which acts as pyroligneous acid upon the meat; and it appears he is likewise aware of the method of preserving meat by keeping the air from it.

These goobies were tightly closed and a kind of pitch or gum was put over it where the opening had been, to prevent the air coming in contact with it.

Some other Indians arrived in the afternoon. To my astonishment, I recognised in one of these a vaqueiro, or herds-man, whom I had seen at Fort San Joaquim. I learnt that oppression, and want of even the trifling pay which appears was only promised to him, but which he said he never received, had induced him to fly; he had since settled in the neighbourhood, married, and had two children: He spoke uncommonly well Portuguese. and he told me that the Brazilians had carried him away, when he was quite young; they kept him at the lower Rio Branco, and when he had grown up, they sent him to Fort San Joaquim, to serve on the Imperial Farms*

*These farms are in the immediate neighbourhood of Fort san Joaquim, but I doubt that they consist at this period of more than three thousand head of cattle. They are at the

[1] Gobi, a bowl made out of a calabash (*DCEU*).

> present under the administration of the Commander of the Fort, and the herds-men are either enslaved Indians or deported malefactors. A number of these Indians escape annually either to the savannahs or Mountains, or to the Colony of British Guiana. Four arrived only lately, who had effected their escape, one of whom proceeded to the lower Essequibo, the other three were engaged by the Commissariat of Fort New Guinea, near Pirara.[1]

as herds-man; until he found an opportunity to escape their thraldom. In his outer appearance he showed little of the civilization he had received while among the Brazilians; I knew him while at Fort San Joaquim to dress with some attention, procuring his clothes by cultivating small spots of provision grounds during his leisure hours; but as several years had elapsed since he escaped, the clothes which he brought with him were worn out, and he had no opportunity of procuring others; I gladly gave him therefore what I could spare and he showed his gratitude by bringing us next day, a gooby with Tapir fat which proved of great use for culinary purposes.

April 29th. The party who had been sent in search of Hamlet Clenan returned the previous evening without success. Engaging all the Indians I could procure, and not exempting an individual of our Expedition, I divided their number into three parties; and having desired Mr Goodall and Mr Richard Schomburgk, to head respectively one of the parties, I conducted the third, and gave orders to depart in three different directions towards the mountains, with the injunction that each party was to fire every ten minutes, to attract the lost man's attention, if he was still among the living. I had great difficulty to induce the Indians to assist me in this search. As soon as they had understood from the Wapisianas, who belonged to our party, that it was a black man who was missing, they did not feel the slightest inclination to stirr. This hatred of the red man towards the black is remarkable, and does not restrict itself upon the Indians of Guiana, but it is equally prevailing among those of Chili, and Peru. I had the satisfaction of finding Clenan at 2 o'clock in the afternoon; he was almost in an exhausted state, and fear and fatigue had so much operated upon him, that I apprehended in the commencement his reason was gone. He certainly offered a most piteous appearance; his wild looks, his clothes hanging in tatters around him, and his incoherent speech, sometimes laughing sometimes weeping, showed what impression his misfortune had made upon his weak intellect; but what amused our men, when they found that he was otherwise whole and sound, was the remnant of a land-tortoise which he had hanging round his shoulder, and which proved that he had made his breakfast upon it, wisely preserving the remainder for his dinner. He had always shown the greatest abhorrence for this animal, and considered us very likely somewhat in the light of Cannibals, when he was obliged to prepare it for our meals; and his repeated observation was, that he would rather die than eat of it. How faithfully he had kept his vow when in necessity, showed the small remnant of the large tortoise.

I had provided myself with the necessary remedies for restoration in case we should find him exhausted, and which were used with full success. As he was too weak to follow

[1] Fort New Guinea was the name given to the fortification constructed by the British military detachment sent to retake Pirara.

us then to the village, I ordered the Coxswain, and one of the Canoe-men to remain with him, and to come next day to our Camp.

We arrived there at dusk; the others had reached before us, naturally without success, but all were glad that also in this instance no loss of life of any individual was to be connected with our Expedition; Petry's accident was still too fresh in remembrance, as to have forgotten the dark forebodings of the superstitious.

We were now comparatively among the mountains. To the north the Tuarutu raised its summit to about 1800 feet, the hill next highest to it being about 1150 feet above the Takutu. The Tuarutu chain is about ten English miles in its greatest length and forms an irregular mass, enclosing Savannahs and monticules of solid granite. Huge blocks of the same rock lie about in great confusion, and witness of some convulsion and a bare rock resembling Ataraipu, which the Indians call Aikuwé, is the most remarkable feature of this knot of mountains.

A plain where savannahs are interspersed with woods of no great extent and here and there covered with hills which rise from 150 to 200 feet above the Takutu, unites the Tuarutu with the Ossotshuni.

That small mountain chain extends 11 miles in N.E. and S.W. direction. No where have I seen so many granitic and amphibolic rocks as in the Ossotshuni mountains; Uruwai, Wapung or Wapuma, Curushiwini are masses of granite which rise to a height of fifteen and eighteen hundred feet, and white spots of Quartz contrast with these masses, which otherwise present such a gloomy appearance in a most striking manner.

The Wapisianas of these regions relate that the Tobacco grows wild on Uruwai.

To the south of Ossotshuni commence thick forests, and the blue outlines of the Essequibo Mountains break the horizon to the S.S.E.

May 2nd. With a supply of provisions, and the promise of our Indian friends to provide more on our return, we started this morning after nine o'clock, and crossed, an hour afterwards, the rivulet Iwurucatakurin.[1] Its long name did not correspond with its importance as a flowing water; it flows into the Ossotshuni, which has its source in the mountain chain of the same name.

Our way led along that chain which we kept a mile or two on our right, and an isolated hillock of a pyramidical form was our guide; we reached about noon the small river Taramtibawau,[2] certainly the most considerable of those we had crossed to day. Its waters were flowing over large shelves of granite; those parts which were not under water were covered with orchideae nearly to the verge of the stream; among them, distinguished itself the splendid Catleya Superba, with hundreds of flowers; some six inches across, diffusing a most fragrant odour. A species of Monochanthus with a spicy smell, and luxuriant specimens of the Oncidium, I discovered some days previous in the Savannahs; Epidendrum licorantum,[3] Agave vivipara, Gesneria, Cacti, and Melo-Cacti, Clusia &c covered the bare rocks, and

[1] This is not a secure reading, but it is certainly not Turerucatakurin, as appears in the published version (p. 57).

[2] Rio Taramtiba and possibly Igarapé Jacamin of the Brazilian maps.

[3] This is not a secure reading, and it has not been possible to identify the orchid in question.

presented an enlivened picture of tropical vegetation. A few Oncidium lanceanum on some of the neighbouring trees, proved that even this species was no stranger to this region.

We passed the isolated hillock on our left and directed our course towards some eminences, called Wawacunaba, from the height of which we enjoyed a most beautiful prospect over the Savannahs.

The mountain Vindaua[1] (Vindiau in my last map) was the most striking object. From here received the Takutu its first tributary of any consequence; It bore S.S.E. Behind Vindaua we discerned the blueish outlines of a large and high mountain Chain, in which I greeted old acquaintances; they were the Wanguwai and Amucu[2] Mountains, near the confluence of the Yuawauri, or Cassikityu[3] with the upper Essequibo. Further eastward I noted a chain of mountains equal in height to the Wanguwai, which the Indians named Uassari,[4] and I gladly recognised in this name a mountain chain which Humboldt mentions in the VI volume of his Personal Narrative, page 523, but the latitude which he states is at least 40 miles too far north.*

* Their latitude is about 1°40′N. according to some angles I took at the Tuarutu.

They appeared to be high, surpassing the Tuarutu or Ossotshuni, but the distance did not permit me to judge whether they were wooded, or consisted, like the Ossotshuni, of bare granite. The rocks which were scattered in fragments on the summit of Mount Wawacunaba were highly crystalline (v. Specimens N° 27). We entered shortly after [a] wood consisting of numerous palms of the Genera Astrocaryon, Bactris, Euterpa, and Maximiliana; the Wild Plantain (Phenakasperum guianensis) grew most luxuriantly, and reached often a height of fifty feet. The soil was a rich loam mixed with white sand.

A small rivulet, the opposite side of which was grown over with nastus, or wild Bamboo, afforded water, and our tents were soon erected in its neighbourhood, shaded by the princely Carolinea.[5] Some of its fruit burst during the night with a loud crack, scattering the seeds which they contained all around, many of which fell on the tent, making a noise like hailstone.

May 3rd. Mount Vindaua bore in the morning S.26°E; we continued our course along the Ossotshuni mountains, and reached, after two hours march, an abandoned settlement on an eminence. On entering a wood, I noted with pleasure the graceful Mauritia aculeata, the first I saw since I left the Rio Negro.*

* V. Journal of the Royal Geographical Society, vol. x.[6]

[1] Serra Vindaua. In the published version and on the accompanying map it is spelt throughout as Vinudaua, whereas in the manuscript it is invariably Vindaua as on the modern map.

[2] Amuku Mountains.

[3] Kassikaityu River.

[4] Wassari Mountain.

[5] *Pachira*, but not a genus of Malvaceae reported from the area.

[6] 'Journey from Esmeralda, on the Orinoco, to San Carlos and Moura on the Rio Negro, and thence by Fort San Joaquim, to Demerara, in the spring of 1839', *Journal of the Royal Geographical Society*, 10, 1841, p. 258.

As at the banks of that river, so it grew here in clusters; the Wapisiana Indians call it Urukush. We walked afterwards through a forest of Bactris, among which the majestic Turu palm raised its head to a considerable height, and large clusters of its purple fruit, contrasted strongly with the dark leaves of the Bactris.

We met in the forest such a numerous body of Cushi ants (Atta cephalotes) that their number would be disbelieved If I were even to estimate it far under the truth. They were marching in an uninterrupted column about a foot wide, and three to four hundred yards in extent, towards their hill or nest, carrying bits of leaves which they brought from a tree 80 to 90 feet high. Some were occupied in bringing the leaves from the tree to its foot where they were deposited, while others took them up, and carried them to their nests; and although the ground was covered with their numbers, passing and repassing, and carrying loads, perhaps five times their own size, no disorder was observable in their columns. They seize the leaf with their mandibles at one of the ends, and carrying it upright it appears like an umbrella, kept over the body to shade the ant against the sun and weather. Some of the Colonists call them therefore the Umbrella Ants; however they are mistaken in the aim which the insect has in carrying the leaf in that manner; it finds it easier to carry such a burden, and takes less room away, while marching towards their common nest. Here they amass heaps of leaves which would astonish any one unacquainted with their habits.

These ants are the greatest enemy to cultivation, and if their nests are in the vicinity of provision grounds, all cultivation ceases. Many a village, many a provision field has been abandoned, as it proved impossible to raise a vegetable. Cassada, Yams, Batatas, are alike subjected to their destruction.

Their hills are frequently several hundred feet in circumference, and from ten to twelve feet high and so undermined that it proves dangerous to walk over them.

Among the palm trees, grew a large number of Juvia trees (Bertholletia excelsa) the fruit of which is well known under the name of Brazil nuts; they had done bearing, but we found a few of the large ligneous fruit on the ground, which we gladly opened to procure the tasteful nut; more delicate in its fresh state, than when shipped across the Atlantic.

We ascended now hills of 150 to 250 feet in height, their summits were dome shaped and covered with numerous fragments of angular Quartz. I observed a large Fig-tree perhaps one hundred and twenty feet in height, with large tabular excrescences like the Mora tree, (Mora excelsa, Benth.) and leaves similar in appearance to those of the Caladium arborescens, or Muc-mucu of the Colonists.

On one of these summits large plates of mica protruded among fragments of Quartz which our Indians collected eagerly as curiosities.*

*V Geological Specimen T N° 30. The Macusis called it Prata-wè or Prata-ewibo. Prata is silver in Macusi and the additional syllable wè or ewibo refers no doubt to the resemblance of the precious metal.[1]

We descended a hill several hundred feet and coming out of a copse of wood saw two huts before us, which were inhabited by Macusi Indians.

[1] 'Prata' is the Portuguese word for silver, adopted by the Macushi.

Figure 6: Macusi houses in the virgin forest. Picture taken from the copy of Richard Schomburgk, *Travels in British Guiana*, I, p. 336, in the editor's possession.

The Chieftain of the place, whose name we understood, was Tuma-tuma, awaited our arrival, with great indifference in his hammock. He was one of the stoutest Indians I had seen.*

*He allowed me in his good nature to measure him after we got better acquainted, and I add here a comparative measurement of him, and another Macusi Indian, by the name of Mingai of nearly the same height.

	Chieftain Tuma-Tuma of Maripa	Mingai
Common height	5 feet 4	5ft 5in
Long diameter of the head from the superior angle of the Occiput to the Chin	" 10	" 11
Circumference of the head round the forehead above the ears	1 10	1 10.2
Upper extremity from the head of the humerus to the end of the middle finger	2 5	2 4
Lower extremity from the head of the femur to the heel	3 "	3 0.3
Circumference of the upper part of the Arm	1 0.5	" 10.8
D° of the lower, below the Elbow	" 11.2	" 9.7
Circumference of the upper part of the thigh	1 9.5	1 7.8
D° of the calf	1 2.8	1 1.4
Breadth of the Chest between the shoulders	1 9.3	1 5.
Circumference of the Chest	3 3.7	2 11.
D° of the Abdomen	3 4.6	3. 0.
Length of the right-foot	" 10.	" 10.
Length of the right-hand	" 7.	" 6.7

While our guides made him acquainted with the object of our journey, he only gave his hem, without betraying the slightest interest in us, or those who were with us. The females however did not constrain their curiosity in such a determined manner as Tuma-tuma; by that sort of freemasonry which prevails among the fairer sex of God's creation, they soon became acquainted with the Indian women in our train, and begging them into their huts; no doubt received as much information as they desired.

Our tents were erected near the banks of the Watuwau which we had crossed a few days previously as a mountain torrent, it was here dwindled to a small brook, only a few yards wide, and at present forming merely pools. It has its source five or six miles further to the Southward.

The inhabitants of the settlement Maripa consisted, at the time of our visit, of only twenty individuals. A sickly looking person, who had dressed himself according to the civilized fashion, in shirt and trowsers, and covered his head with a red woollen cap, made now his appearance and greeted us. He was some relation of the Chieftain, and appeared to pride himself not a little, upon his being dressed like ourselves. From whence he had got his suit of clothes at this distant spot, was for some time a riddle to us; but as we afterwards learnt that he was a great trainer of dogs, of which there were a number equal to that of the inhabitants in the village, we conjectured perhaps rightly, that he might have received it in exchange for one of his dogs.

There were several other settlements in the immediate vicinity; and as our expected arrival had been announced the day previous by some Wapisianas from Tuarutu, who anxious to carry the news of our visit, had not minded the journey of two days; we saw shortly after a party coming to greet us with their Chieftain at their head. These two important men, the Chieftain of Maripa, and the other of the neighbouring Settlement were a true exemplification of Pharaoh's dream of the fat and lean kine. The new comer was a tall lean man with a ghastly expression in his countenance which was much increased by having lost one of his eyes; he told me through our interpreter, that he would order some bread to be made for us, and would show us if we wished, the way to the Sources of the Takutu, where he had been only lately.

I had understood from some of the Indians at Tuarutu, the Sources of the Takutu were at Mount Vindaua; however we learnt here, that this river receives only its first tributary from Vindaua, and has its sources further to the south and east of it.[1]

The Coxswain had sprained his ankle, and expecting that he should be able to walk if a day's rest was afforded him, we delayed our departure.

The reserved manner of Tuma-tuma gave way before four and twenty hours were elapsed; I rather think he mistrusted us; our large number, and the constant fear of those poor beings of the Brazilians or Caraiwa; sees in every stranger of white complexion a Man-stealer. He came to my tent and inspected what appeared remarkable to him; In spite of his size he walks briskly, and before he became so stout, he must have been a handsome Indian. His eyes were rather oblique, and there was a peculiarity in the orbitar process, the horizontal diameter of which is generally with the Macusi Indians from 2.3 to 2.8 inches, while in his it measured only 1.5 inch. His wife was much younger than himself; she had two children, the youngest about five our six months of age, and was again far advanced in pregnancy.

I bought here a young tiger Cat; it had been caught only a few day's previously and possessed all the wildness of its race. It was too young to judge with certainty, but I think it was either a specimen of the Felis pardalis, or the Felis macroura of Prince Neu-Wied.[2] The number of Tiger Cats or Jaguars (Felis onza) must be enormous in this neighbourhood; Men, Women and Children wore tiger teeth round their necks, to which they ascribe talismanic power.

I possess two canine teeth of a Jaguar, which had been killed only a few days previous to our arrival, near the village, which measure three inches and half in length, and three inches in circumference, round the thickest part of the root. As there are no Cattle within eighty or a hundred miles of this spot, they must naturally live entirely upon Aguris, Peccarys, and Deer. The Macusis pretend that they attack man, and told us many wonderful stories of hair-breadth escapes.

We saw several baskets of Brazil Nuts in their huts, and were told that they were very numerous about Vindaua. It appears that tree prefers stony soil, and a moderate height above the sea. I have never seen the Bertholletia at a greater height than 1500 to 2000 feet above the sea, and scarcely lower than 400 feet (on the banks of the Guidaru). The

[1] According to the reference map, a branch of the Takutu, called the East Takutu, rises on the northern slopes of the Serra Vindaua, and another branch, the South Takutu, on its southern side.

[2] See Maximilian, Prince of Wied (1782–1867), *Beiträge zur Naturegeschichte von Brasilien*, 5 vols, Weimar, Im Verlage des Landes-Industrie-Comptoirs, 1825–33, II, pp. 361–79. The *Felis pardalis* is more commonly known in Guyana as the ocelot.

distribution of this useful tree is however extensive. It is found on the banks of the Amazon, the Orinoco, the tributaries of the Essequibo, and as far east as Berbice. Its Geographical range is very likely between the meridians of 57° degrees and 68° west of Greenwich, and the parallels of 6° South, and 4° North; the most northern parallel where I have met it was the third.*

* Near Pukassanti; a Carib Settlement on the banks of the Guidaru in 3°4′ North Latitude.

May 5th. Our number having been increased by the lean Captain, who was to form our guide, and four of his attendants, we started this morning at eight o'clock.

We had to pass the settlement of our lean friend, and had occasion to admire the neatness of his hut. In lieu of walls, it was closed in with the bark of a tree; and the interior was cleaner than the generality of Indian huts. The Utensils of the chase, Bows and Arrows, and the blow pipe were properly arranged, and we found the women had been working on some large earthen pots, which if we consider they were made without the potter's wheel, merely with their hands, deserved every praise.[1]

After it had been settled by means of the Cudgel, and the shrill voice of the Squaws, which dogs were to accompany their master, and which to remain at home; we continued our march, and entered soon after thick forest, which we did not again exchange for Savannahs until our return. Our course was mostly in a South east direction, through a forest, very likely not trodden before by European feet.

As it was merely a hunting path, its traces were soon lost, and we had to trust to our guide's Indian sagacity to wind his path through the wood, which excelled in tropical exuberance. I observed several trees in our path, which I had not noted before; amongst others a gigantic tree with excrescences like the remarkable Yaruri or Massara,*

* The Yaruri is a new species of Aspidosperma (Mart).

the trunk of which from five to six feet in diameter, and to the first branches, perhaps fifty feet high, has the appearance of being fluted, or as if it consisted of numerous slender trees. The present species differed only in its trunk being not so much fluted, and the tabular projections of the lower part being much larger than those of the true Yaruri. The Macusi called it Itsha, and if the seeds which I collected at its foot, belonged really to it, as I was told by the Indians, who by the bye eat them after being boiled, this tree belongs to a different genus and a different order than the Yaruri, which Martius ranges under Apocynaceae.[2]

We followed the dry bed of a river. Among some leaves which the wind had thrown down from the neighbouring trees, and which were being heaped up at a bend of the river's bed, I observed a yeast-like matter; curious to see what was under it, I pushed my stick into it, when a hollow grunting noise issued from it. To my astonishment I was told, that it came from a frog. How the animal makes this whitish foam, which would have filled half a bucket, and for what purpose? remains a riddle; there was no spawn

[1] The standard pottery technique in the region was, and where practised still is, by coiling. Long sausages of clay are made, these are coiled round on top of each other, and then moulded together by hand and smoothed over.

[2] The genus *Aspidosperma* belongs to the family Apocynaceae. There are numerous species of *Aspidosperma* reported from Guiana. According to Mennega et al., one of these is called Musara in Carib, two of them are called Yaruru in Arawak and Warao, and Yarula or Paddlewood in Creole. See Appendix 2. It is not known which tree Schomburgk is referring to here.

visible; and the yeast-like matter, or foam might have been compared to anything else, but to the whites or mucus in which their eggs swim after the spawn is emitted. I was really sorry that I did not succeed in procuring the frog, or at least seeing it, as all our search among the dry leaves proved useless. The Wapisianas called it Pari, and the Macusis Truwé.[1]

We passed several wild bees' nests constructed in hollow trees; these insects have no sting, and their honey or Mapa, as I observed previously, has an acidulous taste. If some unlucky passer-by knocked against the tree or their entrance, which is frequently funnel shaped, and constructed with a sort of mortar made of earth, they appeared in legions, and settled upon us, and although they do not sting, they prove very troublesome, chiefly if they get in the hair which they seldom fail to do. What a scampering over bush and stick, where by accident or in frolic, the winged colony had been disturbed! The Indian runs generally on such occasions with his head bent downwards, and tries to get out of their reach in the quickest manner possible. We attempted to follow his example but generally knocked our heads against the branches, which the other knows carefully how to avoid.

We had been marching four hours without finding any water; the discovery of a Hog Plum tree (Spondias lutea)[2] with numbers of ripe fruit, was therefore highly welcome. It appeared a flock of Peccaries had been regaling themselves when our arrival must have driven them to flight; some of the hunts-men followed, but returned unsuccessful. The Hog plum tree is by no means scarce in Guiana, it forms a high wide branching tree, the wood of which would prove useful.

We stoped at four o'clock near the banks of a dry river, much fatigued, and thirsty to the extreme; We had not found any water in the course of our whole day's march; and although the Indians were digging holes in the ground, it was without success.

The followers of our guide, allowed themselves scarcely any rest before they were assiduously occupied in erecting for their Chieftain a hut of palm leaves. One acted as Cook, and grilled upon a stick, part of a Powis, which had been shot that morning. A plate of delf, no doubt in the possession of which, our guide prided himself not a little, was then carefully taken out of the basket or rutha and put before him. He dined in solitude; only after he had finished, and the plate had been carefully put away, his attendants eat the remainder.

May 6th. Fortunately a short hour's walk brought us this morning to the Takutu; and although the water appeared stagnant, and was covered with a greenish film, our thirst was too great as to prevent us from drinking of it.

The bed of the Takutu was only ten or twelve feet in width, and its waters were merely collected in pools, without flowing in an uninterrupted stream. The colour of the water is almost black, from which circumstance the Wapisianas call the Takutu, Butivan-uru or black river. It would certainly be a misnomer at its lower part, where the water, before it is joined by the Mahu, has almost a blueish colour. Its course through the ochreous & clayey Savannahs changes its colour to a muddy white, and during the dry season to a blueish colour. I have not been able to ascertain the origin of the name Takutu; may be, it is synonimous to black river.

[1] The frog in question may be one of the Guyanese species of Leptodactylid which are known to produce a foam as a form of protection for eggs and larvae. I am grateful to Dr Barry Clarke, of the Natural History Museum, for this suggestion.

[2] *Spondias mombin*.

We followed the bed of the river for several miles upwards. An Indian who mounted a high tree, saw some small hills in the direction of N.80°E. distant a few miles, but no mountains were visible. A small stream, now perfectly dry, joined the Takutu from the right bank; and shortly after another from the left, likewise dry. After the junction of the latter the Takutu dwindles to a rill, bordered by high trees, and thickets of wild Bamboo (Rappu in Macusi). The Vault of heaven is scarcely visible through the arches which the Bamboo, and numerous branches of trees, form along its banks, and consequently every Astronomical observation is rendered next to an impossibility. I recollected however a fine rocky platform, which we had passed, and where the river was wider than generally,

[*margin*: The platform consists of Spec^{s} N° 31 and N° 32 is lieing in large blocks upon it.]

being extended by the passage of the rocks. We returned to it, as the most eligible place for my observations. The place was mean while cleared from all underbrush, and a fine tree having been selected, the mark of the survey was engraved upon it, and the sources of the Takutu claimed in the name of her Majesty.

This spot is in 1°5′N. Latitude,*

*This is the mean of the meridian passages of several north and south stars, but not the mean of the circum-meridian altitudes, which have not been calculated as yet.

and nineteen miles west of Pirara.

The Takutu flows from here North-east, and receives from Mount Vindaua (Approximately in 1°55′N. Latitude) a tributary of nearly its own size; it adopts then a north western course through Savannahs interspersed with wood, and having passed the Tuarutu mountains to the east of that chain, it receives the waters of the Watuwau, a river nearly of its own size. Its course is now a point or two to the east of north through bare Savannahs and its tributaries merely Savannah rivulets until it is joined by the Mahu in 3°35′ North latitude and 24 miles West of Pirara; After the conjunction of the two rivers, they adopt a south western course, and receive the river Zuruma or Cotinga at their right bank. They ultimately fall into the Rio Branco a few hundred yards above the site of the Brazilian boundary fort San Joaquim in 3°1′46″N Latitude.

I calculated the whole course of the Takutu from its source to the junction with the Rio Branco at 200 miles. The last 50 miles it makes a retrograde course (namely Southwest its course having been north) towards the Rio Branco,

May 7th. Starting at an early hour, the return path was much earlier accomplished than our way thither, and I found yet an opportunity to determine on my arrival at Maripa the difference of longitude between our Camp at the sources of the Takutu, and that village by horary angles of the sun; only twenty four hours having elapsed since I had taken similar angles at the Takutu; the difference of longitude (21^{s}.58 in time) can be the more depended upon.

The intensity of the magnetic action manifested itself at Maripa, by one hundred oscillations in 2^{m} 51^{s}.27 at 88°F. by needle L*(a)* and in 3^{m}.56^{s}.89 at 85°.8F. by needle L(*b*).

The thermometer under our tent kept generally in the morning at 6 o'Clock at 70°F., and rose between two and three o'Clock to 98°F., it commenced then to sink gradually to 84° at 6 o'Clock in the afternoon, and stood from 72° to 74° at nine o'Clock in the evening.

Figure 7: Near the source of the River Takutu. The figure seated, wearing a hat, is almost certainly Robert Schomburgk. Watercolour by Edward Goodall, reproduced by permission of the British Library (Add. MS 16939/10).

The latitude of Maripa (mean of meredian altitudes) was 1°54′ 37″N., its difference of longitude, 24 miles West of Pirara.

May 8th. I found myself under the disagreeable necessity of leaving the Coxswain at Maripa, until he considered his foot so far restored that he could follow us. Being amongst friendly Indians, and amidst plenty of provisions, I considered it a much more prudent plan that he should await the cure of his sprained ankle, than commence the tedious and fatiguing march we had before us.[1]

Our guide to the sources of the Takutu, the one-eyed Captain promised to accompany us with several of his followers as far as Tenette, which we gladly accepted, as it was requisite to take from here, and Tuarutu, the provisions for the whole of our return journey to Pirara; we knew from experience that we could not procure any provisions from Tenette.

Our party amounted to nearly fifty individuals, including Women and Children, besides a number of dogs, which would have out-numbered many a pack of a foxhunting Squire. There were twenty five of the Canine Race, which when they commenced in unison their barking noise, was enough to make one momentarily deaf.

The finest among their number was however a dog from the Taruma nation, with which I was so much pleased, that I induced the Owner to part with it for a gun. Its name was Jewanana. The Woyawais and Tarumas are considered the best trainers of dogs; and these animals constitute a kind of merchandise or Article of barter, between them and their neighbours. This dog was of an uncommonly large size; and as it appears a well marked variety, I have added below a more detailed description; I have only to regret that it was emasculated, a custom with the Tarumas follow, under the supposition that they grow fat under it.*

*The Taruma dog: well formed something between a greyhound and pointer; head broad, muzzle long, moderately pointed, ears erect, neck long; tail long, carried erect, slightly arched, the hair on it short; the chest rather narrow; the belly drawn up; eyes brown, pupil black; prevailing colour white with large black patches, ears and head black round the eyes towards the angle of the mouth; point of muzzle white; root of tail black, otherwise white to the point; hair short. Its length from the tip of the nose to the point of the tail 4 feet 3.3 inches; of which the tail was 1 foot; height 1 foot 11 inches; girth of the body 2 feet; girth near the hind feet 1 foot 6 inches; muzzle from the superior angle of the head to the tip of nose 9.3 inches; length of ears three inches; space between ear and ear 4.7 inches; circumference of the muzzle just below the eyes 8.7 inches; fore foot from head of shoulder 1 foot 10 inches; hind foot 2 feet. The dog from which the measurement was taken might have been two years of age.

We reached our old quarters at the Tuarutu mountains at eleven o'Clock in the morning of the 9th of May; and as I considered it necessary to ascertain here the position of some of the neighbouring mountain groups by trigonometrical operations, and to provide myself with a larger stock of provisions, our departure was delayed until noon the 11th of May.[2]

[1] The coxswain, Henry Chessman, made his way back to Pirara to arrive there only a few days after the main expedition. Neither he nor Richard Schomburgk, who suffered from constant bouts of fever, made it to the sources of the Takutu.

[2] Although Schomburgk does not mention him, Hamlet Clenan was found here having made a full recovery from his ordeal (see Richard Schomburgk, *Travels*, II, p. 75).

May 11th. I resolved to cross from the Takutu to the Rupununi, while Mr Richard Schomburgk, who was then suffering under a tertian Ague, returned with his party direct to the Cursato Mountains. We had marched about four miles over Savannahs interspersed with copses of wood, among which I saw numerous specimens of the beautiful white orchidea, which I found first near the Watuwau; when We crossed after one o'clock the river Warimiwau, just a little below the junction of the Paipaitshiwau. I discovered here a tree with large purple fruit which resembled in taste our common Cherry with a slight acidulated flavour. The Indians call it Zuruaku, and I would have considered it a Eugenia, or a genus allied to it, if the presence of stipulae did not argue differently; its leaves are opposite, there are small stipulae at the base of the leaf.

The further we advanced to the east (our course having been mostly north east) the scarcer became those copses of wood which had hitherto almost prevailed over the Savannahs; and about three o'clock, we entered again open Savannahs, with only here and there a Curatella tree, or some Malpighiae and numerous Mauritia palms. The highest summit of Tuarutu bore then N.43°W. and a small mountain chain called Tshuna, along the western foot of which we understood the Takutu was flowing, bore E.b.N.

It was nearly six o'clock before we reached the Takutu; here about eighty feet wide, and flowing over numerous blocks of granite. We found it not difficult to step across without wetting our feet, so numerous and large were the rocks which impeded its course.

May 12th. After I had taken a set of horary angles of the Sun for the determination of the difference of longitude and engraved upon a tree near its right bank the mark of the survey, we continued our march.

Our huntsmen had brought in the morning a couple of deer; these were the first, which they had procured since we left Au'uruparu. It is remarkable that at the Tuarutu, and Ossotshuni mountains, the deer of the Savannah is not to be found, although there is Savannah ground between them.

The Savannahs consisted now of undulating ground, the height of the summits being from forty to fifty feet; they are covered with fragments of angular quartz rocks, which rendered our path very fatiguing. The Macusi Indians call these rocks Wata-Yeku, and the small nodules of Clay Iron-stone which sometimes replaced the other, Mari-Yeku. These nodules are sometimes pisiform*

*Geologl Specimens T N° 37.

and have a black shining surface. A species of grass, which is very common in the Savannahs, is called "Vannah" in Macusi, and the spots which [are] covered with [it] "Vandai", whether this is the origin of our adopted word "Savannah" I will leave the Etymologists to decide; however, since we know it to be of American extraction, the coincidence is strange. The word "Ite" expresses Savannah of any description.[1]

We experienced at noon one of the strongest whirlwinds I had witnessed upon Savannahs; it whirled leaves, sticks and whatelse it could seize, several feet in the Air and raised a columm of dust of two or three hundred feet in height. The Macusis called it "Uranau", and I understood they are very frequent here. The peculiar position of the mountains may give rise to them. The column of dust took a S.S.E. direction. The way

[1] The *OED* only states 'perh. a Carib word'.

in which the Macusi pronounces the word Uranau reminds me of "hurican," likewise an adopted word of our European languages.[1]

A poor goatsucker attempted to escape its influence, but being seized was whirled for a considerable distance in the Air. I had been frequently amused during our monotonous walk over the Savannahs with the strange manners of the Caprimulgi or goatsuckers. When they saw our large column approaching their resting place; they generally cowered down, and ducked their heads to pass unnoticed. On approaching them nearer, they watched every step, and according to the side which one turns, under the supposition that the bird in its stooping position cannot look around, it turns so swiftlv that the Indians say it has a pair of eyes at the back of its head. Only when within a few feet, it takes to the wing. The Indian has the greatest superstition with regard to this bird, and would not kill it for any price; they say it keeps communication with the dead, and brings messages to their conjurors. Even the Common people on the Coast share in a great measure this superstition, and hold the bird in great awe. Its nocturnal habits, the swiftness and peculiarity of its flight, and its note which breaks the silence of the night, no doubt have contributed to the fear which Indians and Creoles have for the Wacurai or Jumpy bird.[2]

The small chain Tshuni,[3] consisting of hills from 400 to 450 feet high bore this evening from our Camp, South, distance about six miles.

We had stopped rather at an early hour; as our guides told us we should find no water before darkness came on, if we started further. It is advisable at all times to pay attention to what the guides observe with regard to water, as I know from experience, that anxious to continue the march, I did not mind once or twice their advise, and had to pay for my imprudence by suffering the horrors of thirst.

Our tent was erected near a small river, and in the vicinity of some large pools of water, cold and nice of taste. These natural reservoirs are a great blessing for those who have to traverse the arid Savannah, where the traveller is exposed to the direct rays of the Sun, and suffers the more of thirst. Numerous temporary huts are therefore generally in the vicinity which are from time to time repaired, or renewed as the circumstance may require, by those who pass the road, and use them as night quarters. Our Camp was this night in 2°19′N. lat. and about 5 miles W. of Pirara.

The Thermometer stood, at 2 o'Clock in the afternoon in the shade of the tent and surrounded by trees at 96°F., at three o'Clock at 87°F., at nine o'Clock in the evening 79°F. The Stars were clouded and undefined.

May 13th. We broke up our Camp and started at 6 o'Clock; the morning was clouded, and a strong breeze (force 5 to 6) blew from the N.E. The Thermometer 71°5. The hills continued to change by turns with low ground; and rendered our march very monotonous.

These Savannahs present many inequalities; they consist of ranges of hills, more or less of a North and South direction, and the aspect of numerous Mauritia Palms convey the hope to the weary traveller, when seeing them in the distance, to meet at their foot a refreshing drink of water; or he fancies to see it flowing in the gullies on the sides of

[1] The word 'hurricane' is recognized by the *OED* as being of Carib origin.

[2] This is a secure reading, but it should probably be 'Jumby' or 'Jumbie', a word denoting the comical or magical qualities of an object (*DCEU*). In the published version it appears as 'Sumpy' (p. 67).

[3] Perhaps Mount Shuna.

the low hills, which he observes to have been torn by torrents of water; in lieu of which the ground under the Mauritia is as dry as the surrounding Savannah, and the gullies present only red ochreous clay, denuded of grass, and covered with the "debris" of quartz rocks.

Our monotonous march was interrupted by a fox hunt; the dogs started a Savannah fox (Waé-ré in Macusi) and gave them a famous chase. They followed rennard under full cry; the Indians shouting and encouraging their dogs; when overtaken, he defended himself with great obstinacy.

That fine large dog "Jewanana" seized him by the neck, and worried him, and he was lieing apparently lifeless on the ground, when unexpectedly he jumped up, seized the nearest dog by the nose, which under most piteous howling tried to disengage himself from such an attack, but in vain; until the master of the dog killed the Waé-ré with a stick. The fox of the Savannahs is somewhat smaller than the European fox, it is of the colour of our hare, whitish under the throat, and along the back to the tip of the tail, which is pointed black, runs a black line; the feet are darker below the knee-joint; the neck below the ears is of a reddish brown; the head from ear to ear lighter; along the snout a little darker; the eyes are brown, with a dark blue pupil.

I have given here this description as the Waé-ré appears to be the same species as Azara's Agouara guazon (the Canis jubatus of Desmarest.)[1]

We had reached the highest ground between the two rivers – the Takutu and Rupununi – at half-past eight in the morning. I considered it one hundred and fifty feet above their level; and as the distance of the Takutu is from here 12 miles, and of the Rupununi merely 6 miles, it will be observed that the rising of the ground is more gradual from the Takutu, than from the Rupununi.*

*I ascended some of the hillocks in the neighbourhood and took a number of angles. It appeared to me that the line of division between the basins of the two rivers runs here in a N.N.W. direction.

Our course was N.56°E.; at the horizon in the E.N.E. I observed the Granitic hills Tamboro,[2] on the summit of which I was in March 1838, and found thus an opportunity to connect the present survey with the former. Further eastward we saw the Carawaimi mountains,[3] their western declivities inhabited by Atorais, the eastern by Tarumas. We entered at noon the Wapisiana Settlement Cau-urua, consisting of five huts. The inhabitants were not numerous, scarcity likewise prevailing at their settlement; and the greater number had gone visiting those of their friends who were more plentifully provided with the staff of life. A dwarfish old man, the smallest Indian I have seen, made us a long speech, which Sororeng interpreted as conveying to us his regret that they could not give us a supply of provisions; but as we had enough, and did not intend to stay longer than the next morning, I told him to console himself in that regard. Towards evening several others of the inhabitants came in, who had been fishing, but with little

[1] See Félix d'Azara, *Essais sur l'histoire naturelle des quadrupedes de la Province du Paraguay*, 2 vols, Paris, Charles Pougens, 1801, I, pp. 307–16, and Anselme Gaëtan Desmarest, *Mammalogie, ou description des espèces mammifères*, 2 pts, Paris, Veuve Agasse, 1820–22, pp. 198–9. It is probably *Canis thous*.

[2] Tamboro Mountain.

[3] Marudi-Karawaimentau Mountains.

success as it appeared. Their whole body was painted black, some of the figures, exhibited labyrinths, others grecques.*

*The Macusis call these figures imeun-casa, those however which we find upon rocks and appear to be the handicraft of an extinct nation, "ta'-emongkong."

Among the females was one who distinguished herself by her fine figure and handsome features.

We had erected our tents among a number of Parica trees (Mimosa acacioides Benth.)[1] the seeds of which are used by several tribes of Indians along the Amazon and Rio Negro, (namely the Uaupes, Puros, &c.) in the same way as the Otomacs and Guajibos at the Orinoco use the bean of the Acacia niopo. They are pounded to powder, burnt, the smoke inhaled; or the powder is put into the eyes, and ears, which produces a state of intoxication, bordering on madness which lasts for hours, and during which time Indians have no command of themselves or of their passions.*

*Compare Journal of the Royal Geographical Society Vol. X. Humboldt's Personal Narrative Vol. 5 p. 662. It appears after all that the chemical properties of the Mimosaceae are the stimulant powers, which exhilarate, and inhaled in superfluity maddens the Indian, although Baron Humboldt doubts it, and ascribes it to the effect of the Calcined lime which the Otomacs mix with it.[2]

Fowls are the only animals which the Indian of Guiana domesticates, and has in large numbers around his hut; but he raises them only for his diversion, as he makes neither use of their eggs, nor of their meat; among those which we saw here, struck me the greater number perfect white ones, and I admired particularly the beauty of two Cocks, of pure white and of large size.

Towards evening I walked to the Rupununi which in an east north east direction, is about a mile and a half distant from Cau-urua. I found it dwindled to the size of the Pirara; its water was dark coloured, and its bed studded with rocks. The Indians said that its source was distant from here, in a south by east direction, about a day's journey, and that it had its source upon the Savannah among Mauritia palms. It is remarkable that the water of the Takutu, and Rupununi is at their upper course black, and at the lower whitish; the same may be observed of the Demerara, and I have little doubt even the gigantic Orinoco has black water near its sources. The riddle is far from being solved according to Physics to what this property is to be ascribed; which I have traced in all rivers of Guiana near their sources. Baron Humboldt is inclined to restrict this peculiarity to the rivers between the parallels of five degrees of north latitude and two degrees of south latitude.*

*Personal Narrative Vol. V. p. 188.

The waters of the Demerara, the Barima, &c., in a far more northern latitude, are as black as those of the Rupununi, and Takutu near their sources; it would be remarkable if that property, were only peculiar to Guiana, taking that territory in its greatest extent, and including the former Spanish, and Portuguese Guiana.[3]

[1] *Anadenanthera peregrina*.

[2] Schomburgk ('Report of the Third Expedition into the Interior of Guayana', p. 182) was right and Humboldt wrong. Some species of the family Mimosaceae contain powerful hallucinogens.

[3] Black water, which occurs all over Amazonia, is usually the result of dyes released by decaying vegetable matter.

I found the latitude of Cau-urua 2°28′25″N. and the difference of longitude about 1 mile west of Pirara. The direct distance between the Takutu and Rupununi is therefore in a S.W. direction 20 miles.

The Thermometer showed under the shade of the Parica trees, at three o'Clock P.M. 90°.3 F. at seven o'Clock 80°; and nine o'Clock 77° F. on our departure next morning it showed at 6 o'Clock, 73°.5 F.

May 14th. We crossed the rivulet Cau-urua – from which the village has received its name – and took the pyramidical summit of Mamette as our guide over pathless savannahs. The Settlement having been only lately erected their paths of communication were not as yet established with the neighbouring Villages. Saeraeri, and the strange rock Dochlopan formed again the striking object in the landscape, and the dome shaped mountains of Cursato rose in the western horizon, and were greeted as old acquaintances. We crossed, at ten o'Clock, the creek Canaru[1] which flows into the Rupununi.

The fragments of Quartz which covered the Savannahs appeared more numerous, than on the previous days, sometimes huge blocks of the same formation pierced the soil. Our path was fatiguing in the extreme, and our feet blistered and injured by the sharp-pointed rocks. The greater part of us Europeans, were either without shoes, or that necessary article was in such a condition, that it did not afford more the desired protection against the Quartz rocks.

I found sometimes specimens of a motley appearance,*

*V. Geol. Specimens T N°. 34 a to d.

at other times it had a violet cast, like Amethyst; the white quartz possessed considerable transparency; however, specimens like the Scottish Cairngorum appear to be entirely wanting in Guiana. The semi-translucent rose-red and milk-white Quartz, are the most prevailing.

We arrived at half past one o'Clock at a Settlement consisting of two huts, built on the declivity of the Pinighette Mountains. The highest summit of this little chain, is about 8 or 900 feet high, and of a pyramidical shape. The huts had such a slovenly appearance, that I gladly returned to the foot of the mountains, and ordered a place to be cleared among the high grass for the erection of our tents. While thus occupied, I saw one of our men and an Indian starting backwards with horror depicted in their faces; a large rattle snake which the Indian saw coiled up under a bush, they were just on the point of cutting down, was the cause of it. Mr Goodall had been sitting near the bush for some time, unaware in what dangerous neighbourhood he was.

We had met only an old woman with some Children at home, however towards evening, several others arrived from their field. I recognised among them with pleasure some of the people who had been taken at the slave hunt in 1838, and whom I released from Brazilian Slavery while passing Santa Maria in the lower Rio Branco in April 1839;*

* V. Journal of the Royal Geographical Society Vol. X.[2] Those whom I brought with me, consisted of two old men, five women and two small Children; a party of seven were to follow one the next day, and these sixteen individuals are the only ones, who out of the number of forty I saw carried away, ever returned to their home.

[1] Kunaruwau River.

[2] 'Journey from Esmeralda', p. 263.

they recognised me as soon as we met and showed me their gratitude by many a friendly token. I was sorry to learn from one of the men, that his wife and five Children who had been taken in August 1838, were still retained in Slavery at Pedrero on the Rio Negro, and he begged me to use my endeavours to see them restored to him.

The traveller, who passes these regions, meets too frequently tokens of the cruelty of the Brazilians towards the natives, committed in the execution of that horrible practice "man-stealing" against which the Holy Bible pronounces the penalty of death,+

+Exodus XXI verse 16, Deut. XXIV verse 7th.

but which nevertheless is carried on through the whole Brazilian Nation. M. Auguste de S^{t}. Hilaire[1] in his late travels in Brazil makes some pithy remarks, which I cannot pass without quoting, as it proves the extent of this barbarism, and shows how energetic our philanthropic Government ought to go to work, to give at least to those of the Indians who are settled within the British Boundary, full protection against Brazilian Atrocities.

"The children of the Botocudos have become a kind of merchandize; abusing a law which gives six years of the natural life of the natives to those who take them for the purpose of civilization, mule drivers proceed to the frontier of Jiquitinhonha, and for an axe, some sugar, or a little Brandy, induce the Parents to part with the Children, under the promise to return them when instructed in our religion, and when they knew how to work. These unfortunate Children are conducted out of their country by their Barbarian purchasers, and resold for fifteen or twenty millereis in the different villages of the province. Then occurred in Brazil, what happens on the Coast of Africa; tempted by the price which the Portuguese paid for the Children, the Chieftain of the Botocudos waged war upon each other for the purpose of having Children to sell".[2]

The Portuguese pretend that it would prove impossible to civilise the Botocudos, unless they were distributed amongst the Colonists; but without speaking of the great injustice and barbarism of such a dispersion, which would only be a reestablishment of Slavery among the Indians, it is evident that such a measure would only tend to extinguish their tribe ("Voyage dans les provinces de Rio de Janeiro et de Minas Geraes, par Auguste de St. Hilaire, Paris 1830 Tome II p. 218.").[3]

How much more cruel is the barbarism of those who in a great measure have depopulated the Takutu, Mahu, and other neighbouring rivers! While those of whom M. de St. Hilaire speaks in his journey, only conducted Children in Captivity; those of the Rio Branco, and Rio Negro spared neither sex nor age, and tore forcibly the dearest family ties asunder. Do not think European reader, who prides himself to stand upon the acme of civilization, that those ties are less tender among the "Stoic of the wood, the man without a tear".[4] I have seen the anguish of the father, the despair of the mother, and thought that every tear of the latter; every curse of the former, were registered by the Avenging Angel!

[1] Auguste François César de Saint-Hilaire (1779–1853), French botantist.

[2] Auguste de Saint-Hilaire, *Voyage dans les Provinces de Rio de Janeiro et de Minas Geraes*, 2 vols, Paris, Grimbert et Dorez, 1830, II, pp.145–6. The Botocudo Indians lived in Eastern Brazil in the inland region approximately behind Rio de Janeiro and São Paulo.

[3] Saint-Hilaire, *Voyages dans les Provinces*, II, p. 219. Although this passage is not in quotation marks, it is a word for word English translation of the original French.

[4] 'A Stoic of the woods – a man without a tear.' Thomas Campbell (1777–1844), 'Gertrude of Wyoming', Part I, stanza 23.

The appearance of the sky had changed. Although periodically clouded for the last fourteen days, it changed now to a uniform purplish grey, and portended the approach of the rainy season. I procured some horary angles of the sun in the afternoon; but my wish to determine the latitude could not be executed, the sky being too cloudy.

The water which flows from the south western side of the Pinighette mountains, goes into the Rupununi. On the north western side has the Sawara-auuru its source, which, as previously observed, affords a Portage between the Takutu and Rupununi. The direction of the small chain is about N.E. by N. and S.W. by S., and their extent two miles.

The Settlement afforded a pretty prospect over the Savannahs, and I noted with surprise the natural Pyramid of Ataraipu, which even at that distance overtopped the mountains in its neighbourhood, and was a striking object in the landscape. It bore nearly N.E. by E., and was about 45 miles distant from Pinighette.

May 15th We passed between the Pinighette and Manette mountains, and now traversed comparatively level savannahs. The small river Sawara-auuru turns towards Saeraeri.

We halted at the rivulet Paiwiyau[1] or Duck River, where we refreshed ourselves with a delightful drink of cool water. The mountain Duruau, the highest summit of which may be 2500 feet, bore at half past ten N.15°E. distant a mile; this group is uncommonly rugged and steep, and forms an angle, one of its sides having a S.W by S. direction, the other standing at W.N.W. A number of groups, only divided by small passes, draw from Duruaru in a W. by N. direction towards Ursato our Cursato; Manoa is the highest next to Duruau.

We stopped near Mount Pauisette, opposite the southern point of Cursato: Although only a few miles from Tenette, we were so fatigued that we could not accomplish the distance.

May 16th We arrived at Tenette at ten o'Clock; and as we had been day after day on the march, since we left Tuarutu, it was necessary to give ourselves a day of rest. On our arrival at Tenette, I saw the impossibility of returning in our Canoes, the Takutu having fallen still more since we left that place; I abandoned them therefore, and I succeeded in engaging the necessary number of Indians, in addition to our regular crew, to assist in carrying our baggage overland to Pirara.

May 17th I repeated the experiment of ascertaining the magnetic force, and found the result differed only .92 by needle L (a) and .24 by needle L (b) from the one which I obtained on the 22nd of April.*

*Time of one hundred oscillations by needle L (a) $2^{m}.52^{s}.08$. } May 17th
Time of one hundred oscillations by needle L (b) $3^{m}.36^{s}.45$. }

Mean of Thermometrical Observations at Tenette.

Period 1842	Forenoon 6 h.	Forenoon 9 h.	Noon	Afternoon 3 h.	Afternoon 6 h.	Remarks
May 16th to 18th	73°·25	78°·67	85°	88°·33	80°·33	clouded May 17th thunder and rain

[1] Possibly Baiewau River.

May 18th It rained and thundered[1] so severely this morning, that it was half past nine, before we could start, and after an uninterrupted march of four hours, and almost under constant rain, we halted on the banks of the river Scabunk or Catu-auuru.

I observed here a splendid tree of the order Labiatae[2] which resembled Hyptis membrancea in its floral leaves, only that in this instance the floral leaves were of a fine rose colour, much larger than those of Hyptis membrancea, and were connected with the Calices; the flowers are in risps and small, and of a blueish colour.*

*Specimens of this remarkable tree have been sent to Mr Bentham, the learned author of 'Gena & Specs Labiatarum.'[3]

May 19th Opposite our Camp on the right bank were numerous plants of the singular Ionidium Itubu of Aublet[4] which diffused its delightful odour; it has an irregular corolla and the labellum is uncommonly large in comparison with the other petals; its size is nearly an inch by seven lines in breadth. This plant which the Brazilians call Poaya da praia, or Poaya bianca,[5] is estimated by the Brazilians the best remedy against dysentery; and the root is often sold as true Ipecacuanha, to which it approaches very nearly in its properties.*

*Kunth's Handbuch der Botanik, p. 564.[6] The Ionidium Itubu grows in abundance round Pirara.

According to St. Hilaire (Plantes usuelles N° 11) the inhabitants of the Rio Grande do Norte, consider the Ionidium Itubu a specific against gout.[7]

The river Takutu meandred through the Savannahs a mile or two on our left hand; there was no path over the Savannahs, and we directed our course to the western angle of the Canuku mountains. We crossed at noon the Sawara-auuru which swollen by the rain of yesterday, ran swiftly over numerous rocks, and rendered our fording it, if not dangerous, by all means apt to lose our footing upon the smooth rocks.

We ought to have stopped here, but anxious to reach Pirara before the 24th of May, we continued our March, and suffered soon from want of water. How deceiving proves in such instances a species of grass, of a light green or rather a blueish colour, which in the distance, and deceived by the mirage, has entirely the appearance of pools of water!

We halted near Mount Curatawuiburi, which forms the most western point of the Canuku Chain; the Indians found some brackish water with which we were obliged to satisfy our thirst.

[1] P. 189 ends here and the next page is numbered 180, rather than 190. This duplication continues for the last ten pages of the manuscript.

[2] Labiatae is a synonym of Lamiaceae.

[3] George Bentham, *Labiatarum Genera et Species: or, A Description of the Genera and Species of Plants of the Order Labiatae*, London, James Ridgway & Sons, 1832–6.

[4] The Guianese species is *Hybanthus calceolaria*. Ipecacuanha is a Tupi-Guaraní word for a plant with emetic properties.

[5] Poaia is another word for ipecacuanha.

[6] Karl S. Kunth, *Handbuch der Botanik*, Berlin, Duncker und Humblot, 1831. Karl Sigismund Kunth (1788–1850) took over from Bonpland the preparation of the botanical volumes arising from Humboldt and Bonpland's travels when the latter's efforts were deemed unsatisfactory.

[7] Auguste de Saint-Hilaire, *Plantes usuelles des Brasiliens*, Paris, Grimbert, 1824, no. xi, p. 4., under *Ionidium ipecacuanha*.

May 20th Started at half past six, and rounded the Canuku Mountains the northern side of which we followed now at a distance of a mile or two; we passed the rivulet Maripa outé (Cucurit river) and where we issued from the small forest of Cucurit palms which girted its banks, we passed another monument which the Brazilians have erected as a silent witness of their barbarism. Here stood once the Macusi Village "Iquari"; it was surprised by night by a descimento, or slave hunt, the houses set on fire, and all the inhabitants then in the village, were conducted into slavery; not one has returned to tell his relations how they fared and where they were led to.

Marks of the incendiary; blackened pots, and broken pottery, is all that has remained of the village, except the tradition of the misfortune, and the silent curse of those who lost their relations. One of our Indians lost his sister on that dreadful night, and I saw the working of his muscles when he related the circumstance.

The village near Ilamikipang, where I spent several days in 1838 to look for the Urari and see the preparation of that poison, was abandoned; several of its inhabitants had died, and the others had left for fear that their sojourn at that place, was unpleasing to the evil spirit; the same was the case with the Macusi Settlement Quariwaka, where at that period I found ten Indian huts inhabited; not one was left; only the bare ground denoted that there had been once habitations. However at a short distance from the site of that village, we saw three huts, built by some of the former inhabitants of Quariwaka; they called their settlement Curata-kiu.

In one of the houses we noticed the apparatus of the most famed Urari boiler of the Macusi tribe; unfortunately he had paid nature's debt; and his numerous utensils, consisting of filters, sieves made of Palm leaves and fibres, pots &c. were lieing unused in the corner of the house, in which he himself was buried. Opposite the village is a remarkable rock of granite, with a rounded piece of white quartz in the middle, which shines when the sun reflects its rays upon it, like the Kaira in the mountain chain, which bears its name from it; the Macusis call this rock Uruqua-ka-epping.

May 21st. We reached, towards evening, Awarra, having accomplished that day the greatest distance we had made during our present journey, namely about twenty miles. If it be recollected that we had to march over open Savannahs, and a path covered with fragments of conglomerate rocks, under a noon-tide heat of 120° F., it cannot he wondered that every individual was glad to sling his hammock on our arrival. As many of the Indians, who accompanied us resided at Awarra, or in the neighbourhood, it caused a great concourse of people, anxious to see their relations after an absence of two months.

We learnt here that the Military boats with a supply of provisions for the Detachment at Fort New Guinea, near Pirara, had arrived two days previous, and we were most anxious to proceed on; under the expectation of finding letters from absent friends; but our limbs refused their duty; however we started next morning before day light, and reached Pirara shortly after eight o'Clock in the morning. A large package of letters and a file of Newspapers, was a most welcome present which awaited me on my arrival. After an absence of five months from Demerara, information from Europe, or even from the Colony is an enjoyment, which only he who is transplanted in a wilderness like ours, can appreciate sufficiently.

We had been absent two months from Pirara, and although exposed to the extreme heat, and constant fatigues, no serious sickness had taken place amongst our party, if I except the accident which occurred to the Canoe-man Henry Petry, who certainly under Mr Fryer's attention had improved, but nevertheless was found still lingering and far from being perfectly recovered from his wound.

With the exception of a Thermometer which was broken, while being carried with the other baggage over a fall in the Takutu, the instruments which I had with me, did not suffer any injury and the two Chronometers (Arnold N° 6062 and Frodsham N° 389) presented a good mean rate, which gives me great confidence in the difference of longitude determined between Pirara and the sources of the Takutu.*

*The difference of longitude between Pirara and Tenette was found to be $29^{s}.13$W. by horary angles of the sun and between Tenette and Pirara or by remeasurement $29^{s}.23$W. for the determination of the latter difference I used the mean rate of the Chronometers deducted from horary angles on our departure (March 26) from Pirara on our return to that village (May 22nd).

We were sufficiently recovered from our fatigues to do every honour at our Command at this spot so distant from all civilization, to the birthday of Her Most Gracious Majesty. This was the third which I celebrated at this village, and where the twenty one guns fired by our small battery told the surrounding Indians that the Paranaghiri had occasion to celebrate that particular day to their 'hearts' content.

[****]

The military supply boats, whose arrival is referred to in Schomburgk's report, set out on their return journey to Georgetown around 30 May. Among those who went with them were the missionary Thomas Youd who was to make a visit to England,[1] Henry Peterz who, against all expectations, had survived although not fully recovered, and Hamlet Clenan, who had been dismissed for theft. Schomburgk also sent Nicholas Reuter and Hendrick Chessman to Georgetown. The latter, fully recovered from his sprained ankle, had made it back to Pirara from Maripa shortly after the main expedition had returned.

The reason for sending these two men to Georgetown was to obtain new supplies for the next stage of the boundary survey which involved tracing the course of the Surumu and then Cotingo as far as Mount Roraima and then following the Cuyuni downstream to the mouth of the Akarabisi, thus linking up with the survey conducted the previous year. It had been hoped that this stage of the boundary survey could get under way before the end of August as both the Surumu and Cotingo were full of rocks when shallow and would be much easier to ascend when still swollen from the rainy season.

The rainy season proved a fairly trying time for both the boundary commission and the military detachment. There was very little they could do to keep themselves occupied. The savannahs round Pirara become inundated which made any movement difficult. Furthermore, life in the camps was far from comfortable. Not only were insects very bad during the rainy season, but any higher ground became a refuge for many creatures, including numerous rattle snakes. The boundary commission's supplies arrived on 22 August and with them came the order for the military detachment to

[1] Youd never made it back to England; he died at sea between Georgetown and Barbados early in August.

withdraw; Brazil and Great Britain having agreed to the neutrality of the Pirara region until such time as the boundary had been settled.[1]

The soldiers departed on 1 September, and took with them Nicholas Reuter who had been dismissed for theft. Both he and Hendrick Chessman were found guilty of pilfering the supplies that they were bringing up for the boundary commission, but Chessman whose role as coxswain was vital to the success of the expedition was allowed to continue in service. Reuter never reached Georgetown as he died on the way.[2]

The boundary commission got under way during the second week in September. We have only a very sketchy account of the journey from Schomburgk. There is reference to the journal that was kept of it, but, as with the others, no trace of it has been found. Light does not record having received a report; nor is there a record of his having transmitted one to the Colonial Office. What may well have happened is that Schomburgk did not have time to prepare the fair copy he mentioned, between his arrival in Georgetown on 23 January 1843 and his departure for Pirara on 14 February, and never got round to doing it after his return from the Corentyne in October that year. Accordingly we are dependent for a description of what happened during the course of the next part of the boundary survey on a letter from Schomburgk to Light,[3] written soon after the former had arrived back in Georgetown. This is reproduced here.

Georgetown
Demerara 23d January 1843

Sir

I have the honour to report to your Excellency my arrival in Georgetown.

I informed your Excellency in my letter dated Pirara the 25th August a.p,[4] that it were my intentions to proceed to Roraima, and from thence towards the headwaters of the Cuyuni, in order to close in with the survey where I left off in July 1841.

I have this safely executed, and while the detailed relation of this journey for the first time performed by a European, belongs properly to the journal, kept during the period, I beg leave to mention here only such occurrences, as relate more strictly to the mission, for which I have the honour to bear Her Majesty's Commission.

In the plan of operations which I laid before your Excellency, before I left Georgetown, I observed that I considered Mount Roraima, and the other Sandstone mountains in its neighbourhood, of such importance, that a stay of some time would be required to determine this point with every geographical precision at my command. This mountain ridge is the culminating point of the different groups which send tributaries to the three great rivers of the northern part of South-America, namely the

[1] The agreement that a neutral zone should be declared had been reached in Rio de Janeiro that February, but it took until August, six months later, for this news to reach Pirara. The boundary line was not settled until 1904 after the two countries had agreed to submit the matter to arbitration (see Appendix 1).

[2] They had stolen from the commission's supply of wine and rum by driving nails through the seams of the crates to break the bottles, and catching the liquor as it drained out the bottom (Richard Schomburgk, *Travels*, II, p. 110).

[3] TNA, CO 111/204, Letter of Schomburgk to Light, 23 January 1843. This document is sparse on details but Richard Schomburgk provides a full first-hand description of the journey as far as Roraima (*Travels*, II, Chaps V–II), and a second-hand account, obviously obtained from his brother and perhaps from the missing journal, of that from Roraima to Georgetown via the Cuyuni (ibid. pp. 271–7).

[4] TNA, CO 111/195, Letter from Schomburgk to Light, 25 August 1842.

Amazon, the Orinoco, and the Essequibo, and here unite therefore the natural boundaries of British, Brazilian, and Venezuelan Guiana.

The river Takutu, from its sources to its junction with the river Cotinga,[1] has been provisionally assumed as part of the southwestern boundary of British Guiana (v. Parliamentary papers relative to British Guiana ordered to be printed 11th May 1840).[2]

I informed your Excellency in a former report (dated Pirara May 30th 1842)[3] that I had executed the survey of that part of the boundary which extends from the sources of the Takutu to the mouth of the Mahu; it became now my duty to continue the Survey to the sources of the Cotinga at Mount Roraima.

Our canoes having been hauled over the portage for that purpose, the party under my command left Pirara on the 11th September;[4] and arrived at the junction of the Cotinga with the Takutu, we encamped. Under the supposition that Colonel de Matheo, the Brazilian Commissioner,[5] who, as I was informed, was then expected at Fort San Joaquim, might have arrived by this time, I considered the propriety of calling on him, and to pay him my respects as a proof, that it is the desire of Her Majesty's Government to settle the question of the relative boundaries in the most amicable manner. I left therefore the Expedition at the mouth of the Cotinga, and took only the requisite number of paddlers with me, and proceeded to Fort San Joaquim, where I was received with great kindness and attention by Capt Leal,[6] the Commandant of Fort San Joaquim, but was sorry to learn that Colonel de Matheo had been taken ill on his way up the Rio Branco, and had returned to Santarem for medical aid. I left therefore the Fort the following morning, and had the pleasure to welcome Capt Leal a few days after at our Camp.[7]

The magnetic and meteorological observations which are carried on in Her Majesty's observatories on 21st and 22d September, were attended to likewise by me; the 24th of September approached therefore before we left the mouth of the Cotinga, and ascended that river towards its sources. We were only two days advanced, when I received information from Fray José, with the name of whom your Excellency is already acquainted, that he wished to pay me his respects, and desired me to stop in order to allow him to come up with us the following day. He had been travelling day and night, since he learned at his mission of our arrival, and when I considered his age, and weak constitution, I knew the more to appreciate this mark of attention. His conversation turned upon the boundary question, and breathed the wish, as on a former occasion, that every thing might be settled amicably. He showed me several letters, which he had

[1] Schomburgk identified as the Cotingo, the river from its junction with the Takutu , and considered the Rio Surumu a right-bank tributary. On the reference map, the main river is the Surumu and the Cotingo is one of its left-bank tributaries.

[2] See Parliamentary Papers.

[3] TNA, CO 111/195, Letter from Schomburgk to Light, 30 May 1842.

[4] A certain amount of the commission's property was left behind at Pirara with one of the German crew members, Tiedge, to watch over it.

[5] Colonel João Henrique de Mattos. For an account of his trials and tribulations, see Rivière, *Absent-minded Imperialism*, *passim*.

[6] Captain Antonio dos Barros Leal. For a description of him and his activities, see Rivière, *Absent-minded Imperialism*, *passim*.

[7] A description of this remarkable entertainment which included not only Leal but a number of other local Brazilians, is given by Richard Schomburgk, *Travels*, II, pp. 126–8, and is reproduced almost in its entirety in Rivière, *Absent-minded Imperialism*, pp. 142–3.

Figure 8: Valley of the River Cotinga. Picture taken from the copy of Richard Schomburgk, *Travels in British Guiana*, II, p. 144, in the editor's possession.

received from his Superior, the Vicar-General of the province, which lauded his amicable conduct towards "the English", and desired him to continue the same, and to prevent every possible rupture.

Fray José is considered to have political influence, and I know from experience, that he is generally esteemed in the province of Rio Negro.

Your Excellency will therefore observe, that I have made it my duty to keep up the best understanding between the Boundary Expedition, Fray José, and Capt. Leal, upon which I flatter myself the more, since they were both prejudiced against me, before we became personally acquainted.

We continued the ascent of the Cotinga, and passed its junction with the Zuruma[1] on the 4th of October. After entering the large chain of mountains, which is marked as the Sierra Pacaraima in maps, we were obliged to abandon the canoes for the present, and to continue our journey over land; a task not without its difficulties in consequence of the mountainous nature of our road, and the number of Indians, that were required to carry our baggage, as those people are only able to carry a trifling burden if compared with Europeans and Africans.

[1] Rio Surumu.

We found great scarcity prevailing among the Indians of these regions, partly owing to an uncommon drought, partly to an internal feud among the Arécunas in which several were killed, and which had rendered the mind of the others, and their places of abode unsettled.

I selected our chief Station at the left bank of the Cukenam,[1] from whence Roraima was about fifteen mile distant. This point was not only astronomically determined, and the situation and height of the most remarkable mountains ascertained by trigonometry, but a series of magnetical and meteorological observations were entered into, which I hope will prove of great interest. As I stood unassisted[2] during any of the scientific and geodesical inquiries, which I considered it my duty to enter here into, December approached before I had procured the necessary data for the future construction of the maps.

The scanty information which I had been able to obtain, with regard to my future journey across the mountains to the river Cuyuni, foretold difficulties of no common nature; I came therefore at once to the resolution to order the expedition to return to Pirara, and to await my arrival there.

As I speak now from experience, it would have proved nigh to impossibility, to carry our baggage and the greater part of the instruments across the mountains and through the thick forests which I have traversed, nor do I think that the constitution of the other gentlemen, who were attached to the Expedition, would have empowered them to accomplish the task. I undertook it therefore alone, and confided the command of the Expedition to Mr Fryer, the gentleman whom I engaged in consequence of his medicinal knowledge, and who has been paid hitherto out of my own means. I have the honour to enclose herewith his instructions for your Excellency's information.[3]

Mr. Fryer left our camp near Roraima with his party on the 4th of December, and I started next morning, accompanied by the Coxswain of the Expedition, three canoemen, and the necessary number of Indians to carry the indispensable instruments and our baggage. Mr. Fryer returned to Pirara, while the party under my command followed the Sandstone ridge of Roraima in a northwestern direction, in order to trace the tributaries which the river Mazaruni receives from the west, and to fall in with the ridge of mountains that divide the rivers which flow into the Mazaruni, from those which empty themselves into the river Cuyuni.

The river Carimani or Carimang[4] is the largest of the western tributaries of the Mazaruni. It is succeeded by a ridge of Sandstone mountains, from which flows the river Wénamu[5] northward into the Cuyuni.

I consider that Her Majesty has undoubted right to any territory, through which flow rivers that fall directly, or through others, into the river Essequibo. Your Excellency is

[1] Río Cukenan on the Venezuelan map; Kukenaam on the Guyanese. The expedition had crossed the divide between the Amazon and Orinoco river systems; the Cukenan is a headwater tributary of the Caroni, which flows into the Orinoco.

[2] There must be some doubt about this claim since Fryer, the assistant surveyor, was present throughout this stage of expedition.

[3] According to Richard Schomburgk (*Travels*, II, p. 221) the two parties were composed as follows: with Robert Schomburgk, Hendrick, Adams, two coloured men, Sororeng the interpreter, and some further Indians; with Fryer, Richard Schomburgk, Goodall, Stöckle, Moore (of African descent) and a large group of Indians.

[4] Kamarang River on the Guyanese map; Río Camurán on the Venezuelan.

[5] Wenamu River on the Guyanese map; Río Venamo on the Venezuelan.

well aware, that the river Cuyuni falls a few miles above the penal settlement into the Mazaruni, and both rivers after their junction empty themselves, at Bartika Point, into the Essequibo. Upon this principle the boundary line would run from the sources of the Carimani, towards the sources of the Cuyuni proper, and from thence towards its far more northern tributaries, the rivers Iruari and Iruang,[1] and thus approach the very heart of Venezuelan Guiana. These rivers are of less importance to Great Britain, but as a maritime power, the possession of Point Barima is of great importance, and relinquishing the claim to the territory, watered by the upper Cuyuni, and its northern tributaries, the Iruari or Iuruario and Iruang, Her Majesty's Government acquires additional grounds to impress the claim of Point Barima, the Dardanelles of the Orinoco, as it has been lately styled by the Venezuelans.[2] Upon these grounds, I considered it unnecessary to proceed further towards the sources of the Cuyuni, which according to Indian information are about three days' journey farther to the west, and close to those of the river Apangwau,[3] one of the great tributaries of the Caroni.

I traced therefore the Wénamu, and after a march of fourteen days (not including numerous delays which were unavoidable) from the time we left our Camp near Roraima, we reached the cataract Imaparu, from whence we had to continue our route by water. We procured here some bark-canoes, slight skifts,[4] (as the name employs[5]) made merely of the bark of trees, and which are so light that they can be carried over land, when necessity requires it. Such a navigation is always connected with danger, nor was my party allowed to pass without all accidents. One of the small canoes sunk; fortunately no human life was lost, but of the cargo which consisted of our kitchen-apparatus, and the little stock of Salt, so precious to a traveller through these moist regions, nothing whatever was saved, and I was obliged to adopt for the remainder of the journey the Indian fashion of eating.[6]

We entered the Cuyuni on the 4th of January, and after having procured the necessary stock of provisions and engaged two Indian pilots, to guide us on the river so full of dangerous falls and rapids, and numerous islands, we commenced to descend the Cuyuni on the 7th of January, and continued the survey along the right bank, until I met the mark, engraved on a tree, where I had left off in July 1841; accomplishing thus the whole line from the sources of the Takutu to Point Barima on the Atlantic Ocean.

The deprivations and difficulties we had to contend with during this journey, can only be judged by him, who has been travelling in a country so sparingly inhabited as

[1] Río Yuruari and Río Yuruan respectively.

[2] This may seem an exaggerated comparison given that just over a decade later the Crimea War was fought to prevent Russian naval access from the Black Sea to the Mediterranean via the Dardanelles, but in the last years of the 19th century Britain and the USA nearly went to war over the British Guiana/Venezuelan frontier.

[3] Presumably the Río Aponguau, but this river is a long way from the source of the Cuyuni. On the other hand certain of the Cuyuni tributaries do rise close to the banks of the Caroni. As Schomburgk states, a boundary that took in the whole of the region drained by the Cuyuni and its tributaries would have encompassed a large area of Venezuela. Interestingly enough, during the negotiations with Venezuela over the boundary in 1887, the British, at one point, did make a claim that included this territory.

[4] Presumably 'skiffs'.

[5] Presumably 'implies'.

[6] This could refer either to the fact that Indians rarely used salt in cooking or to the necessity of eating with the fingers.

Guiana. We were repeatedly put upon small allowances of Cassada-bread, and in one instance did not taste a morsel for thirty six hours; nevertheless I can almost without exception, attest the readiness and good conduct of the men who accompanied me from Demerara, and the entire confidence and willingness of the Indians who joined me partly at Pirara, and partly near Roraima.

The accompanying list of Mr. Goodall's drawings and sketches, will serve as a proof to your Excellency, how industrious he has been since I had last the honour to report proceedings; and I have great satisfaction to attest his amicable conduct in general. This refers likewise to Mr. Fryer, who as your Excellency is aware, accompanies the Expedition for the sake of his medical knowledge, and who, as his salary is paid by me, receives only a small remuneration for the numerous fatigues and deprivations, he is obliged to undergo.

It has been impossible to me to calculate the numerous elements which I possess for the construction of the maps; nor can I present at this time to your Excellency, a fair copy of the Journal which I kept during the journey.

Your Excellency will please to recollect that deprived of an Assistant, not only the conduct of the Expedition to the smallest details, but likewise all observations and scientific inquiries fell solely upon me, and occupied me more or less day and night.

Some of the instruments have suffered from the roughness of the road, but none materially, if I except three common Thermometers and one Mountain Barometer.

In order to bring the survey which has been entrusted to me, to a final and satisfactory close, it remains now to trace the country between the sources of the river Takutu and those of the river Corentyn; and to descend the latter to its embouchure into the Atlantic. This might be accomplished, as far as I can foresee, in six months from the time I leave Demerara.

While I can not help to wish most earnestly, that I may receive your Excellency's orders for the completion of the Survey of so interesting a portion as remains yet to be explored, I can not deny the knowledge of the peculiar situation of the Expedition under my command, as hitherto Her Majesty's Government has only given orders to pay half the contingent expenses, connected with so arduous an undertaking; and if your Excellency had not come forward, and advanced meanwhile the other half which the colony refuses to pay as was contemplated, the object of my mission would have failed in all commencement, in lieu of which I stand now near its accomplishment.

Comparatively a small sum will be required to effect the execution of the plan as originally matured; and Her Majesty's Government will then not only possess a firm foundation upon which to base future negotiations for the settlement of the boundary, which could not have been procured without the actual exploration of the territory, but I trust that my researches during that period have proved likewise of some interest to science in general.

If necessitated by circumstances Your Excellency should have resolved to withdraw the timely assistance, the Expedition has hitherto received from your Excellency, and thus render the continuation and conclusion of the Survey for the present impossible, I beg leave to observe that it would be necessary for me to proceed with such despatch as circumstances permit to Pirara, in order to conduct the Expedition back to Georgetown. A number of valuable instruments have been left in Pirara, and among others the reflecting Telescope which only arrived at the time of our departure in Pirara, unfortunately too late to make then the necessary use of it.

Neither Mr. Fryer, nor Mr. Goodall possesses the necessary experience to pass the numerous and dangerous falls of the Essequibo with every chance of success. There is an other circumstance which I beg leave to bring under your Excellency's consideration, and which urges my return to Pirara. I ordered the reflecting Telescope already mentioned, for the determination of meridional distances by the occultation of Stars, and the eclipses of Jupiter's Satellites. It arrived unfortunately at a period, when the rainy season, and our departure for Roraima, rendered all observations at Pirara impossible. The expense of this instrument would not only prove useless, if by proceeding to Pirara an other opportunity was not afforded to me, to fill up this great desideratum, but the series of astronomical observations would remain incomplete.

The large Canoe of the Expedition is now in Georgetown, and the Indians and Canoemen who accompanied me to the coast, are willing to return to Pirara; indeed the Indians have been promised at the time they were engaged, to be sent back to Pirara.

It is therefore my firm persuasion, that considering the necessity of returning to Pirara, the completion of the Survey as originally contemplated, would not cause a much greater expense than circumstances require me to incur in order to conduct the expedition back to Georgetown. The necessary stores, provisions, articles of barter &c for the maintenance of the Expedition for the next six months, would scarcely amount to more than twelve hundred Dollars, which sum does not include Salaries, and wages to the Canoemen, the latter of whom I intend to reduce to a Coxswain and three Canoemen.

I have the honour to be Sir your Excellency's most obedient and very humble Servant

Robert H Schomburgk

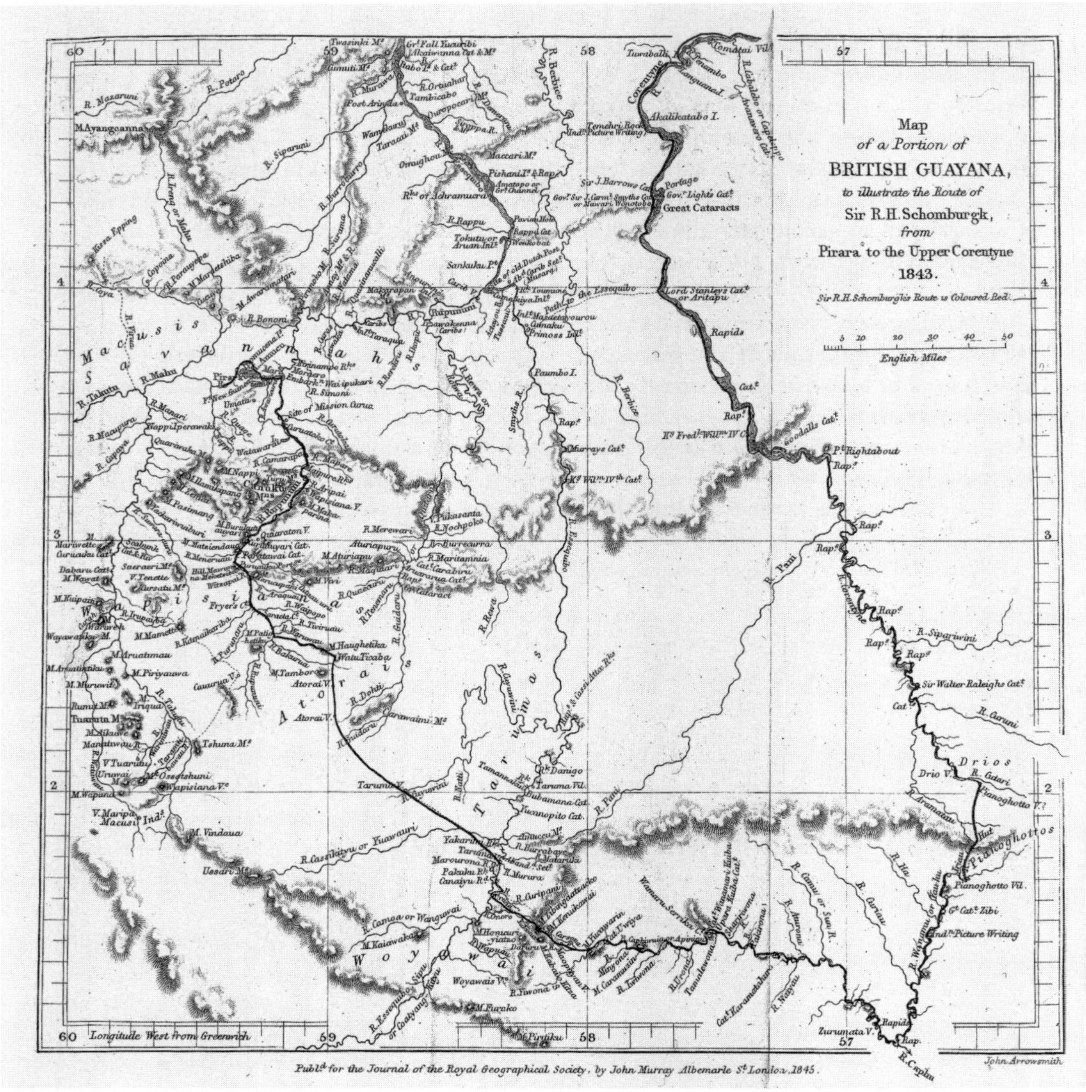

Map 3: Map of a portion of British Guayana to illustrate the route of Sir Robert Schomburgk, from Pirara to the Upper Corentyne. Reproduced by permission of the Royal Geographical Society.

CHAPTER 6

Dutch Guiana and the Eastern Frontier: 1843

Schomburgk stayed just over three weeks in Georgetown before setting out again for the interior. The reason for the haste was that there was a strong chance that the survey of the boundary between British Guiana and Dutch Guiana would be cancelled; as it would have been if the ministries in London that were involved had not been so dilatory. The Treasury had written to the Colonial Office on 30 September 1842 recommending that the survey be discontinued if the colony remained unwilling to contribute its half-share to the cost. The Colonial Office did not get round to responding until 5 January 1843 when it agreed that the Dutch Guiana boundary should not be surveyed. The correspondence between the two departments continued in a desultory manner, but nothing had been done to direct Governor Light on this, when a despatch from him dated 15 February arrived, stating that he had given Schomburgk, who had departed for the interior that day, permission to proceed with the survey. Faced with this *fait accompli* the Treasury backed down and agreed to pay.[1]

The manuscript account of Schomburgk's survey of the Corentyne is held in the archives of the Royal Geographical Society, shelfmark JMS 6/42. It is entitled 'Journal of an Expedition from Pirara to the Upper Corentyne and from thence to Demerara, executed by order of Her Majesty's Government, and under the Command of M^{r} now Sir Robert H. Schomburgk KRE. Ph.D. &c &c.' It is divided into three parts, (a), (b) and (c), all with their own subtitles, which are respectively forty, fifty-six and forty-one pages long. The manuscript is in good condition, legible and with few editorial amendments; the changes that have been made are mainly in the same handwriting as the main text and have thus been accepted in the transcription. It was received by the Royal Geographical Society on 9 December 1844, nearly six months after Schomburgk arrived back in London. It was published in the *Journal of the Royal Geographical Society* (15, 1845, pp. 1–104) under the same title and only slightly altered and shortened.

JOURNAL OF AN EXPEDITION FROM PIRARA TO THE UPPER CORENTYNE AND FROM THENCE TO DEMARARA, EXECUTED BY ORDER OF HER MAJESTY'S GOVERNMENT, AND UNDER THE COMMAND OF M^{r} NOW SIR ROBERT H SCHOMBURGK KRE. Ph.D. &c&c.

JOURNEY FROM PIRARA BY THE RIVER RUPUNUNI TO THE WAPISIANA VILLAGE WATU TICABA

In furtherance of the objects of my mission to British Guayana I left George Town on the 15 February 1843 with the intention of joining the other gentlemen of the

[1] See Rivière, *Absent-minded Imperialism*, pp. 150–51.

expedition at Pirara, to which place they had proceeded from Roraima while I investigated the regions north of that remarkable mountain chain.

I ascended the Essequibo on this occasion for the fourth time; the route therefore offered no novelty, if I except the appearance of that remarkable comet, which we observed first on the 8th March, and which remained an object of wonder alike to us and to the Indians.[1]

Without having met with any accident on ascending the falls and rapids, I reached Waiipucari[2] on the 24th March, and had the pleasure of finding my companions in excellent health & spirits after an absence of about four months.

Several of our large boats were still at the confluence of the Pirara with the Mahu and as they were required for the ascent of the Rupununi, and could not be brought by water in consequence of the rains usually expected in February not having fallen, I made the requisite arrangements for having two of them brought over land, though the distance in a straight line was twenty three miles, and this would be considerably increased by the detours necessary for avoiding high ground. Pasico,[3] the Chieftain of the Macusis about Pirara undertook to bring them which he successfully accomplished.

Before we left Pirara we had the grief to see it partly burnt down, through the negligence of a child. I fortunately rescued from her blazing hut, a poor old woman who, though she knew there was gunpowder in it, seemed determined to save a few articles belonging to her absent grandson. The remains of the former Catholic Chapel, and the houses of the late Mr Youde, next fell a prey to the flames, and it was only with the greatest difficulty that the further progress of the fire was arrested.

The preparations for our departure being completed and a number of Indians engaged to accompany us, on 30th April I bid adieu to Pirara; Nor was it without feelings of regret that I cast a last glance on the hut which for so many months had been my residence, for humble as was its appearance when judged of by the scale of European constructions I contrasted the comforts I had enjoyed under its shelter with the sufferings of the many nights I had passed under the canopy of Heaven. The ruins of the Missionary's house, near which my road lay as I commenced my journey, its blackened walls attesting the cause of its destruction, and the remembrance of its former inmate whose first arrival at this spot I had witnessed, and under whose benevolent and judicious conduct I had seen the seed of our Christian religion sown in the heart of the Savage, but not permitted to ripen into the perfect harvest before it was cut down – all combined to make me sad. The large wooden cross, which, when the Brazilians possessed the village had been erected in front of the chapel & which, as a protestant I could not view in the same light that a Roman Catholic would, still brought relief to my feelings; it spoke of hope, the hope that there might yet dawn a better day, for the poor abandoned beings who just initiated into our religion were again left to themselves and to nature.

As I proceeded to the embarkation at Wai-ipukari recollections crowded upon me; with what various feelings had I, at different times, trodden the same path, from the

[1] C/1843 D1 or the Great March Comet.

[2] Yupukarri. The Rupununi landing of the Pirara portage.

[3] Elsewhere also Basico. For a portrait, see Goodall, *Sketches of Amerindian Tribes*, p. 42.

time when I first traversed the Savannahs in 1835 in company with Lieutenant (now Captain) Haining,*

*Since this has been written poor Haining died in Jamaica.[1]

to the present moment! And had the cause of religion, had humanity been advanced during the long interval of eight years? Alas no, the ruins of Pirara, scarcely a hut inhabited, the regular paths which traversed the village during the Missionary's residence among the Macusís, overgrown with rank grass, no human being visible to greet us with his hearty Tombowai – all replied in the negative.

At Wai-ipukari, I found the bustle of preparations for our departure. The Indians from the Kanuku mountains, were all encamped round our tents, to see us once more before we departed. The boatmen were putting the craft in order, and the hoarse voice of the Coxswain was evidence of the free use he had made of it during the day. Upon examination of what had been done and what yet remained to do, it was clear that more than one day would elapse before we could start on our further ascent of the Rupununi. In order to provide for the multitude without consuming our salted provisions, hunting parties were immediately despatched in quest of game.

The Inlet, on the banks of which we were encamped, abounded in Kaimans; on land they are too timid to be dangerous, and the animal is so well aware of the disadvantages under which he labours when on Terra firma, that on the slightest appearance of man he immediately plunges into the water – once there, he is the most dangerous animal in the rivers of Guayana. I was anxious to possess a perfect skeleton of one of these monsters, and accordingly, Naripo, the Kaiman slayer, a fine Macusi who had settled near Wai-ipukari, was armed with my rifle for the purpose. He promised to do his best to shoot one before we started, and he kept his word, for the following day he arrived with the intelligence that he had succeeded; on hearing which, we all started for the sea-like expansion of the Awaricuru, near which Naripo had erected his house and shot the Kaiman. The animal was still in the water but fastened with a bush-rope or liana round the neck to a tree. With the assistance of several indians and the greater part of our own men, it was hauled ashore. Life was extinct and we had nothing to fear though when looking at his horrible jaws fringed with teeth and involuntary shudder almost came over me. Its whole length was twelve feet three inches; (the largest I have ever seen, measured sixteen feet, but there may be larger ones), the head was eighteen inches long and the circumference passing over the eyes was twenty inches, the girth of the body where most slender was four feet five inches. We placed it in a position which appeared to us the most natural, and our artist took a drawing of it, while Naripo with no small vanity related how he had shot it. He went, he said, with his little son, a boy about eight years of age, to the inlet of Kirahagh, and having tied a fowl to a long string threw it into the water as a bait. The fluttering of the bird in his attempts to escape soon attracted a Kaiman to the spot, when it received the contents of the rifle, but apparently with little effect, for though it sunk it soon rose again & made a second rush at the fowl which had been again thrown into the Water. Naripo fired a second time and little Danappé now loaded his fowling piece with slugs. The ire of the monster being raised the dainty bait was withdrawn, and the Macusi merely splashed the Water with his foot,

[1] See Vol. I, p. 44, n. 1. According to *The Army List*, Haining died in Sierra Leone on 21 July 1844.

holding firmly against a tree on the margin, when the Kaiman rushed furiously towards him, and received the shots of both the father and son. Seven discharges were necessary before they proved successful; several slugs had penetrated the skull and one of the balls had passed through the eye. The drawing finished the Kaiman was hauled on dry ground and a kind of cage work of large sticks built over it to protect it from the Vultures and other carrion-loving animals; this precaution, however, was of no avail, for when Mr Fryer returned about a month afterwards to secure the skeleton it was not to be found.

I had selected the rainy season for ascending the Rupununi as at that period I would be able to reach a higher point with our canoes than I could at any other, and the river was now rising rapidly, still were we not ready and the 3d of May approached before we could start. At length the corials being reported ready we took leave of our Indian friends. In the number were many who had accompanied me on my previous journeys when I first landed among them eight years ago; these pressed my hands, a custom with which they had since become acquainted, and gave me their Tombowai or farewell. We did not start under a good omen, if such superstitions were entertained by any of us. The Corial which I occupied, the Louisa, ran foul of a sunken stump on leaving the inlet, and one of her planks was stove in. We could scarcely keep her from sinking before we reached the bank of the river. Our Indians were however alert and, after an hour's delay, the damage was repaired and we were once more *en route*.

4th May. What a night we passed, it appeared as if the floodgates of heaven had been opened. Anxious to make every progress, it was almost dusk before we began to pitch our tents the previous evening and we had not got them up when the thunder storm commenced: All Sounds were overpowered by that of the falling rain; even the thunder was scarcely audible and announced itself only by the vivid flashes of lightning, which as they shot along illuminated groups of canoemen and indians, seeking shelter as best they could, and trembling with fright & cold. As for keeping a fire alive, it was out of the question, and our tents were no protection from the rain that fell in torrents. We were much amused by my brother who having fallen asleep in spite of the uproar, awoke half dreaming, and as the rain fell in heavy drops upon him, and had already saturated his hammock, fancied the river, on whose banks we were encamped had so much swollen in the course of the night that its waters already reached our tent, and his efforts to escape a watery grave were really ludicrous. I estimated the quantity of rain which fell that night at from 5 to 6 inches. The Thermometer stood this morning at 65 Fahr. & the wet-bulb thermometer at 64°.7, a certain indication of the Atmosphere being overcharged with moisture.

At ½ past 8 we passed the mouth of the Curotoka,[1] which a few miles higher up is inhabited by Macusi Indians. This stream meanders partly through Savannahs and has its source in the Eastern continuation of the Canuku mountains.

Two hours afterwards, we passed the first rapids which the 'Louisa' did in fine style, but the large canoe had to be hauled over by main force. At a short distance above this point is the site of the former protestant Mission.[2] While the people were occupied pitching our tents, I strolled up the stream and again had occasion to admire the

[1] Kuratoka River.

[2] Urwa. Youd withdrew here from Pirara in 1839, fearful that the Amerindians were too exposed to Brazilian incursions. See Rivière, *Absent-minded Imperialism*, Chap. 4.

beautiful Palm, the Curua of the Macusis. It appears to be a species of Maximiliana Minsignis,[1] and if so is certainly the handsomest of that interesting family which though social in its habits is only found in regions far asunder. Time had wrought great changes since the Mission had been abandoned. Vegetation in Tropical exuberance, had overpowered what man had planted, and the cotton plantations and provision fields which under Mr. Youd's direction afforded so gratifying a prospect, presented now only a few straggling shrubs of a Eupatorium,[2] numerous young Trumpet trees Cecropia[,] Phyllanthus[3] and various sedges, which being the first occupants of the soil before man had cultivated it again sprung up when he relinquished his labour. The Missionary's house and the building in which he was wont to instruct the Indian in the Christian religion & the English language, had fallen in and added the gloom of their ruined aspect to the forlorn appearance of the once flourishing Mission, and yet scarcely three years had elapsed since circumstances had obliged Mr. Youd to withdraw.

Seeing marks of recent footsteps I followed their direction through clusters of Plantain trees, which, in consequence of their rapid growth had withstood the throng of ranker vegetation and attested by their height, in some instances from forty to fifty feet, the fertility of the soil. Indeed, the Indians consider those regions where the Curua grows, the most fertile of the forest. Of the numerous indian habitations, which I had seen here on a former occasion, only two miserable huts remained and the pools of water which were standing in them, were sufficient evidence that their roofs were not water-tight. I saw no human beings in them; but some wood shavings, a gun with its lock taken to pieces to be cleaned, and a spindle with cotton plainly indicated that people had recently been there and that, alarmed at my approach, they had fled into the wood. I went in search of them, but without success, and when I returned to the huts I found there two women and a man, the former spinning cotton the latter cleaning the lock of his gun as though they had never been disturbed and evincing no surprize at my appearance they did not even look up from their work. A few words spoken to them in their own language soon inspired confidence and they told me that hearing me approach and knowing by the sound that he who came wore shoes they had concealed themselves till they had ascertained who it was. I purchased some ground provisions of them and gave the females some glass beads after which I returned to the camp.

May 5th. The first large fall of the Rupununi, which is a little above the late Mission, was passed without accident and we continued our journey at half past eight in the morning.*

*The rocks which form the fall are of compact Basalt.

The river Avaricuru by means of which and the small river Quatatta, Pirara is reached within a distance of two miles, has a second outlet about a mile above the Fall, and eleven miles from the lower (outlet?). Such divisions of a river, although frequent near the Coast, are but seldom to be met with in the Interior.

Three miles higher up and on the left bank of the Rupununi, a small channel leads to a lake-like expansion of water called by the Macusi Indians Watawarai, to which they

[1] This is not a secure reading. Perhaps *Maximiliana insignis* (*Attalea insignis*) is intended.

[2] The species of *Eupatorium* found in Guiana have been transferred to other genera of the family Asteraceae.

[3] *Phyllanthus* is a genus of the family Euphorbiaceae.

resort for fishing, as this inlet is well stocked with the finny tribe. We halted at half past three o'clock in the afternoon and were then in sight of the Canuku Mountains. Thick opaque clouds enveloped their summits and deepened the dark green tint of the gigantic trees which clothed their sides. Our tents were scarcely pitched when the clouds dissolved & the rain poured down in torrents during the whole night.

May 6th. Nothing could exceed the beauty of the morning. The sun rose partially covered by fleecy clouds, as if afraid to contend for the supremacy of the day against the dark hovering masses to the West, but when its rays fell upon the rain drops on the foliage the scene was one of fairy land. The majestic Mora (Mora excelsa) refreshed by the late rains was covered with a succession of leaves whose tints, varied with their age, passed from yellow through red to dark green. While the splendid Petrea[1] with racemes of bright blue flowers sometimes from twelve to eighteen inches long, trailed from bush to tree, and with the orange coloured Combreta[2] gave variety to the surrounding scenery.

We were now surrounded by mountains through which the river had forced its way, and proceeding onwards arrived, by one o'clock, at the small river Aripai on the right bank of the Rupununi. A settlement of Wapisiana Indians was established at a short distance from its mouth and as I was in hopes of procuring here a fresh supply of provisions, we halted at that early hour. Immediately on landing I proceeded to the Village, which is only half a mile from the river. The yell of the numerous dogs informed the Indians of my approach, and when I came in front of the first huts I found the inhabitants all assembled. An old Indian stepped forward, and greeting me in tolerably good Portuguese said he was the Chieftain of the people I saw before me, and who appeared to be all young men. One by one now came up to me and gave me their hand with a good-natured smile. The Chieftain's wife a tall woman, who although middle aged was still good-looking also spoke broken Portuguese, and dropping her courtesy seized my hand which she attempted to kiss, which I prevented.

On entering the hut I was accosted by my name, and a young female whom I remember to have seen at Senhora Librada's in 1838 welcomed me. She spoke the language well enough to inform me that dissatisfied with her employer she had left her and come to this village, the Chieftain's wife, a Paravilhana by birth being a relation of hers.

I perceived no difference in the construction of the huts of this village from those we had seen at the Macusi settlements. Some of the women were occupied in preparing a Laba (Coelogenus subniger, Desm.[3]) which appeared in such excellent order that it tempted my appetite. Three or four men were lounging in their hammocks, each of whom had a large calabash of yellow plums (Spondias lutea L. Maropi in the Macusi language) which they were so eager in devouring as to have no time even to cast a glance at me. Large heaps of firewood piled up for use in the winter, when the torrents of rain render the wood in the forest unfit for burning showed a foresight I had but seldom met with among the Indians.

The old Chieftain requested I would accompany him to his provision grounds, which were adjacent to the houses. He showed me the pride of his heart, spots of Cassada (Janipha Manihot), some, as he said six moons, others four moons in the ground. (The

[1] *Petrea* is a genus of the family Verbenaceae.
[2] *Combretum* is a genus of the family Combretaceae.
[3] Anselme Gaëtan Desmarest (1784–1838), French zoologist.

Indians divide the time from rainy season to rainy season into lunar months or moons.) There were his yams and there his tobacco plants, and arborescent cottons of a size that would have astonished a planter from the South of the United States, and attested the fertility of the soil, which was a black mould slightly mixed with sand. The old man could not sufficiently expatiate on the fertility of the soil around his village, in proof of which he pointed to the graceful Curua-palm towering high above the adjacent forest trees. We saw the trunk of a large Silk-cotton tree (*Bombax Spec.*) which if it had remained standing in his fields, its wide spreading branches would no doubt have rendered the ground under it useless. It had accordingly been cut down and the labour of the operation may he judged of when I state that the trunk was twenty two feet in circumference.

The forests about Aripai abound in cedar trees (*Icica* of Aublet), so called from their resemblance in smell & the colour of the wood to the real cedar (*Larix cedrus*), though they belong to a quite different family.*

*I consider this tree to be undoubtedly Aublet's *Icica altissima* though in one respect it differs from his description of that species, as it has from 9 to 12 pair of leaflets while Aublet says it has but from 3 to 4. He describes the white Cedar and speaks of a species which he considers to be merely a variety and the wood of which he says is of a red colour. This latter is probably the red cedar I found at Aripai but which is decidedly specificly different. It may stand provisionally as I.Camiau.

This tree often attains a height of from 60 to 70 feet and canoes made of it are considered more durable than those made of any other wood. It is called by Wapisiana Indians Camiau and by the Macusis Paranguai. The colonists esteem it much for furniture particularly for drawers as its aromatic odour keeps away the insects.[1]

I had scarcely made my way back when the whole village returned my visit. Our baggage was examined with eager curiosity and the sight of the objects we had brought for barter excited great ecstacy and the promise that all the provisions that could be spared would be immediately brought. I observed here a strange custom among the young girls, and adopted by them perhaps till womanhood; they wore their hair short with the exception of a lock, which hung from the crown of the head down towards the neck.

The following day being Sunday I resolved on remaining here. The season had set in with uncommon rigour; torrents of rain fell during the night; heavy thunder & frequent showers alternated during the day time, with sun-shine and blue sky. The river was rising, almost visibly, and the current ran, where the river turned sharply round, with the swiftness of a rapid.

May 8th On awaking this morning I found my hut under water & might have stept at once from my hammock into the Canoe, but as we had anticipated this, the baggage had been previously secured. The river had risen ten & a half feet in thirty six hours, and still continued to rise. Previous to our departure, the Indians brought us about thirty cakes of Cassada bread, for which they demanded mock coral glass beads, and others of a white colour, the latter being the most esteemed by the Wapisiana ladies; the men requested for their share knives, hooks, &c. A little girl with very pretty eyes but by no means an intelligent countenance particularly when seen in profile, appeared

[1] For a discussion of cedar trees, see p. 23, n. 1, and Appendix 2.

not to possess any provision fields to enable her to offer Cassada bread, but anxious, nevertheless to possess some of the favorite beads she brought under her arm a white hen and a number of young chickens in a small basket which she set down at my feet. I gladly presented her with what she so much desired and for which, no doubt, she would willingly have sacrificed her pet, which I returned to her with the young brood; an act of generosity so unexpected that she first looked with astonishment at the beads then affectionately at the hen and chickens, then with smiles of thankfulness at me.

The following are the meteorological observations at Aripai

Period & number of observations	Instruments	Mean	Maxima	Minima	Remarks
From May 6th to May 7th. 11 Observations	Barometer	29in.667	7th 10h a.m. 29in.706	7th from 4 to 6 p.m 29in.606	The weather very changeable; fine in the Afternoon.
	Attached Thermometer	77°.72	7th 12h 80°.78	7th 6h a.m. 72°.14	
	Detached Thermometer	77.80	81°.2	72.50	
	Wet bulb Thermometer	76.72	7th 5h p.m. 79.7*	72.10	

*The detached thermometer stood at the same time at 80°.2.

Having brought the Corials to our tents, we steered through the rows of trees; certainly a novel species of navigation. The river was, in some places running from five to six knots an hour. The Louisa was a fine boat and her crew some of the ablest men, still there were places where with all our advantages we only advanced inch by inch. The other boats had to be hauled along by means of the branches on the river's banks. Accumulating flakes of white foam came floating down the river, which at those parts where it was narrowed in by the mountains, almost covered its surface, and gave it the appearance of a northern snow-field. Great was its contrast with the verdure and the many-coloured flowers that adorned its banks, particularly where a beautiful *Combretum aurantiacum* (Nov. Spec.)[1] immerged its drooping spikes into the stream below so bright its colours that their reflected image was visible even in the muddy waters of the Rupununi. This beautiful shrub common to almost all the rivers of Guayana appears to bear a constant succession of flowers, but at this season the banks were covered with them, and together with the *Petrea Macrostachya* (n.sp) and the thick garlands of pink flowers of several species of *Bignoniaceae* presented a most splendid appearance.

In the early part of the morning the heavens looked gloomy, they presented an expanse of a dull blue tinge with here and there a few interspersed Cirro strati. At 10 o'clock the appearance of the clouds became highly interesting. Thick bulky masses of rain clouds with a blue sky intervening indicated an approaching storm & offered a fine subject for a painter. The cry of the Toucans, that sure prognostic of rain, told us what we had to expect. It is remarkable that the Buradi or Carauwui (Ramphastos Toco) and other allied species generally commence their cry on the approach of rain. If therefore, their disagreeable voices assailed our ears at any other time than in the morning or

[1] *Combretum fruticosum*.

evening, when we allowed them to set up their cry as in the regular course of things, we generally looked out for our cloaks or made other preparations to avert the threatened wetting. Our Toucans very seldom proved false prophets. Noisy as is the Buradi, it is surpassed by a species of Hawk called by the Macusi, Callau-callau; by the Warraus, Yacka-tata (*Ibycter teucogaster* Vieill.[1]) a bird of prey which appears to be spread all over Guayana, and which has been honored by the Colonists with the name of Bull-dog. It is always found in societies of ten or twelve, occupying the loftiest trees along the banks of the rivers where they act as warders. Scarcely do they spy a boat in the far distance, when they set up their Yacka-tata; a cry which when the whole company join in the alarm, proves almost stunning.[2]

The prospects on ascending the Rupununi where it has worn itself a passage through the mountain chain, are by no means uninteresting. In one place the mountains approach the immediate banks of the river and the stream bathes their feet, in another they surround it like an amphitheatre or recede at the distance of a mile or two; but no where do we find a bold cliff overhanging the water. It seems as if the river followed the windings of the valley & gliding from mount to mount without abrading any portion of their granitic foundations. Wherever the current has washed away the clayey soil large rounded blocks of gneiss are visible. At one of those spots where a high mountain turns the river boldly to the east, I enjoyed a most beautiful prospect. Not only the banks of the river but also the steep side of the hill, were almost covered with trees of that species so splendid in appearance, which I first discovered at the river Scabunk, a tributary of the Tacutu,, but where the absence of flowers prevented me from ascertaining its character; here they were in profusion.

Among those plants which bear bracts, that organ is generally situated immediately below the Calyx; but in this instance the bract sometimes occupies the common situation, but is more frequently attached to one of the segments of the Calyx, a peculiarity common to the genera Calycophyllum, Mussenda, & Pinckneya.[3] This tree belongs to a new species of the first of these, and is of singular beauty. Its large pink bracts, almost clothe the tree in that beautiful colour, and it is only upon a near approach that one can discover the shining green leaves and the spikes of small flowers of a velvety blue. Let the reader call to mind the splendid aspect afforded by our rose bushes in full blossom, then let him imagine the great garden of nature in Guayana, clothed in tropical exuberance, and then among the luxuriant productions of a genial sun & fertile soil, trees from forty to fifty feet high presenting a mass of leaves of the colour of the rose from the deepest carmine to the faintest blush, and he may then form some idea of the picture I now witnessed. The wood of this beautiful tree is extremely hard, of a yellowish brown, and very bitter, and I have little doubt possesses febrifuge properties. The Macusi Indians call it Dehpoyeh. I have already observed that it is a new species, and, according to the custom of botanists frequently to name their own discoveries after the Patrons of the Science of Botany, or other distinguished individuals, I have requested the permission of Lord Stanley,

[1] Louis Pierre Vieillot (1748–1831), French ornithologist, author of *Analyse d'une nouvelle ornithologie élémentaire*, Paris, Deterville, 1816.

[2] The Red-throated Caracara (*Daptrius americanus*).

[3] These are all genera of the family Rubiaceae.

who, during the time the plant was discovered by me, was Secretary of State for the Colonies, and by whose authority the expedition was undertaken, to make this splendid tree known to the botanical world by the name of Calycophyllum Stanleyanum.*

*The description of this beautiful plant was first read at the meeting of the British Association at York, and has since been published in Sir William Jackson Hooker's "London Botanical Journal, Vol.III.[1]

There were no palms in the vicinity of the spot where we intended to rest for the night, and as the sky threatened rain our Indians had to find a substitute for Palm leaves, with which they usually cover their temporary resting places. The huts were accordingly covered first with the branches of trees & then with a thick layer of fallen leaves, which at this period covered the ground several inches thick, and thus rendered impervious to the rain. A pair of those pretty monkeys called Marmousets in the colony (the red handed Tamarin, or *Midas rufimanus*[2] of Geoffroy[3]) gamboled from tree to tree near my tent. I was rather surprised to meet this pretty species of Midas so high up in the interior, as I had always considered their abode to be near the coast region, where they occasionally visit the plantain walks*

*A plantain walk is a piece of ground allotted for the cultivation of the *Musa paradisiaca* and *Musa sapientum* or the Plantain & Banana of the tropics.

in such number & feed upon the ripe fruit that they commit great injury. They are easily caught in traps, but seldom survive their confinement above a few months.

May 9th. It appeared the river had reached its maximum height yesterday as from 2 o'clock in the afternoon of that day to 6 o'clock this morning, it had fallen upwards of five feet. We found the large boat (42 feet in length) resting with the stern in the fork of a tree and the bow on dry land; certainly the fault of her coxswain who had comfortably taken his rest under her awning while the water was running rapidly off beneath her bottom. It required the whole strength of our crew to get the boat out of this unwonted situation.

We entered, at about 11 o'clock, between two large rocks that part of the river, where, on ascending, a continuation of rapids and cataracts commence forming obstacles to further progress; but the lovely prospect from this point made me forget the dangers and difficulties which were to be surmounted. These natural portals of gneiss form a conspicuous feature in the view. On the rivers right and from its very brink the Savannahs stretch away; towards the SouthWest rose high mountains; while the Peak Burkutuau-yari, with its granitic cliffs formed the background of the picture. The whole landscape appeared "Couleur de rose." The splendid Calycophyllum Stanleyanum gave a marked & peculiar character to the prospect.

We, in the 'Louisa' halted before noon near Mount Burkutuau-yari; the other boats did not arrive until past two in the afternoon. In the meanwhile I walked with Sororeng, one of the Indians who accompanied me to London in 1839, and who since 1837 had served me as interpreter, to a village of Wapisiana Indians, which was called Kuiraton. It

[1] See Schomburgk, 'Description of a New Species of *Calycophyllum* from British Guiana', *London Journal of Botany*, 3, 1844, pp. 621–3.

[2] Probably *Saguinus midas* which has orange or yellowish hands and feet.

[3] Etienne Geoffroy Saint-Hilaire (1772–1844), French zoologist.

is situated about a mile from the river on its right bank. Here it was that, in January 1838, on my return from the sources of the Essequibo I was detained for a fortnight waiting the arrival of one of my boats in order to join our party at the lower Rupununi.[1] I found on my arrival that no other craft was to be had but a small corial, just large enough to hold two persons. I was therefore obliged to send the Coxswain and an Indian to Curassawaka at the lower Rupununi to bring up one of the boats of the expedition. The interval of fourteen days is indelibly impressed upon my memory from the excruciating pains of Tic doloureux[2] and Neuralgia which I suffered in consequence of exposure during our journey to the sources of the Essequibo.

The huts had been removed and of the five dwellings which composed the settlement in 1838 only three were standing. The greater part of the inhabitants and among them their Chieftain was absent in the provision fields and I found only two men and a few women at the place. One of the latter appeared to have just recovered from smallpox, and another was still suffering under that baneful disease. I learned to my great regret that many of the inhabitants had fallen victims to the malady which had been introduced in 1802 from the Colony.[3]

I left a message for the Chieftain to come and see me on his return, and as there still remained a few hours of daylight after my arrival at the camp I measured the height of the Peak Burukutuau and found it to be near 2076 feet above the Savannah,*

*Buntens barometer indicated the height of our camp on the Savannah to be 407 feet above the sea, the summit of Burukutuau would therefore be 2483 feet above the level of the sea.

and a set of altitudes of the Sun, West of the Meridian gave me the Chronometric difference of our camp one mile west of Pirara (59°21′W of Greenwich). On a former occasion, I had ascertained the latitude of Kuiraton to be 3°1′40″N. which would give for the peak 3°1′N. While we were still occupied with our measurements the Chieftain of Kuiraton made his appearance, but merely waving his hands he passed on to the camp where I afterwards joined him. He was a stranger to me, a fine looking man & rather tall in stature for an Indian. He bore himself quite upright & a piece of cotton cloth thrown over his shoulder in form of a Toga, added to the dignity of his mien. One of his followers was entirely painted over with Lana,*

*A black pigment, prepared from the fruit of the *Genipa Americana* & *G.caruto*.[4]

with the exception of the face which was painted with Roucou†

†The Roucou is prepared from the red pulp which covers the seed of the *Bixa Orellana*.

so that the nether parts of his figure presented the appearance of a Negro, with which the red colour of the face formed a most ridculous contrast.

As our journey over land was to commence at Watu Ticaba I was anxious to send a messenger there to desire of the Chieftain that a *quantité* of Cassada bread might be

[1] See Vol. I, p. 253.

[2] A severe facial neuralgia with twitching of the facial muscles.

[3] In the published version (p. 12) the date has been changed to the more probable 1842. In fact there were several occasions on which smallpox was reported from the interior of British Guiana in the first half of the 19th century.

[4] *Genipa american* = *G. caruto*.

prepared for us, the more particularly as our friends in Watu Ticaba possessed well-stocked provision grounds while in and near Kuiraton, they appeared to endure great privations. I was also desirous of procuring a guide as I was perfectly unacquainted with the Rupununi above Kuiraton. I succeeded in both objects. Our new pilot was a young man who had been for a short time with M^r^. Youd, before that zealous missionary had been obliged to withdraw from Pirara & Curua.

May 10th The Thermometer stood this morning at 6 o'clock at 82°3′F. The Wet-bulb thermometer 79°F; the difference, amounting to 3°3′ proved that there was less moisture in the air than on the preceding day.

Previous to our departure we received the visit of upwards of forty Indians of both sexes and of all ages; they were conducted by their Chieftain and looked with much curiosity at our baggage.

Among the visitors was an old woman whom I immediately recognized from having seen her on my former visit in 1838 when her appearance was so shocking that I then compared her to a living skeleton and thought she could not have survived the next month; my astonishment was therefore great indeed at seeing her among the other visitors and still strong enough to walk the distance of a mile for the purpose of satisfying her curiosity. Her appearance was rendered more pitiful by numerous white spots of different sizes which covered her body. These patches occupied chiefly the abdominal region, but there were also some on the face extending over the Chin and under the jaw towards the neck. They were scaly and, no doubt, the consequence of disease of the cuticle. This ghastly object was strongly contrasted by the appearance at her side of a young mother with her firstborn, an infant of fourteen days old and on which she lavished her caresses.

We fired our small cannon and cohorns[1] before we started, to the great delight of our Indian friends. The reverberation of the sound from the near Burukutuau-yari and the distant mountains, at first loud, then dying away into an almost inaudible murmur, produced a sublime impression.

Our toil now commenced; rapid followed upon rapid, and as we had but inexperienced hands for our crew, our ascent was not without danger. We passed in the course of the day the Curuayair or crooked falls, the Ruru-ruru, the Tremetre, Trekutara-tepau, and several other falls. The hills Matziendaua form apparently on the left bank of the Rupununi, the South-eastern extremity of the Canuku chain. At their foot flows the small river Méneruau; they extend North-north-west and are connected with the high mountains Awarre-tequi and Burukutuau. Opposite Matziendaua, the river Catuau-uru joins the Rupununi on its right. This river has its source in the Savannahs South-east of the Tarucupani which forms the southern angle of the Canuku chain. Burukutuau bears from its mouth North 27° East.[2]

It was late in the evening before we had overcome the difficulty of ascending the fall Paratawai, and as we had to encamp on the open Savannahs without even a copse of bushes for shelter, we earnestly hoped that the night might pass without a storm. Our

[1] Cohorn or coehorn is a small mortar for throwing grenades and was named after the Dutch military engineer, Baron Menno van Coehoorn (1641–1702).

[2] None of these names appears on the reference map although they must all refer to sites within the fairly circumscribed area where the Rupununi flows through the Kanuku Mountains.

Indians proceeded with their hammocks to the Matziendaua Mountains, which were thickly wooded at their foot.

May 11th. The heavy work at the falls and rapids, the frequent rains & short provisions combined to indispose our followers to accompany us further. On calling over the list, in order to share out allowances, we found that four Macusis had decamped during the night. If I except one of their number, who had appeared at least to be well disposed to give his assistance the other three had proved themselves the laziest & most indolent of our crew, so that had it not been for the bad example and the fear that others might follow, I would have cared little for their desertion. I therefore addressed those who remained in their own fashion and told them that if any one was desirous to leave, he might do so at once, but that he should not deceive me by stealing away in the night like a thief or night-murderer (Kanaima). Not one came forward to avail himself of the permission, and we continued our course with reduced strength.

During the dry season the river must be almost overgrown with Guava bushes (*Psidium aquaticum*, Benth). At this time, with their tops above the water they assisted us in hauling the canoes over small rapids; besides which they break the swiftness of the Current and their flexible branches allowed the boats to pass over them.

Shortly after eight o'clock we arrived at Paruauku, the portage mentioned by Horstman, Santos and others, and by means of which the Saruau-uru is reached during the rainy season. It is a low Savannah which stretches West-south-west towards the Saerere Mountains, on the eastern side of which (not the western as I had been erroneously informed) the Suruau-uru flows. The Eastern Peak of the Saerere Mountains bears from hence South 73° West, distant about ten or twelve miles; the highest point of the Cursato Mountains South 65° West and Burukutuau, which we had left the previous day North 20° East; its distance being merely five miles. There are several isolated hills in the neighbourhood of this ancient portage, which has been known for the last hundred years, it being in April 1740 that Horstman traversed it, no doubt the first European that ever set his foot in these regions.*

*See Humboldts Personal Narrative, Vol. V. p. 594.

Santos followed in 1775 and Barata in 1793. I gazed with interest on the ancient trees that fringe the river near the embarkation. Perhaps Horstman had encamped here, dreaming of the rich treasures which he fancied were buried in the mountains he saw towards the north, and little suspecting the difficulties which the transport of his boat across the Savannahs would cause him and his crew on the following morning.[1]

Some small hills approach the Rupununi a little above the Portage, and on the left bank are some hillocks which the Wapisianas call Mawunnama-meketsiba, the translation of which is eye-hills. I could not ascertain why they were so named, nor could I myself see any reason for the apellation.

At about one o'clock we were opposite the most southern point of the Canuku Mountains, which our Indians called Tarucupani. The small river Witzapai[2] joins the Rupununi on its right and above the junction is a rapid, which insignificant as it appeared gave us a great deal of trouble to ascend. A small cluster of trees which we

[1] As already mentioned, Humboldt was wrong about Horstman who crossed from the Rupununi to the Mahu by the Pirara portage. Barata did cross via the Sauriwau portage but made no mention of the Paruauku.

[2] Probably Wichabai (*Manilkara bidentata*) which is the name of a current small settlement.

approached after three o'clock in the afternoon was too inviting for us to pass them without pitching our camp under their protecting branches. Trees are so scarce in these Savannahs, that we were obliged to carry the poles for our temporary huts with us in our canoes.

May 12th. It was fortunate we had Yesterday halted in a spot somewhat sheltered for a thunderstorm came on with hurricane-like fury that would doubtless have swept away our tents, had they been pitched in the open Savannah. Our crew had some difficulty in securing the canoes from injury, while others were constantly employed during the rain in baling out the water. The Thermometer stood in the morning at 70°F. and the Wet bulb Thermometer showed only a difference of half a degree.

The Rupununi which the Indians call Camoyepaugh, or Sun River has not much diminished in breadth. From the entrance among the mountains to the Savannahs it has varied only between 250 and 300 feet. The splendid *Calycophyllum Stanleyanum* which imparted so peculiar a character to the Landscape, has now vanished from the banks of the river and is replaced by the scarlet Elizabetha (Elizabetha coccinea). It was bearing numerous seeds, and the velvety appearance of the seed-pod, of a rich crimson, was not without its attractions, more particularly since the late rains had called forth the budding leaves whose white colour contrasted strongly with the crimson legumen.[1]

There are very few among the genera of tropical trees which like the beautiful Elizabetha put forth leafbuds enveloped with leguments resembling those of Liriodendron and covered with a viscous juice. The scales of a hard texture drop off when the delicate pinnated leaf, of a whitish colour when it bursts, first appears partially folded up. It expands in the course of a few hours, but too weak to sustain its own weight until irrigated when it assumes the position of the older leaves. The light by degrees colours it green; but the various shades of the leaf during its transition from white to green together with the crimson seed-pods give the tree a remarkable appearance. It is probable that the dry weather which precedes the Tropical rains may operate in a manner analogous to that of the winter of the northern hemisphere on our deciduous trees, in rendering torpid the vegetative powers; but it is remarkable that its effect should be limited to so few genera, which, like the Elizabetha, produce leafbuds, nor is that tree at any time entirely deprived of its foliage. While the delicate white leaves make their appearance, others of a yellowish green and the old perfectly formed foliage of a dark green colour, cover the tree. I know only of a few trees, and among these a Bignoniaceae, a species of Erithrina (*E. corallodendron*)[2] and the *Spondias Cirouella*[3] of Tussac,[4] which shed their leaves entirely, but even then, these genera are covered with flowers: indeed the Bignoniaceous plant is so clothed with bright yellow blossoms, that the trees appeared at a distance like yellow hillocks.

Our crew spied out with much pleasure a quantity Guanas on the branches of the trees which girded the river, a delicacy highly prized by the Indians and not despised by us Europeans. Many of the females had eggs of which there are frequently from eighteen to twenty four in the ovarium, perfectly formed, and somewhat larger than

[1] Just as this word and 'leguments' in the next sentence are secure readings, it is probable that 'tegument' and 'teguments' were intended. In fact, this is how the latter appears in the printed version (p. 16).

[2] This species is only found as a cultivated plant in this area.

[3] *Spondias cirouella* = *Spondias purpurea*.

[4] François Richard Tussac (1751–1837), French traveller and botanist.

the egg of the domestic pigeon. I confess I am semi-savage enough to find them very delicious.

We halted above a rapid near the mouth of the river Kamaikariba at half-past two o'clock. The large canoe, though manned with fifteen paddles, hove in sight only towards six o'clock in the evening. I watched them coming up the rapid. When the impediment was nearly overcome they attempted to cross the stream diagonally by force of paddling; a feat which we in the 'Louisa', with six paddles had performed successfully; but the Coxswain and bowman not acting in concert the strong current seized the boat and hurried her broadside down the rapid to the great consternation of the crew: the next moment she was driven against a rock. The greater part of the people jumped overboard and swam to the bank. The boat balanced for a moment but our interpreter Sororeng and some of our indians by timely assistance prevented her from upsetting. Mr. Goodall & Mr. Fryer[1] who were under the tent of the boat were described to me as having turned rather pale, while I freely confess, that though witnessing the accident from the shore I felt the greatest uneasiness. As it was, the boat had shipped a great deal of water, and had to be brought in shore with all dispatch to prevent her from sinking. My brother, in the mean time had been enjoying himself hunting an ant-eater (*Myrmecophaga jubata*),[2] which he and the Indians had espied drinking at the river's bank; and jumping ashore had chased along the savannah. It was unfortunately a female; I have never had the good fortune to see a male of this remarkable animal, although in the course of my journeys in Guayana more than twelve females were procured.

May 13th. It had rained nearly the whole night and till after eight o'clock this morning, While they were loading the Canoes I ascended a small elevation and looking southwestward saw the flood at the distance of a mile or two come rushing over the Savannahs. I returned with all dispatch to hasten the loading of our Canoes. The river had swollen considerably in the course of the night, and had invaded the camp of the Indians.

The great Cataract of the Rupununi[3] the Cutatarua or Iruan of the Wapisianas the Coroni*

*The word Iruan in the Wapisiana, and Corona, in the Caribisi signifies respectively "the fall" *par excellence*, there being no larger one nor any like it in the Rupununi.

of the Carabisi was before us[. W]e had heard the noise of its falling waters during the past night, but as the river had greatly risen, we found it less dangerous than we had anticipated, and passed it without much difficulty. We had to unload nevertheless, and as there was a little sunshine after the dismal night, we profited by it and the large granite ledges to dry our baggage. The river near the Cataract was about eight feet higher than when I saw it in 1835, but the trunk of a large tree that was lying on the highest rock of the fall, and which had been carried there by the flood on some former occasion proved that it had yet to rise from 14 to 15 feet before it reached its maximum height.*

*On the return of Mr. Fryer and my brother, about ten days afterwards the trunk had been carried away, a proof that the water had risen above the rock. This will serve to give some idea of the extent to which the flat savannahs which border the river, are inundated.

[1] The rapids in question are probably those known today as Fryers Fall.
[2] The giant ant-eater.
[3] Kutata Fall.

We started from the upper Cataract at One o'clock but our progress was very slow, in consequence of the swiftness and strength of the current. The river Phurnaru or Purnaru[1] joins from the S.S.W. on the river's left bank, a small stream when I visited it in 1835 but now nearly as large as the Rupununi itself. Opposite the junction the river Waipopu[2] also comes in.

I halted at three O'clock, at the fall Sarata,[3] and dispatched my Coxswain and some able hands, to assist the other boats which did not arrive until six o'clock. Large blocks of granite near which we encamped afforded an extensive view, and I ascended to their summit to witness the sunset, a spectacle so rare during the rainy season. The fiery orb surrounded by golden clouds was sinking behind the Cursatu mountains, whose bases were enveloped in mist, while their summits were gilded with the sun's parting beams. The eastern peak of Saerere rose above a sea of vapours, and the remarkable mountain Dochlopan out of the reach of the sun's rays, stood like a sombre rock in the ocean. To the north black clouds piled upon each other, and capped by the highest pinnacles of the mountain masses, spoke of the storm that was raging there and contrasted finely with the calm & placid picture in the west. At the same time the moon, then at its full, rose in the east, and of an apparent size of which we in Europe can have no idea. So true it is that in the tropical regions of the Western hemisphere not only the earth and its productions but also the heavens and their phenomena differ essentially from what we have been accustomed to from our childhood.

May 14th. The river had continued to rise during the night and we were now able to row over the savannahs in a straight direction avoiding the serpentine course of the river and its increased current. The expanse of waters resembled an extensive lake. The snow-white Egrette (*Ardea Egretta*) in great numbers; the American Stork (*Ciconia Americana*),[4] the stately Jabiru (*Mycteria Americana*), Cormorants, (*Carbo Brasiliensis*),[5] and large flocks of the Spurwing Plover (*Charactrius Cayanus*);[6] enlivened the surface, above which the tops of trees, granite blocks, clothed in tropical vegetation and here and there a small spot of elevated ground, alone remain visible. We saw the Savannah Deer, hemmed in by the water on one of these small spots. They cast an anxious look at our approaching boats, then turned round to ascertain if there were no means of escape, hesitating to take to the water and stamping impatiently with their fore feet. The buck at length plunged into the water; but the doe resorted to a ruse. It doubled down in the grass. Two Indians noiselessly left my canoe and swimming with one hand while with the other they held their loaded guns above their heads, approached the island. Arrived there they stole gently towards the spot where the Deer lay hid. The Easterly wind informed the crouching animal of the approaching danger, for though the sight of the Savannah Deer is dull, its scent is remarquably sharp; it suddenly rose pricked up its ears, looked for awhile steadily in the direction whence the scent came, and then endeavoured to escape, but the Wapisiana had already fired with unerring aim the poor deer sprung twice into the air, and fell lifeless on the ground shot through the head.

[1] Possibly the Kunaruwau River.
[2] Waipopo River.
[3] Sarata Falls.
[4] The Maguari stork (*Euxenura galeata*).
[5] Probably *Phalacrocorax olivaceus*.
[6] *Hoploxypterus cayanus*.

Whenever the water on the Savannah became too shallow for us we were under the necessity of regaining the river, which we had some difficulty in doing having to cut our way with cutlasses and axes through the thickets that fringed the stream. Myriads of ants, driven by the waters from the savannah had taken refuge on the bushes and trees and these now assailed us on all sides and inflicted merciless bites and stings. That species which like the bee and the wasp is provided with a sting is fortunately scarce, for the pain which they cause surpasses in violence either that of the bee or wasp.

I have already observed that I dispatched a messenger over land to Watu Ticaba, on leaving the Wapisiana village to inform the Indians of our approach, and to desire them to bring a supply of bread to the mouth of the river Warriwau or Awarra.[1] We were now approaching that stream, and discovered on some high ground which rose above the surrounding Savannahs several human beings, and two or three temporary huts. At some distance from this group, stood an individual apparently dressed according to the European fashion and holding an umbrella over his head. My glass proved that I was correct; the distance did not allow me to discover the colour of his face but I saw that he wore upon his head something like a military cap, that his loins were girded by a belt, that his bearing was upright and that he stood apart from the rest who remained at a respectful distance. My curiosity was much excited as to who this mysterious personage might be, but as soon as the canoe touched the ground, this strange personage, shutting up his umbrella, and walking directly up to us announced himself as the Captain or Chieftain of Watu Ticaba. The rules of etiquette were of no avail, and unable to contain myself I burst out into a loud laugh at the wonderful metamorphosis of my Indian friend. Imagine reader, a thin-boned being, his face adorned with an eagle nose and an uncommon large mouth, clad in a pair of sailor's trousers, which had once been white reaching to his ankles and fitting him à la Oliver Twist, and having round his loins a piece of blue cotton cloth from which depended a naked sword (in which by the bye I recognized an old acquaintance), his head covered with a red woollen cap ornamented with a large yellow tassel, and carrying in his hand a blue cotton umbrella, (upon which he seemed greatly to pride himself), and you may perhaps form some idea of Captain Wayapari in his full dress.[2] He had inherited the sword from his brother Captain Sirruai of Eischalli who bore it with him as a faithful companion, when he guided us in 1837 to the sources of the Essequibo. Poor Sirruai, I was sorry to learn had paid the debt of nature, and thus it was that the sheathless sword had descended to his brother Wayapari. As for the umbrella, we learned that Wayapari journeying lately by means of the portage at Primoss to the Corentyne had observed that article in the possession of one of the woodcutters at the lower part of the river and had become so enamoured of it, that he procured it at any price. Since it had been in his possession, the handle had been broken but his ingenuity had substituted in its place the forked branch of a tree. His followers were robust and well made men of the Wapisiana tribe; they had brought down some baskets of bread and several bunches of plantains, which greatly delighted our crew who had for the last two days been upon short allowance.

Our tents were soon pitched and protected against approaching rain; fires now blazed in all directions surrounded by groups of swarthy Indians warming themselves or cooking their plantains on the coals.

[1] Awariwau River.

[2] See Plate 31 in Goodall, *Sketches of Amerindian Tribes*.

We had for some days past suffered from a minute sand-fly (Simulia Spec?) which from sun-set til sun-rise, inflicted upon us acute bites that left marks of a scarlet hue for several days after. I was quite unacquainted with this nocturnal species, as the sand-flies we had hitherto met with, had proved troublesome, only during the daytime, and vanished at dusk, nor was their bite so piercing as that of the nocturnal species. They were this night so numerous that they compelled us to discontinue our astronomical observations. The size of this little insect scarcely exceeds that of a small pin; and as it attacks not only the face and hands but gets into the hair and inflicts its bites upon the scalp, we suffered from it more than I can describe.

The meteorological observations at the mouth of the Warriwau (432 feet above the sea) were as follows –[1]

Date	Hour	Barometer	Thermometer			Remarks
			Attached	Detached	Wet bulb	
May 14th	p.m.					The weather fair
	4.30	29.575	86	86.4	79	but partially
	5	" .570	86.90	87.6	80	clouded.
	6	" .571	77	77	75	At 6 P.M. a
	6.30	" .590	77	77	75	shower passed
	7.15	" .602	76.82	77	75.2	over

May 15th. We recommenced the ascent of the small river Warruwau which was running with uncommon velocity; and as its course was much impeded by trees, we had to overcome some difficulties, before we reached the spot from whence the path leads to Watu Ticaba. Arrived there our boats were unloaded and the baggage divided into convenient portions for being carried overland.

May 16th. I had given orders for starting early in the morning but the rain descended in torrents, and it was nearly noon before it abated sufficiently to allow us to proceed on our journey. Several Indians had arrived to assist us, and Wayapari dressed in his gala suit, with his umbrella over his head, led the van. We had not proceeded far when a heavy shower overtook us and Wayapari disrobed himself, and shutting up the umbrella stood the shower with Indian fortitude. We had sometimes to wade up to our arms through the water and every rill had become a torrent. Indeed if it had not been for the Chieftain who walked before us and searched for the shallowest places to ford the torrent, with [i.e. we] might have met with serious accidents. As it was an Indian boy nearly lost his life, while crossing one of the streams just above a Cataract: he carried a basket which he had slung round his neck to prevent the contents getting wet. The torrent swept him away towards the Cataract, he sunk, and not being able to rid himself of his burthen could not stem the current, and was given up for lost, when his brother, who at that moment arrived jumped into the water and fortunately seizing him, dragged him half-dead to the shore. Well aware that I could not swim, I grasped with all my strength the pole which assisted me in wading and I kept as close as possible to Woyapari.

Our march was fatiguing in the extreme the sun was rapidly approaching the horizon, and as yet nothing was to be seen of Watu Ticaba. The party to which I belonged had

[1] At the foot of each column, in different handwriting, there is a figure. They are respectively, counting the three types of thermometer readings as a column each, 5, 4, 5, 4, 5, 5, 6. It is not clear what they indicate but they are possibly typesetting instructions concerning the width of each column.

been marching and wading for the last six hours without resting, we were all very tired and indeed the Indians dropt off one by one and loitered behind. We were now ascending some rising ground, when our guide halted, and raising his umbrella which he had shut up and carried under his arm during shower and sunshine, his young son commenced blowing a strain upon a fife or flute made of bone, a certain indication that we were near a village.[1] I now observed five huts and a great stir among the inhabitants, the sound of another whistle reached us from the village, and a ghastly figure of a Negro perfectly naked, except the waistcloth, came out of one of the huts and met Wayapari whom he respectfully saluted and then placed himself behind him took no notice of us or of any other person of our suite. Thus we entered Watu Ticaba, the locality of which had been changed since I was here in 1837. The former chieftain of the village having died they had buried him according to their custom in his hut, burnt down their dwellings and established themselves further to the Southeast.

The strangers house (the Tapoi of the Macusis) had been put in order for our reception, and that portion of our luggage which had arrived before us was placed upon large beams to prevent it coming in contact with the moist ground. The Chieftain now bid us welcome and the influential men followed his example, while the women & children remained at a respectful distance. I counted fifty-eight individuals of all ages and of both sexes. They did not differ from the Wapisianas we had met on our journey to the sources of the Takatu. We had now full leisure to examine the strange figure of the old negro who came to meet Captain Woyapari on our arrival. His gaunt figure, which appeared all the more meagre and ghastly for being clad only in nature's garb, his close woolly hair bleached by age, his ears standing out immeasurably from his head, altogether gave him, when he began to caper and dance for the amusement of others, the appearance of a lascivious satyr in the train of Bacchus. His history which he told me in sufficiently intelligible Portuguese was as follows – He had escaped from the Rio Negro and after having been for some time among the Wapisianas, he married one of their females, by whom he had a son. This individual, who we saw the next day and who was called Wannéhré, might have been about 25 years of age; he had all the striking peculiarities of the negro, being merely in colour somewhat lighter than his father, but in all other respects the negro characteristics were strongly marked, and indicated that in him the father's blood predominated.[2] He had married a Wapisiana woman, by whom he had two fine-looking boys; their hair and countenance generally resembled those of the Indians, though the hair, in one of them, was not quite straight, but might be called wavy, like that of the Teutonic race, and was of rather a dark colour. Their skin was somewhat darker than that of the generality of the Wapisianas. Bastin the elder might have been eight years old and his looks were very intelligent. His forehead was rather arched, and though his lips were by no means curled up as in the Negro still it might be observed that they were somewhat thicker than those of other children of his age of pure Wapisiana descent.*[3]

*In order to enable the reader to compare the offspring of a negro man & an Indian woman, with an Indian of pure descent, I subjoin the measurements of Wannéhré and of Akarighur, the latter an Atorai of nearly the same age with the former –

[1] Certain types of Guianese Indian flutes are made from the tibia of a deer, although, in myths, the bone is often human.

[2] See Plate 67 in Goodall, *Sketches of Amerindian Tribes*, for a portrait of this individual.

[3] Plate 66 in Goodall, *Sketches of Amerindian Tribes*, is possibly a portrait of this boy.

	Akarighur an Atorai	Wannéhré a Cobb of mixed blood	Bastin the son of the latter
Estimated age	20 Years	25 Years	8 Years
Common Stature	5ft 8¼ inch^s	5ft 8 inchs	4ft 0in1 tenth
	ft. in. tenths	ft in tenth	ft in tenth
Circumference of the head round the forehead & above the ears	1. 10. 3	1. 10. 0	1. 8. 1
Long diam^ter from the occiput to the chin	0. 10. 1	0. 10. 0	0. 9. 6
Length of nose	0. 2. 2	0. 2. 0	0. 1. 5
Breadth of nostrils	0. 1. 7	0. 1. 9	0. 1. 2
Circumference of the neck	1. 1. 6	1. 2. 3	0. 10. 1
Length from the point of the shoulder to the tip of the middle finger	2. 6. 0	2. 6. 1	1. 9. 0
Length of Humerus	1. 2. 0	1. 2. 0	0. 9. 7
Length of Ulna	0. 11. 3	0. 11. 5	0. 7. 5
Circumference of upper part arm	0. 11. 1	0. 11. 0	
D° at the wrist	0. 6. 4	0. 6. 2	
Length from the division of the thighs to the sole of the foot	3. 4. 0	3. 3. 5	2. 2. 1
Length of the foot	0. 10. 0	0. 10. 3	0. 7. 5
Breadth of do	0. 4. 0	0. 4. 0	0. 2. 2
Breadth across the shoulder	1. 5. 0	1. 6. 0	0. 11. 0
Circumference of the same part	3. 1. 3	3. 5. 0	
D° of the Pelvis	2. 7. 5	2. 5. 1	
Height of Vertebral Column	1. 9. 0	1. 9. 0	1. 1. 2
Circumference of the upper thigh	1. 8. 0	1. 8. 7	
D° of the Calf.	1. 2. 0.	1. 5. 0	

I had resolved to send M^r. Fryer back to Demerara, with the boats and such of our instruments and baggage as were not indispensable, because we had before us an arduous undertake, which rendered it impossible to carry any superfluous baggage with us. My brother had likewise resolved to return with his collections to the coast, as the journey overland through pathless regions, could add nothing to his collections and was likely to endanger what he had already collected. I had ascertained that the provision grounds at Watu Ticaba were amply stocked with Cassava, and Woyapari promised to provide us with plenty and to inform the other Indians of the neighbouring villages of our wish. The inundation and swollen state of the rivers restricted his communicating with the surrounding villages. Watu Ticaba was comparatively on elevated ground and the intercourse with the western or left bank of the Rupununi broken up.

May 17th. The whole female population of the village were in motion fetching the Cassava roots, scraping them and making them into cakes, even the little girls assisted and assumed as much importance upon the occasion as if the weal of the whole village depended upon their occupation. Among the females was one who deserves to be mentioned as the belle of the Wapisiana tribe. Although Mayori-Epping was mother of a fine girl, her youthful appearance did not bespeak it. Her figure was very small, her height being only four feet and a half, but so symmetrical that she might have served as a model; the greatest breadth of her hand was two inches and eight tenths. The Wapisianas marry very early. Mayoriepping did not appear above fifteen years of age, and in the neighbouring house there was a young mother with two children, the eldest perhaps two years the youngest four or five months old, while she herself did not appear more than fourteen. During our stay in the village I had frequent opportunities of admiring the equanimity and the industry with which, so young, she attended to her maternal duties and the comforts of her husband.

The females puncture and stain the skin round the mouth in an eliptical form; this practice although occasionally followed by other tribes is much more general among the Wapisianas. This figures are punctured in the Cutis without affecting the cuticle; but their grand art consists in painting with pigment, a practice common to all the tribes of Guayana.

I had an opportunity while here of seeing two females of the mixed race. Their father a Negro, also from the Rio Negro, joined the Wapisianas about forty years since, in his language and manners he was far superior to the other black man, and his dress though plain, was clean and neat. When he paid me his first visit, he even sported a white pocket-handkerchief. I recollected having seen him some years since at Pirara. He was now accompanied by his two daughters the issue of his intermarriage with a Wapisiana female, the one a grown up woman, the other a little girl eight years of age, in both the characteristic features of the father prevailed. I had seen the elder with her father in Pirara, she was then stout and apparently in excellent health, now she was thin and looked sickly. The father told me, when I asked the cause of this change, that when she accompanied him four years since on his visit to Pirara, she possessed a great many glass beads which she ostentatiously wore round her person. This treasure in the eye of the Indians was coveted by a Macusi but reluctant to part with it, she refused to give it away, on which, said the old man, the Macusi in revenge had administered poison to her. It is not likely, however, that this was the true cause of her indisposition, but as the disease commenced soon after her return from Pirara, and the Macusis are held in the worst odour as slow-poisoners or Kanaimas among the Wapisianas, every evil from the aching of a tooth to the most serious disease is ascribed to their malignity.

May 20th. Early this morning the number of inhabitants was increased by a young stranger. We had seen the mother walking about only a short time previously with her first child about eighteen months old in her arms, and scarcely fifteen minutes afterwards, she delivered herself without any assistance, and was now sitting on the ground with the child in her lap while her husband was occupied in constructing a small hut of palm leaves inside the large hut, over her head. The other women did not come near her; a female after labour being considered by the Indians, as the Jews looked upon them when in that state. A fire having been lighted and some calebashes filled with water, put within her reach, she was left to herself and the care of her husband who kept his hammock almost the whole time from the period the little hut which was to separate her from the community was finished.[1] The child was rather small but nearly as white as a European, the hair on its head rather thick, the face round and though the nose had not the flatness of the Mongolian race, and was more distinctly formed and projecting; it was uncommonly thick towards the nostrils, the openings directly downwards; the hands & feet were very small, the nails well formed. I presented some cotton cloth for covering the newborn child and some glass beads for its future adornment, with which the young mother seemed much pleased.

[1] This childbirth ritual, whereby the father and mother of a newborn child are very restricted in what they may do and eat, is widespread among Native Amazonians and is known as the couvade in English.

Figure 9: Young Wapisiana Mother *aet* 14. It seems highly likely that this is a portrait of the young woman referred to in the journal entry for 17 May. Watercolour by Edward Goodall. Reproduced by kind permission of the Pitt Rivers Museum, Oxford (1884.117.80).

May 21. M[r] Fryer and my brother left us this morning, we felt the parting, the journeys of both parties had dangers in prospect, they had to descend many a difficult cataract before they could reach the Coast-regions & M[r] Goodall & myself to direct our course towards regions perfectly unknown.

This day the quietness of the village was disturbed. One of the Wapisianas thought he observed during the night while stepping out of his hut, a Kanaima or night-murderer stealing towards the village, who on seeing him made his escape across the Savannah. This betokens harm to one of the inhabitants of the Village; as the Kanaima will execute his revenge (for the perpetration of which he stole near the village) at an earlier or later period.

May 22[d]. The husband of the young woman who was delivered the other day, continued to occupy his hammock, for according to their superstitious belief any labour which he might perform at this particular period, would have an injurious effect upon the child; his diet is much restricted for similar reasons; if he were to eat meat or plaintains, it would prove hurtful, not to the eater, but apparently, by sympathetic influence to the child.

The Vaqueiro or herdsman I first saw at S[t.] Joachim in 1838 and met afterwards near the Tuarutu Mountains paid me a visit to day. He had settled in the neighbourhood of Watu Ticaba, in consequence of some dispute with one of the Macusi Indians at his former residence. I found him meek in appearance, and quite dejected and soon learnt that he considered the Macusi, with whom he had the dispute had injured him by secret charms and passion.

I was rather astonished when I saw he refused to take the refreshment from me, which I offered him, and give as his reason that his wife having lately borne him a child, it would injure the health of the baby if he were to eat or drink anything presented by a white man.

It has frequently been observed with what stoicism Indians endure pain. I had an opportunity of observing it again to-day. I do not know how I have acquired the honour but I stand in high repute among the Indians as a great physician. A boy upon whose head a palm tree had fallen, was brought to me to have his wound dressed. It appeared he had lost a great deal of blood and it was still flowing. I had the head shaved round the part injured and styptics applied during which operation, he did not give the slightest indication of the pain he must have suffered. He was not more than seven years of age and, in appearance rather a weakly child which made his fortitude the more remarkable.

May 24th. This day did not pass without a bottle of Champagne being drank to Her Majesty's health. It had been reserved for this purpose and our numerous followers partook in some measure of the festivities. The bottle whose contents we emptied with the best wishes for Britain's youthful Queen, we intend to bury where the flagstaff now stands on which on this occasion the Union Jack was hoisted, putting into it an account of when & why it was emptied of its sparkling beverage. Will civilization ever extend to the poor benighted beings who now surround us, so as to render it likely after the present generations shall have passed away, that the plough or the hoe may bring it again to light? The hope is but slender and reluctant as I am to despair, the conviction is forced upon me that the Indian race is doomed to extermination. Six years have scarcely passed away since I wandered to this spot on visiting the sources of the Essquibo. We left the settlement Eischalli Tuna, and passed on our route to the Taruma Indians three villages of Astorai or Atorayas and one of Torais, the latter containing the remnant of that sister tribe of the Atorai nation. The villages have vanished, death has all but extirpated the former inhabitants, and I am informed that of the true Astorais only seven individuals were alive. From the accounts I received in 1837 I estimated the number of Astorais and Taurais at two hundred including the descendants of mixed marriages, and of that number, about sixty are now left.

The measles, so dreadful to the Indians has twice decimated the Atorais, and at the commencement of the present year, the smallpox brought from the Colony to Pirara ravaged from thence southward as far as these poor people. Their belief in the secret influence of the Kanaima, who has only to breath upon his victim in anger to send him to an untimely grave, operates as banefully as that species of witchcraft called Obeah practised among the negroes, which operating upon their superstitious fears is frequently attended with disease & death. Nor is it is the Aturais and the Taurais alone, whose rapid extinction is thus going forward; similar causes are operating over the whole indian population of the Colony. The village of Wapisiana Indians called Eischalli Tuna, from which I started in 1837 is no longer in existence, and of its then inhabitants only one female and three children are now living. Many of my former acquaintances among the Taruma Indians are now buried, and I have already alluded to the rapid decrease of the Macusis. But the most affecting picture that now presented itself among the many Indians assembled around us was Miaha, the last remnant of the once powerful tribe of Amaripas. Singled out by destiny to be the sole survivor of a nation, she wanders among the living. Parents, brothers, sisters, husband, children, friends, and acquaintances are all gone down to the silent grave; she alone still lingering as the last memorial of her tribe, soon to be numbered, judging by her faltering voice and tottering steps, with those of whom tradition alone will record that such a tribe existed. Alas, a similar fate awaits other tribes; they will disappear from those parts of the earth on which Makunaima, the good spirit, placed them,*

*This is a common expression among the Indians, always accompanying it with the assertion that consequently, neither Portuguese, nor Spaniards nor English have a right to remove them.

and which since the arrival of the European, has become the vast cemetery of the original races.

Miaha appears to be about sixty years of age; her hair however is not bleached either by age or grief and would still be called luxuriant if she allowed it to grow, she has a large aquiline nose (a feature of frequent occurrence among the Wapisianas and Atorais) and a low forehead & the general cast of her face is jewish.

The language of the Amaripas differed only in dialect from that of the Atorai. They inhabited the regions about the Tuarutu Mountains near the river Wampuna and as Miaha well recollected when the late Dr Hancock was at the upper Rupununi in 1811, I had a fixed point from which to date my inquiries as to whether the extinction of the Amaripas had been slow or rapid. She told me that at that time their number was not quite so many as two men had fingers and toes, I concluded she meant about thirty five individuals and of that number Miaha alone remained in 1843.

The grass of the extensive savannahs which surrounded the dwellings of the Amaripas will no more be trodden by one of the descendants of this tribe, and ere long the deer alone will range over those thousands of square miles of herbage plains once the bed of a vast inland lake, now the grand burial place of the Amaripas, the Atorais, the Wapisianas & Macusis. Let us hope however, that the poor remnant of these people may be preserved from destruction and that instructed in the Christian religion and relinquishing their unsettled mode of life and superstitious customs they may become happy and useful members of a christian community.

Mr Goodall has sketched the portraits of several of the remaining Atorais and Taurais as also of Miaha, the last of the Amaripas,[1] while I have collected a vocabulary as far as circumstances would permit me, to save from utter oblivion a small part at least of the language of a tribe which with one death more becomes extinct.

We had in the afternoon a grand consultation, for although the weather had abated nothing of its severity, and still rendered it impossible for us to cross the rivers, I was anxious to ascertain how many individuals I could depend upon to accompany me from Watu-Ticaba to the Tarumas. Captain Wayapari summoned the principal men into my hut, and opened the subject to them. Adopting the squatting posture peculiar to an Indian, he addressed himself to a Wapisiana, who though by no means old in appearance seemed to possess great influence, and poured forth such a stream of words that I was astonished at his volubility. His speech was unaccompanied by either gesticulation or strong intonation, but flowed uninterruptedly for nearly half an hour. He to whom the discourse was addressed answered from time to time by the monosyllable hm sometimes varied with ha. The Harague[2] concluded, it was responded to by the Wapisiana, but fortunately at less length. I was silent and all impatience for the issue. Sororeng, our interpreter whose maternal language is closely allied to the Wapisiana did not betray the nature of the Colloquy either by a smile or the motion of a muscle of his face. It would be a violation of all decorum to interrupt the speakers upon whom the eyes of the whole assembly were fixed, some squatting, some standing like statues but all preserving a profound silence.

The arguments of the first speakers being apparently exhausted I expected that some of the other Chieftains, of whom there were four or five present, to give their counsel. I fixed my eye upon an old man who with one finger on his mouth, had such an intellectual appearance, that I regarded him as the Solon of his tribe, and imagined he must be a Demosthenes in eloquence. I awaited his harague with impatience but he merely smiled. Woyapari now addressing himself to Sororeng told him to interpret to me that he himself would accompany us & procure moreover a sufficient number of people to carry our baggage, but that we had to wait for more favorable weather which he thought would take place with the young moon five days hence.

Thus ended our palaver which if it was not marked by those bursts of oratory peculiar to the Indians of the northern part of America, was remarkable to me for the earnestness with which it was conducted and the characteristic countenances of the assembly.

It appears to be against their custom to permit females or children to be present at their consultations. A young girl anxious to offer some cakes of Cassada in exchange for some tinkling ornaments, and too impatient to wait till the conclusion of the palaver, unceremoniously came into the hut, but she was as quickly expulsed as she had entered.

May 25 to 31. Those Indians who have kept aloof from intercourse with the Colonists, show the greatest abhorrence for the use of pork. A strict Hebrew could not reject it with greater loathing than does a Wapisiana. An old man of that tribe, whose children had accompanied us on a former journey to Roraima had permitted their doing so, only on the condition that they were never to eat any viands prepared by our

[1] Some of these are reproduced in Goodall, *Sketches of Amerindian Tribes*, pp. 62–5 and 67.

[2] Presumably 'harangue'. Long speeches are a typical feature of decision-making in Lowland South America where persuasiveness depends on the strength of the speaker's words.

Cook, for fear he might have used pork in their preparation. We met with another instance at Watu Ticaba. A young Indian female was sometimes so far indulged as to receive the remnants from our table. She began to complain and became seriously ill. Her relations immediately ascribed it to her having partaken of our food. She suffered the severest headache, and her uncle who pretended to medical skill, insisted upon bleeding her, an operation which he performed by making three incisions on the side of her head, with the serrated spine of the stingray, for the purpose of opening the temporal artery. He made a bad business of it however, at least the other sapient men of the village did not think the incision high enough & condemned his practice. The young woman however felt no inclination to have the operation repeated, and accepting our remedies soon recovered.[1]

This disgust for pork is very remarkable and is likewise met with among the Indians of North America. Adair,[2] who was forty years among the Indian tribes of North America, gives several instances. The objection does not however extend to the native hog (*Dycoteles labiatus* and *D. torquatus*) which is eaten by the laity indiscriminately, except when pregnant or after delivery. The native conjurers partake but seldom of the native hog, as they consider it injurious to the efficacy of their skill. A candidate for that office does not taste it during his novitiate in this latter case however the prohibition extends to all thick-skinned animals.[3]

The influx of visitors from the surrounding country was very great, and such quantities of provisions were brought to us that twenty additional canoes would have been requisite for its transport. At the same time, to have refused the purchase, for a few trifling objects of barter, of what they had brought from a distance of twenty or thirty miles would have proved a great disappointment to them. Among the articles brought to us were many of those fine pine-apples which I have already had occasion to admire on my first visit in 1837, for their size and exquisite flavour. That variety which is called in the Islands the sugar-loaf pine is very plentiful. The Wapisianas call it Curtoi-ruai or Tapir's head, from its form bearing some resemblance to the head of that animal. Nana is the common term for pine-apple in its general sense.

June 2^{d}. I had planned leaving Watu Ticaba this morning to proceed on our journey to the Upper Essequibo but as one of the Chieftains with his followers failed to arrive at the stated time, our departure was delayed till the next day. We had been weather bound for nearby three weeks, and as a favorable change seemed to have taken place I was anxious to take advantage of it. During the period of our sojourn the weather had been exceedingly variable. The results of our Meteorological observations were as follows.

[1] Native Amazonians are careful about what they consume and its preparation, especially food of animal origin. To put it very simply, this has to do with ideas about the essence or soul of the animal concerned, and its impact on their own. These concerns increase at times of ritual seclusion and sickness when certain foods may be banned altogether. Scarification and blood-letting are quite common remedies for various ills throughout the region.

[2] James Adair (c. 1709–83), born in Ireland but had migrated to America by 1735. He spent most of his life as an Indian trader. He is the author of *The History of the American Indians, particularly those Nations adjoining to the Missisppi, East and West Florida, Georgia, South and North Carolina, and Virginia*, London, Edward and Charles Dilly, 1775. Almost half the book is given over to arguing that Native Americans are the descendants of the Ten Lost Tribes of Israel, and their refusal to eat pork forms part of the argument.

[3] Shamans throughout Lowland South America face limitations on their diet and behaviour. These vary from group to group.

Mean of Meteorological Observations at Watu Ticaba
Height 624 feet above the sea.

Period	Instruments	Forenoon 6h	Forenoon 9h	Noon	Afternoon 3h	Afternoon 6h
[May 17th to June 2nd][1]	Barometer	29.369	29.416	29.417	29.364	29.358
	Attached Thermometer	74.64	77.49	83.47	84.06	79.43
	Detached Thermometer	74.80	77.36	82.69	83.55	79.08
	Wet Bulb Thermometer	73.41	75.54	79.41	78.94	76.40

The highest indication of the thermometer in the shade was 85°.1.F. the lowest 72°.F., The black bulb thermometer exposed to the sun rose on the 21st. of May at ½ past twelve to 125°, not surrounded with black paper, but equally exposed to the sun, 113°F. The wet-bulb thermometer under similar circumstances 93°F. The thermometer in the shade 88.F.

The evaporation which during the twelve hours of the day amounted in Pirara to 320 grains out of 1000 grains, which were exposed in a cylindrical glass vessel*

*I have followed here Dalton's[2] plan, as the more rational. The glass vessel was 2.63 Inches in diameter and 1 inch deep. It was generally placed 4 feet above ground.

in the morning was only 278 grains. From the 18th of May to the 27th, the prevailing wind was North. It then veered Eastward between E. by N. and N. by E.; its force, except during thunderstorms, seldom amounting to more than Force 2 of Captain Beaufort's table.

The latitude of Watu Ticaba deduced from thirty circum meridian altitudes of α and γ *Ursae Majoris* and α *Crucis* was 2°32′3″ North; the meridian distance, east of Pirara 20′52″ in arc, consequently 58°59′8″ West of Greenwich.

[****]

Part (b) of the manuscript, numbered pages 1–56, is entitled:

JOURNEY FROM WATU TICABA TO THE FROG INDIANS AND THENCE BY THE RIVERS CAPHIWUIN AND WANAMU TO THE PIANAGHOTTOS

June 5h.[3] It was nearly nine o'clock before we left Watu Ticaba this morning, and as several of the Indians who promised to accompany us had not arrived I was under the necessity of leaving a great part of our provisions behind. We followed a different path from the one taken in 1837, and shortly afterwards arrived at a small settlement consisting partly of Wapisianas and Atorais. The huts were in a most dilapidated state, fissures and holes in every direction, quite large enough to afford the loungers every facility to contemplate the canopy of heaven without leaving their hammocks; but howsoever wretched the huts, the principal one was provided with two large troughs for Paiwori, besides a number of jars on a gigantic scale, made of clay, intended to be filled with favorite beverage.

[1] The manuscript is blank but the column has been filled with these dates in the published version (p. 30).
[2] John Dalton (1760–1844), a British scientist who formulated a law of evaporation.
[3] A mistake for June 3rd. It is corrected in the published version (p. 31).

The Chieftain was sick in his hammock, and presented a ghastly appearance[;] indeed all the inhabitants looked sickly, and I was informed that the smallpox had carried many to the grave.

June 4$^{th.}$ Our last night's voyage had been on the verge of the savannahs; but we this day entered the forests. After a few hours' march we crossed the first rivulet, which was flowing into the Guidaru.[1] The ground was undulating and in many places covered with blocks of Chert and granite which extended to considerable distances in a North and South direction.

The forest abounded in that remarkable tree the Murre-Marri of the Macusi Indians, Aublet's *Couratari Guianensis*, the woody capsules of which covered the ground. They are of a very remarkable form, sometimes four or five inches long & somewhat three-cornered.*

*The seed is a *Samara* or winged fruit which is attached in three grooves to the sides of a hard and fleshy, somewhat triangular body which fills the woody capsule, and is, near the summit, convex, protuberant, and slightly furrowed, so as to close the capsule firmly. At the time the seeds are mature, this fleshy body shrinks up, and dropping out of the capsule, sets free the winged seeds, which are carried by the winds in different directions.

We found them very useful as cases for preserving insects and other curiosities in, which were thus effectually secured from accidents.

Our march was this day a short one, for at 11 o'clock we reached a settlement, and as it was the last inhabited one we should find previous to our reaching the Tarumas on the Cuyuwini, our guides and carriers could not be induced to continue our march further. A circular hut of small dimensions appeared the only inhabitable house. Not less than six hammocks were slung in it and all of them occupied, and the heat was intolerable. Several blowpipes and materials for preparing the arrows, were hanging from the roof ready for use. The thrifty housewife was assiduously occupied in scraping the Cassada roots for preparing bread for the numerous visitors. Her long hair was, for the occasion tied up in a queue with uncommon stiffness so that with every bend of her body, while occupied in scraping the Cassada root that peculiar ornament was in violent motion; at one time describing circles, at another, at another sticking out almost at a right angle from the head to which it belonged. It was so ridiculous that I could not refrain from laughter.

The hut was thickly surrounded by Arnotto bushes (*Bixa orelhana*). At some distance from the hut I observed the remnants of a large fire, and Sororeng, the interpreter, told me that the people had lately burnt here one of their dead. The Atorais are as far as I know, the only tribe in Guayana who place the dead body on a pile of wood and burn it; the ashes are afterwards buried.

We erected our tent at a short distance from the hut, near the provision grounds. The towering stems of some Palm trees which are called Manica by the Colonists, and which I consider to be a species of Euterpe really astonished me by their height. I had one cut down and it measured without its feathery foliage eighty two feet; its total length could not have been less than 100 feet & in spite of such a height, the circumference of the trunk near the base was only nineteen inches.

[1] Kwitaro River.

The mean of a number of circum-meridian Altitudes of the star γ Ursae Maj. and α Cucis, gave me 2°18′24″N. as the latitude of our camp. Our meridian distance was 21 miles East from Pirara. The Thermometer stood at half-past seven o'clock in the evening at 74°F. The wet bulb thermometer 72°.9 F.

5th. June. We started at 8 o'clock and soon afterwards crossed the small river Dohté one of the largest tributaries of the Upper Guidaru.*[1] We again met with numerous blocks from 14 to 15 tons weight which extended as on the former occasion North & South and soon stood at the foot of the Carawaimi mountains[2] which we had to traverse. I estimated their height here about 1000 feet. Their highest summit more to the Eastward of our path is about 2000 feet. On the southern foot flows the river Guidaru.

It rained very heavily in the afternoon and we were glad to pitch our camp at an early hour, near one of the smaller branches of the Guidaru.

6 June. A march of half an hour through thickets of bamboo and melastomas, brought us to the Guidaru, here about twenty feet broad & consequently our present path traversed it much higher up than in 1837.

About four miles further southward we came upon an abandoned Atorai settlement, the former inhabitants of which were all dead except two children who are now with their distant relations near the Rupununi. We had to cross numerous swamps abounding in a species of Cocoa (*Theobroma bicolor*?).[3] The large melon-shaped fruits of all sizes, some green, some of a bright yellow, were eagerly collected by our Indians and indeed the pulpy arillus which surrounds the bean has an agreeable vinous taste. There was sufficient evidence that rats, Agouris, Monkeys and Peccaries were as eager after the fruit as our Indians who, however, did not appear acquainted with the fine aroma which the seed itself possesses, which induced Linnaeus to call it the food of gods. We collected a number of seeds of which I counted as many as sixty in one capsule. They afforded us for some mornings the most delicious cup of cocoa I ever tasted.

It is known to botanists that in Theobroma, as in Gustavia,[4] Crescentia,[5] Cynometra,[6] some Swartziae,[7] the new genera Lightia[8] and Alexandria,[9] the remarkable Omphalocarpus[10] &c, the flowering buds, break through the rough bark of the trunk in lieu of appearing at the tender branchlets, as in the generality of trees. It is remarkable that Cacao in its wild state is only found in swampy or at best moist situations. The trees which I observed although of a peculiar growth, almost shrubby, and the trunk less developed than in large forest trees, often attained a height of fifty feet. The capsules were large and contained from sixty to seventy seeds, which were larger than in the cultivated kind but not so thick.

[1] Although a footnote is indicated here, it is not provided in the manuscript.
[2] Marudi-Karawaimentau Mountains.
[3] Although Mennega et al. lists *Theobroma bicolor*, the NMNH checklist does not.
[4] A genus of the family Lecythidaceae.
[5] A genus of the family Bignoniaceae. *Crescentia cujete* is the Calabash tree.
[6] A genus of the family Caesalpiniaceae.
[7] A genus of the family Fabaceae.
[8] Reclassified as *Euphronia*, a genus of the family Euphroniaceae (Vochysiaceae).
[9] Renamed *Alexa*, a genus of the family Fabaceae.
[10] *Omphalocarpum*, a genus of the family Sapotaceae, but does not appear to occur in Guiana.

We passed soon after noon the site of the Daurai settlement where, on my journey to and from the sources of the Essequibo in 1837–38 we had rested. It was now perfectly overgrown with bushes and the spots where their huts formerly stood, could not be reached without using the axe and cutlass. Scarcely six years had elapsed since I found here a settlement of nearly forty persons; two grown up individuals of the number are now all that are known to be alive, both of whom strange to say are singularly marked – the one from his birth the other by accident. The younger was born with only one eye, a large tumor overspreading the place where the other should have been. The elder lost his eye while shooting at a monkey with a blowpipe in a perpendicular direction. The poisoned arrow missed the intended victim and in its descent fell right back into the eye of the Indian. He had sufficient presence of mind to withdraw it instantly, a severe sickness however and the loss of the organ were the consequences. These two individuals have now withdrawn from their former abode, where their wives and children had fallen victims to the unhealthy swamps by which it was surrounded. Some young orphans however demanded their care when they abandoned the village, and these with themselves now constitute the last of the Daurais, soon to be numbered with the Amaripas as tribes only known by tradition among the old men of the adjacent and more populous nations.

June 7th & 8$^{th.}$ On the first of these days we continued our march till two o'clock in the afternoon. It rained almost the whole night and had not ceased when we started on the morning of the 8th. Knowing, however, that we should reach the first Taruma settlement at an early hour, we passed on. The site of the village had been removed from its former position to another near the left bank of the Cuyuwini. We reached it soon after eleven o'clock delighted that our pedestrian wanderings were for the present at an end. The weather had, upon the whole, been favorable to us, for a cessation of rain had enabled us to cross with comparative ease the swampy ground which we had found it difficult to traverse even in December 1837, and which in the rainy season is scarcely practicable.

We crossed a large clearing, their former provision fields but now apparently abandoned for more fertile grounds. Cassada plants (*Janipha Manihot*) from the former cultivation, were were growing up amid a rank vegetation but had nevertheless a strong and healthy appearance proving the great fertility of the soil and that it was not exhausted by former crops.

We found three dome shaped huts, one of large size and very neatly finished with branches of the Murumutu palm (a species of *Astrocaryum*, perhaps the *A. Murumuru* of Martius). The inhabitants were for the most part absent some on visits others hunting. The latter arrived in the evening, they had not been very successful, as a single Coita (Spider Monkey) and a few Toucans, were the sole produce of a whole day's hunt. I recognized only two of my former acquaintances among them; many had died since I last visited them, and some had moved higher up the river.

The few remarks which, on a former occasion I made upon the Tarumas as a tribe*

*Vide Vol. X page 167 of our Journal.

may be here repeated with such additional information as I have been able to procure.

It appears the Tarumas were formerly located at the tributaries of the Rio Negro and the Portuguese missionaries had some Neophites of that nation in Manaos. Disagreement with the other tribes however and other circumstances induced a portion of the

Tarumas to fly. They proceeded up the rivers which have their sources in the Sierra Acarai of the maps, and settled at the head waters of the Essequibo. As to the Neophites who had remained at the Rio Negro, death had committed such ravages among them that Von Martius not aware of the party that was settled at the Upper Essequibo pronounced the tribe extinct.[1] Meanwhile information that a tribe of Indians called Tarumas was settled at the Upper Essequibo, was brought to Demerara half a century ago by Mahanarva, a Carib Chief, who had lived some time at the Cuyuwini but such was his exaggeration that he described them as amphibious and living in cavers under the water, and flying on the approach of man. Mr. Morrison and myself were the first Europeans who visited them which we did in December 1837, and found the first settlement we fell in with about five miles higher up than the present one. I have since seen from a hundred to a hundred and fifty individuals of the tribe, who in stature are little inferior to the Wapisianas but by no means so good looking. The features of their females are by no means engaging and they are so slovenly in their persons as almost to create disgust. The head of the Taruma is rather small in proportion to his body, and the outline of the cheek bone rather hard; in colour however and in the general proportion of their bodies the Tarumas do not much differ from the generality of Indians. The dialect spoken by these people differs from that of the other Indians of British Guayana, and the pronunciation and accentuation of their words are so strange that their language strikes the most unconcerned in philological researches as different from the Macusi and Carib tongues. Their huts are dome-shaped and do not materially differ from those of the Wapisianas. The interior has a more cheerful appearance, as the horizontal or cross pieces of wood that connect the upright in the hut of the Macusi and Wapisiana are dispensed with. The centre of the floor is occupied by the large trough for the intoxicating drink, and the hammocks are slung from the upright central pole to the circular wall, along which runs a kind of grating, occupied by the dogs as their sleeping place.

The Tarumas are great dog-fanciers and are considered very skillful in training them for the chase; but as fire-arms are still scarce among them, few of the dogs stand fire. I have elsewhere*

*Vide Vol. XIII. p. 65 of our Journal.[2]

given a full description of one of the finest dogs I saw among the Indians which came from the Tarumas. We saw here some fine specimens of that variety which resembles Buffon's St. Domingo Greyhound;[3] but none could vie with the dog alluded to, and which I unfortunately lost by an epidemic that prevailed among the canine race in Pirara. The stock of other domestic animals possessed by the Tarumas was small, but I observed a few cocks of our Dunghill breed which evidently were kept more for state than use as they were unprovided with hens. Like other Indians these people neither eat their flesh nor make use of their eggs. Two Trumpeters (*Psophia Crepitans*) were masters of the poultry yard; to their sway four footed and two legged animals alike

[1] See Spix and Martius, *Reise in Brasilien*, III, p. 1304.

[2] 'Visit to the Sources of the Takutu, in British Guiana, in the Year 1842', *Journal of the Royal Geographical Society*, 13, 1843, pp. 18–75.

[3] See Louis Leclerc Buffon, *Oeuvres complètes. Histoire des animaux quadrupèdes*, 10 vols, Paris, 1775–9, I, pp. 327–8.

submitted. If any disaster occurred they settled the dispute by their interference, and all new comers had first to submit to their ordeal viz. a severe pecking and slapping with their wings, before they were duly admitted of their society. We had brought some poultry with us and among the rest a cock which imagined it had as much right to crow and flap its wings as those that were born or had been long at the place. This caused contention and the new comer had to fight it out with its conquerors; but it was ridiculous to see how quickly the trumpeters settled the matter and pecking at our poor cock soon drove him out of the field. They equally attacked strange dogs, which generally ceded to them the palm of victory and sought safety in flight. The Indians had brought three Powis (Crax alector) to the place, a bird much larger than the Psophia; but so little rest was allowed them [by] the latter, that the owner of the Powis was obliged to tie them up.

The inhabitants of the village appeared to cultivate the Sugarcane to some extent. At least large quantities were brought to us for barter. It was of that variety which is called the Otahaiti[1] cane (*Cana de Otahaiti*). As common as the banded and Bourbon cane is near the coast, I have never seen a plant of that description among the Wapisianas and Tarumas. It is known that Cook & Forster were the first who made us acquainted with the Otahaiti cane. Bougainville[2] introduced it into the Mauritius, whence it found its way to Cayenne, and from thence towards the close of the last century to Martinique, Guadaloupe, & St. Domingo.[3] Captain Bligh brought some to Jamaica, and scarcely fifty years have elapsed and we find it spread over the tropical part of America, esteemed alike for its superiority by the planters and the Indians; the provision grounds of the latter being never without it.

The Cane cultivated by the Indians differs neither in size nor sweetness from the Otahaiti cane which is cultivated along the coast and as, no doubt, scores of years have elapsed since it was transplanted among the Manihot and Cassada fields of the interior, it may be considered as acclimated, and it is not likely that with common attention it will degenerate in its quality.

In the provision fields behind my tent stood one of the finest trees I have ever seen in my wanderings. The compound leaf, the small leaflets, and indeed the whole appearance bespeak it to be a Mimosa. Its whole height is 168 feet.*

*The height was ascertained by measuring a small base line, and ascertaining the angles with a sextant and artificial horizon.

The trunk from the base to the first branches 73 feet. About a foot and a half above the ground it measured only 27 feet in circumference, but ascending perpendicularly of almost equal thickness, to the first branches. It appears like a slender column bearing, bearing its finely formed leafy capital. Numerous nests of the Oriole which generally build in families, were constructed on one of its branches, near the summit, sure of being there unmolested by either Monkey or Tiger-cat. The Indians themselves seemed to have some regard for this singular tree; it stood amidst the provision fields,

[1] Tahiti.

[2] Louis Antoine de Bougainville (1729–1811) FRS, led a scientific expedition round the world in 1766–69, the account of which is contained in *Voyage autour du Monde, par la frégate du Roi La Boudeuse, et la flûte L'Étoile; en 1766, 1767, 1768 et 1769*, Paris, 1771.

[3] Today the Dominican Republic.

and while almost all the other trees had been felled by the axe, this was spared, and its gigantic size appeared all the more striking.

We remained at the Taruma place much longer than I wished; but, as on leaving Watu Ticaba we could not procure a sufficient number of carriers for all our baggage, I was obliged on my arrival here to dispatch people for the remainder. The delay was however favorable to more detailed observation. I have mentioned the Indian females as ill-favored and neglectful of their persons. Their hair is uncommonly coarse & wiry and does not fall over the outlines of the head, but stands out as if the individual were frightened or as the imagination of the artist paints Shakespeare's witches. They paint their whole body, including the face, with broad black bands, without order and of unequal thickness, and their peculiar walk, with protruding abdomens, their coarse voice and the peculiar intonation of their language increases the abhorrence which the *tout ensemble* excites.

The Tarumas are famed as manufacturers of Cassada graters which they barter to a great extent with the neighbouring nations, chiefly the Wapisianas, who carry them to the Colony where they receive about a spanish dollar (4/2^{d})[1] for each. These graters consist of a flat board from 2½ to 3 feet in length and from 15 to 18 inches broad, and which is prepared from the soft wood or outer layers of the Purple-heart tree (*Copaifera Spec?*). After it has been properly smoothed and a slightly concave form given to it, the Indian marks it with cross lines, along which he beats one by one, small angular pieces of a hard rock (very compact greenstone), which he calls Tempé, and which is found a few days journey up the Cuyuni.[2] The angular points project about a line and a half out of the wood. The board being covered with these stony asperities, he takes the milk of a tree called Hennicarro which he colours with Roucou or Arnatto (*Bixa Orelhana*), and spreading it equally over the board, it serves as a glue in fixing the rocky fragments, besides giving a varnish to the whole. It is afterwards painted fancifully according to Indian taste, and exposed to dry in the shade, when the grater is completed. It is not possible to complete a grater in less than five or six days, including the fetching and preparation of the materials. Nevertheless he exchanges it for a common knife with the Wapisiana, who carries it four or five hundred miles and sells it for a dollar. Time is however of no value to the Indian, and the inclination to work at that particular job, is his sole inducement regardess of the little reward of his labour.

13$^{th.}$ June This was one of the hottest days we had had since we left Pirara, the forenoon was cloudy, the sun scarcely making its appearance till after twelve o'clock. At ½ pas[t] one o'clock the Thermometer under the tent, open and surrounded by bushes stood at 91°.2 F. The Thermometer exposed to the sun at 132° and the black-bulb Thermometer 136° above which it would have risen if the length of the tube would have permitted it.

14$^{th.}$ The weather was similar to that of the preceding day, till evening when a white fog rose which increased in intensity till midnight.

15 On the morning of this day the rain commenced and continued uninterruptedly till seven on the 16$^{th.}$ occasionally coming down in torrents. On the following day I

[1] 4s.2d. or approximately 22p.

[2] This could be a slip and Kuyuwini intended. However, the region was criss-crossed by extensive trade networks and the Taruma may well have obtained this raw material from the Cuyuni.

resolved in spite of the weather to proceed without waiting for the remainder of the luggage from Watu Ticaba.

Notwithstanding the rain and clouds I had been enabled to procure 28 Circum-meridian altitudes of North & South stars which gave me for the Latitude of the settlement 2°4′30″N. The Chronometer (Frodsham's) made its meridian distance 39'56″ (in arc) East of Pirara.*

* Pirara		59°20′ W. of Greenwich.
Taruma Place	–	39′56 East of Pirara.
		58°40′4″ W. of Greenwich.

Sixty five observations of the Thermometer and Barometer during our stay gave the following results

Instruments	Mean	Maxima	Minima	Remarks
Barometer	29.270	12th at 11^{h}.30 a.m. 29.316	14th 6 a.m. 29.191	
Attached Thermometer	79.68	13th at 1.30 a.m.[1] 90.8	10th 10 a.m. 70.3	
Detached Thermometer	79.92	91.2	70.5	
Wet bulb Thermometer	76.92	83.7	69.6	

The greatest range during one and the same day was 19°.3 F for the Thermometer and 0^{i}.118 for the Barometer.

The following Meteorological Table will give the other data of my observations at the usual hours–

Period & number of observations	Instruments	Forenoon 6^{h}	Forenoon 9h	Noon	Afternoon 3^{h}	Afternoon 6^{h}
9th of June to 17th	Barometer	29^{i}.230	29^{i}.268	29^{i}.279	29^{i}.226	29^{i}.225
Inclusive	Attached Thermometer	71°.60	76°.46	82°.24	81°.93	76°.07
65 Observations	Detached Thermometer	71.90	76.62	82.25	81.95	76.18
	Wet Bulb Thermometer	70.13	75.08	78.16	77.96	74.72

The evaporation amounted on the 12th June from $6^{h.}$ a.m. to $6^{h.}$ p.m. to 252 grs. of out a thousand grains exposed to the sun.

I vibrated the magnetic needles on the 13th June. The result of 100 vibrations at 88½° F was $2^{m}53'08$ pr needle La, and $3^{m}42'$ at 90 pr needle Lb.

The settlement is entirely surrounded with wood. The nearest Savannah being about 30 miles to the North of it and the small river Cuyuwini here only 100 feet broad flows at a distance of a few hundred yards. Its height above the sea, according to Bunten's barometer … .[2]

June 18th We this morning commenced our journey by water in two small Corials and six bark canoes, the whole party consisting of twenty three individuals. The river was full to overflowing, the current, nevertheless, quite sluggish and the water muddy. The few trees and shrubs that were in blossom along the banks of the river had only one or

[1] A mistake for p.m.

[2] This sentence is incomplete in the manuscript but in the published version (p. 38) it is completed with 'was 745 feet'.

two flowers of a white colour and of great fragrance. It is remarkable that white flowers, under the tropics possess the greatest fragrance. The Sawarai palm the Murumuru and the Uassa (Astrocaryum Jauari; A. Spec? & Euterpe) were the most frequent palms, occurring in small groups. The beautiful *Ionopsis violacea*[1] with its large paniculated scapes of nodding flowers *Brassavola Martiana*, *Zygopetalum rostratum*,[2] the common *Epidendrum* variegatum,[3] alike abundant at the shores of the sea as in the interior at a distance of six hundred miles from the ocean, here and there a *Brassia* and a peculiar orchideous plant (of which, from its large clusters of stems, I expected some striking and handsome flowers when I first visited the Cuyuwuni in 1837) were the genera of that order which I found in blossom. I was much disappointed in the last of these, their large clustering stems producing only heads of small white flowers, nestling as it were in a gelatinous substance. I have seen something similar in a highly interesting plant, the genus and affinities of which, I have not been able as yet to make out to my satisfaction. This gelatinous substance is so abundant that it flows off as soon as the stem which bears the clusters of flowers is shaken.*

*Aublet's figure of *Serapias Caravata* (t.320)[4] resembles this plant. I observed it likewise to my astonishment on the high mountains in the Island of Grenada.

June 19th & 20th Since we left the Taruma place the rain has descended in torrents. It is peculiar that about 10 o'clock in the forenoon severe showers set in and last for an hour or two, when they abate and the weather becomes variable. The river had overflowed its banks so that we were compelled to pitch upon a sand bank a few feet above the water as the only possible spot for establishing our night quarters. But no sooner had some of our Indians jumped upon it than they began capering about as if suffering from St Vitus's dance and immediately retreated. The whole place was swarming with that species of ant which our creoles call the Yagerman, and of whose unmerciful bites we had already had sad experience. The day was far advanced and it was doubtful whether we should find another convenient place for our camp. We therefore determined to contend with the ants for the possession of the place. A fire was soon lighted and by means of large firebrands and Caabashes of Water, we broke their colums and eventually succeeded in driving them away from as much ground as we required for our tents, and we passed the night in peace.

June 21st to the 23rd We entered the Essequibo at half-past nine o'clock in the morning of the 21st. of June and its bed being full to overflowing the current which I found in November 1837 to be scarcely more than half a mile pr hour, now ran at the rate of three miles and rendered our progress very slow.

With the best intentions of making the meteorological time observations this day, scarcity of provisions obliged us to push on in order to gain a village of Taruma Indians. Indeed on the 23rd we had to paddle on from six in the morning till half-past six in the evening without permitting even a short repose to our people. On arriving we found

[1] Unidentified species of orchid.

[2] *Zygosepalum labiosum*.

[3] *Epidendrum vespa* or *Prosthechea vespa*. Dunsterville & Garay list twenty-one synonyms for this plant (see *Venezuela Orchids Illustrated*, 6 vols, London, Andre Deutsche, 1959–76, II, p. 154).

[4] See Aublet, *Histoire des plantes de Guiane Françoise*, II, pp. 816–18 and IV, Tabula 320. It is the Orchidoid orchid, *Elleanthus caravata*.

the greater part of the inhabitants absent on a fishing expedition; but as one of our bark canoes had started from our last camp seven hours previous to ourselves it had reached the settlement earlier than us, and had sent for the absentees who were hourly expected.

The place consisted merely of two huts, one in the form of a common shed, the other circular and perhaps from forty-five to fifty feet in diameter, rising, not like the generality of huts in a dome shape, but pyramidal. It was open above to allow the smoke a free passage, and overtopped by a smaller roof to protect the inmates from the inclemency of the weather. A tree taken out of the ground with all its roots, but of which the branches were cut being fixed to the uprights at about five feet from the ground supported the smaller roof at its upper end while the roots served as pegs to hang up divers household utensils, caps &c. The tree was stripped of its bark and painted according to the Indian fashion. The interior of the hut did not differ otherwise from the general construction; numerous dogs were lying peaceably on the gratings, and to my astonishment did not set up that furious barking which had so tried our patience during our sojourn at the former Taruma village.

I observed large baskets full of the leaves of the *Bignonia Chica* with which the Indians prepare a red pigment called Caraveru by almost all the Guayana Indians. The leaves are dried in the sun and at the first exposure, after having been plucked from the vine which produces them, they show the abundant feculent substance which they contain. This colouring matter is more known in the United States than in England and is used as a dye. The Artist of our Expedition, Mr Goodall considers the pigment equal to madder which is rather an expensive colour. The Caraveru might form an article of export if it were sufficiently known, the more so as its preparation is extremely simple.

I allowed our crew a day of rest. The weather had been very unfavorable during our journey, and those among us who wore clothing had no opportunity except at night to change their wet garments for a dry suit.

The Indians of the village offered their services for fishing or hunting. We were told that Pacu and Haimura (*Myletes Pacu* and *Erythrinus Macrodon* vide Fishes of Guayana in Naturalist's Library Vol.I. p. 236 & 254) were very plentiful, and in the course of the day several were brought to us. These two kinds of fish are here baited with the ripe berry of a species of Phytolacca.[1] A bunch of these berries is suspended about an inch or an inch and a half above the water, and should there be any Pacu in the neighbourbood they are sure to be attracted to the spot, and as they rise above the water to seize the dainty morsel the Indian shoots them with his arrows. This fish is equally partial to the ripe fruit of the Lana (*Genippa Americana*, G. Caruta &c) but as the subtle fish might suspect some danger if the fruit were handled by man, the Indian presses it without touching it with his hands, and puts it into a little basket made of withes, which he suspends as in the former instance. The Indians assert that the Pacu scents the fruit at a great distance; it likewise takes the hook which is baited with the fruit of the Genippa, the Crabwood (*Carapa Guayanensis*) or the acidulous berry of the Casami, a species of *Eugenia*. The Pacu ascends the river only when it is full to overflowing. When the water is low and the rocks that impede their course are nearly above the water and overgrown with different species of Lacis, the fish will not take any bait, as he feeds on the aquatic plants.

[1] A genus of the family Phytolaccaceae.

Numerous pine apples of a superior taste, the orange coloured fruit of a species of Passiflora, which the Colonists in Demerara call Scimitu, and large bundles of sugar cane were brought to us. Some of the cane measured above eight inches in circumference, and the joints were from seven to eight inches apart.

25 June. We started in the morning but the strong current prevented any rapid progress. A circumstance this day occurred which induced me to remark that the chief apprehension I have ever entertained in travelling through the forests and Savannahs of this country, has been of the venomous snakes which lurk in such places. Indeed the repeated instances I have known of death or misery for the rest of life which has followed the bite of one of these reptiles, are well calculated to inspire dread. I was sitting with Mr Goodall under his tent, towards dusk, when I felt something crawling between my feet, and before I had time to see what it was, Mr. Goodall jumped in great terror from his seat crying, "a snake." Although I felt when it crawled along my foot, I could not see it, but Mr. Goodall considered it to have been a rattlesnake. I doubt this, as they are not generally found in forests, but I was none the less thankful that I escaped unhurt; a single motion of one or other of my feet might have induced the snake to inflict its bite.

June 26th The morning was quite foggy, and we could scarcely see a few yards before us. Indeed the Essequibo rather resembled the Thames on a November morning than the tropical river we were navigating. The Thermometer stood at 72°F. the wet bulb Thermometer at 70°.7 F., and the air had a sulphurous smell which I found annoying to my lungs. The rain set in again at ten o'clock and lasted the greater part of the afternoon.

June 27th We arrived at about 10 o'clock at the second Taruma settlement on the Essequibo. Since I last visited it (1837) they had removed the site of the village more to the Southward. We were instantaneously observed, and a large crowd assembled in front of the principal hut. It appeared that our Taruma friends had got better acquainted with fire-arms since I visited them on a former occasion, and one less familiar with their customs might have taken alarm on seeing them place themselves in a line near the bank of the river and fire off several muskets. It was however a mere feu-de-joie. Those who had no muskets had each a calabash in his hand, which was emptied before we touched the ground. "Paiwa, Paiwa, cried our men, we have just arrived in time to partake of a Paiwori feast." And so it was, my old acquaintance Yamoriko, the Barokoto Captain, came staggering toward me and gave me his hand; an example followed by all the rest as well as they could. As I had to go through the ceremony of shaking hands with upwards of fifty[,] babies at the breast not being excepted, I was really tired. Some of the children from five to six years of age, struggled most valiantly to subject themselves to the friendly shake; but when they perceived that the colour of my face was so different from what they were accustomed to see, besides being mustachoed and whiskered they commenced crying most lustily.

The Barokoto who was already a fine looking Indian when I saw him in 1837 had become more portly in the interval. His beard was certainly stronger than I had seen it in any other Indian of pure descent. His two former wives were dead but he had consoled himself with two others, the one *enceinte* the other a young girl perhaps not yet fourteen years of age, who appeared so much attached to her husband that she followed him like his shadow. Their years were certainly disproportionate, for he was fifty or upwards. Such mis-alliances, as we should call them, are by no means uncommon, and the strong frame of Yamoriko made it probable that he would outlive many of the striplings around him. On entering the festive hut we observed a Corial or boat twenty two feet long and three and a half wide, which had evidently been filled with their drink, but which was nearly empty to the dregs. Next to it, stood a large trough about fifteen feet long, two and a half wide and one and a half deep, filled to the brim with Piawori, which was yet to be emptied before the feast should be over. The men were highly painted with Caraveru, and wore on their heads that tasty cap, made of the green feathers of the Parrot surmounted by the snow white plumes of the Harpy Eagle (Cocoi in the Taruma language, Guan, in the Macusi; the *Harpyia destructor*).[1] The ceremony of greeting their acquaintances and the stranger, by inquiring after every individual of the family occupied several hours.

At a later period they resumed their dancing which did not differ much from the same amusement as practised among other tribes. The men had large sticks in their hands, round which a number of seeds were tied that made a rattling noise when the stick was struck upon the ground. Others had merely their bows and arrows and other implements of war and the chase, in a word, every one carried something. They placed themselves round the large vessels that contained their drink, and having sent forth their loud and startling yells, followed by a shrill whistle which they produced by blowing through their fingers, they put themselves in motion, going round the troughs sometimes slowly sometimes quicker, stepping after each other keeping time and bending the whole body at each step. After this had been continued for some time another yell was set up, when the women, who had not been present during the first dance issued from a neighbouring hut led by one of their number having in her hand a maracca or rattle; the others followed in succession, each having her right hand on the shoulder of the one who went before, and all with looks bent on the ground. Some carried their babies, others a puppy or some other object in their left arm. They now joined the men in the dance, forming an inner ring near the troughs and going round them in a contrary direction to that of the men. Their step remained always the same however rapid or slow that of the men might be, and the[y] accompanied their dance with a low monotonous song.[2]

They continued their amusement till far in the night, and in order to avoid disturbance I had removed my tent to some distance. At an early hour in the morning I was woke by loud voices hard by. It appeared that some one of the votaries of the feast had indulged too deeply in his potations, and persisted in preferring the cold wet ground to his hammock to which his better half objected, and after a long debate she induced him partly by good words partly by threats to follow her into the house. The mode adopted among the Warrau Indians in similar cases is admirable. When the men are intoxicated with Paiwori and the persuasion of their wives remains without effect, the women join together and raising the refractory votary of the Indian Bacchus from the ground and placing him with great agility in his hammock, and with a rapidity truly surprising lace him in, where he remains like a mummy or a babe in its swaddling clothes, till he comes to his senses.

Yarimoko[3] was merely a visitor at the village, his own settlement was a few miles higher up the river. The information I received with regard to the Corentyne and the

[1] *Harpia harpyja*.

[2] See Goodall, *Sketches of Amerindian Tribes*, p. 68, for a view of the feast.

[3] This is presumably the same person as Yamoriko referred to above.

possibility of descending that river was quite contradictory. The Barokoto acknowledged ultimately that he had drunk too much the previous night and that his head was not yet clear. He seemed to have some object in view by deterring us from commencing our journey. He told us of water-spirits which dwelt in the Corentyne, and who wrought the destruction of any one who navigated that river, and when he found this had no effect he said the country we had to pass through was inhabited by Indian Sorcerers and tribes of very bad character, and that food was scarce. I treated his observations as they deserved, and in order to impress him with the power of white men I profited by circumstances as the great discoverer of the new world had done before us, and predicted that in the course of an hour the Sun would be partly obscured.[1] He related through our interpreter what I had told him, and a cloudless sky favored me by allowing the partial eclipse to be visible. I showed it to those who were sober, through the telescope of the sextant, well knowing that their relation would lose nothing by being retold. Yarimoko left the village the next morning (June 29th) and before going promised that he would accompany me to the Maopityan or Frog Indians,[2] where we might procure information with regard to the Corentyne, which he called Curitani, and that his people should prepare bread for us, as the Maopityans were short of provisions.

The Tarumas do not like to intermarry with other tribes; indeed the Barokoto Yarimoko, who is the head of one of the settlements is the only exception among those at the Essequibo and Cuyuni. A few more Tarumas are said to live among the Maopityans of whom some of the men have married Maopityan women. This repugnance to intermarry with other tribes must materially tend to keep down the numbers of this tribe which does not exceed one hundred and fifty among whom a great family likeness prevails. It must also be observed that as the number of women is not equal to that of the men, young girls before the age of puberty are admitted to the rights of married life, a practice which must injure their health and prevent their becoming mothers, or if they have children these usually do not survive many days or remain sickly beings all their lives. We saw at the Barokoto's place a young woman not more than twelve years old, or twelve rainy seasons as they term it, with a child at her breast. Another woman with an infantine face had already the sign of puberty, and had been a wife for some years without having had any offspring. This early intercourse and the circumstance that chastity is no virtue among unmarried women, among the Indians must also greatly tend to the decline of the tribes.

The language of the Taruma Indians is not so disagreeable, when slowly articulated, but it is their peculiar mode of speaking it and of uttering the first syllable of each word strongly through the nose that renders it inharmonious. It would take a European years to learn that nasal twang, that starting and raising of the voice that renders their dialect one of the most remarkable among the tribes I have hitherto visited. They possess the sound of the english th and the greek ph. While European languages are distinguished by the brevity of the numerals the Indians require great circumlocution to express a number above five, which being the number of fingers of one hand serves as a radical number. Six, in the Taruma language is thus expressed, "Oshekia akanna urapani

[1] On 29 February 1504, Christopher Columbus, while marooned on Jamaica, predicted an eclipse of the moon; a prophecy that so impressed the local native population that it stopped making difficulties for him.

[2] The few remaining Mawayana live today among the Waiwai and the Trio. They are an Arawak-speaking group.

ahumi" signifying that one finger from the other hand must be added to the radical number. Twenty is expressed, as in other Indian languages, by "one man," namely the number of fingers and toes a man possesses, "Oshe Coarse" in Taruma.[1]

It is very remarkable that they call an eclipse of the moon "Piwa-toto" being a compound of "Piwa" moon and "toto" earth. Are we to suppose that this is accidental, or are they aware that the interposition of the earth is the cause of the eclipse? and whence have they derived that astronomical knowledge?

Well acquainted with the aversion of the Indians to meddle with the dead or to touch their remains, I have experienced an almost insurmountable difficulty in procuring any skulls for the elucidation of the physical character of the different tribes. From what I had observed, however, of the Chieftain at Watu Ticaba, I was encouraged to offer him a large present, if he would give his assistance for the disinterment of one of the Wapisiana Skulls. I succeeded beyond expectation and though the skull which I procured was not quite perfect in consequence of the root of a tree having grown over it, nevertheless, it was sufficiently so for the study of its form. This success emboldened me to put the question to the Barokoto, who, to my astonishment entered immediately into the proposal, and observed that, if handsomely paid, he would give me the skulls of his mother-in-law, and of two of his former wives the elder of whom I had known on my former journey. He came this morning July 5th and requested that while he should go for the skulls, I would remain at the place to avoid suspicion. He accordingly started with Mr. Goodall, my coxswain and the interpreter to the site of the former settlement, and commenced digging where his former hut stood and soon exposed the bones of the younger of the two.*

*Judging from the skull and teeth, she could not have been more than ten years of age when she died, and as Yarimoko told me, she had never recovered after giving birth to her first child.

The elder was buried close to her, and Mr. Goodall relates that he appeared rather affected at seeing her remains, and told him that she had been a very good wife, who had always seen that his comforts were provided for. She was buried in a bark canoe, and with her was deposited a bottle and a drinking cup. On enquiry he told Mr. Goodall that she herself had requested it, that she might not suffer thirst while proceeding to the other world: the bottle still contained some water. In the grave of the younger were some glass beads and a few articles of dress. Next to her was the grave of her mother and of her young child. Near the remains of the child was lying a looking-glass and a broken cutlass; and near those of an old woman were some glass beads. The Child must have been very young as the skull was broken up and could not be removed. The other three were however in good order. As ready as Yarimoko appeared to part with the members of his own family he could not be induced to show Mr. Goodall and his party a grave, where a man was buried. The former he considered his property in life and death; but over the remains of an individual who was not connected with him by the ties of blood or marriage he considered he had no right.

As we were the first white people who paid a visit to the Maopityans I had, in the first place despatched messengers to inform to inform them of our arrival, and to request them to meet us at that place on the Essequibo, where we had to disembark in order to continue

[1] Uncertainty continues to surround the linguistic affiliation of the Taruma. Čestmír Loukotka (*Classification of South American Indian Languages*, Los Angeles, Latin American Center, University of Los Angeles, 1968, p. 150) classifies it as an 'Isolated Language'.

our journey overland. The time had now approached when we might expect to find them at the spot and I gave the necessary orders for our departure the next morning, the 8th July.

I must confess that of all the Indian tribes, the Tarumas appeared to us the most friendly and obliging. We lacked neither fish nor game during our stay. Whenever we found that our larder was getting empty hunting or fishing parties were undertaken, and it was sure to be replenished. That excellent fish the Haimura (*Erythrinus Macrodon*) was very abundant, and one which was caught in a trap measured three feet four inches and weighed thirty pounds and a half.

Forty circum-meridian Altitudes of α and β Centauri and γ Ursae Majoris gave me as a mean 1°43′58″N. for the latitude, and the Chronometer gave the meridian distance 59′ 9″ East of Pirara.*

*Pirara .... 59°20′ West of Greenwich.
2d. Taruma Place
on the Essequibo ... 59′9″ East of Pirara.

———

58°20′51″ West of Greenwich.

The following data are extracted from the Meteorological register

Number of Meteorological Observations from June 27th to July 6th inclusive = 81. The height of the settlement 767 feet above the Sea.

Date	Instruments	Mean	Maxima	Minima	Remarks
1843 June 27th to July 6th inclusive. 81 observations	Barometer	29.248	29th June 9 a.m. 29.341	6th July 6 p.m. 29.166	Weather quite variable frequent thunder, with rain, and severe fogs in the morning. The air generally calm & if there was a slight breeze, it came in general from the E by S. and E by N.
	Attached Thermometer	75°.11	30th June 1 p.m. 80°.60	28th 6 a.m. 67°.28	
	Detached Thermometer	75.22	80.60	67.00	
	Wet bulb Thermometer	73.92	77.20	66.00	

The greatest range, during one and the same day was on the 28th June when the Thermometer varied 12°.8 and on the 1st July the Barometer varied 0i.107.

The evaporation for the 12 hours of daylight was
On the 2nd July* 13 grains out of 1000 grs.
" 3 " 21 " "
" 4 " 17 " "

*27 June.[1]

[1] This is a marginal note but it is not clear what it means. If it is a correction and the reading was taken on 27 June the easiest thing to have done would be change the date. It could mean that an identical reading was obtained on that date.

July 8th. We started this morning at eight o'clock accompanied by several of the men from the Taruma settlement, and the number of our fleet of Bark Canoes was considerably increased in consequence. After we had proceeded for about three miles we halted at Yarimokos settlement where we found a large quantity of bread prepared for us which we bartered for cutlasses, axes, knives, &c. The chief himself accompanied us with his whole household including men, women, children, dogs, parrots and all that was moveable. He himself with his three wives, of whom the youngest had not yet arrived at puberty, occupied a bark canoe for their exclusive use of which he acted as coxswain and his wives as paddlers. I do not think they would have looked upon him with much affection if they had known that only a few days previous he had disinterred the skulls of his former wives to sell them to me and that a similar fate might await their own skulls, if they should die before him and another traveller direct his course to these regions. He wisely kept his dealings in organic remains to himself.

One of the Tarumas of our party, a handsome young man was attacked while en route by a strange disease. His tongue and teeth bled to such an alarming degree that towards nightfall he was sinking rapidly; Our small medicine chest was, unfortunately, not provided with a styptic and I was really glad that I succeeded towards evening in arresting the profuse bleeding with warm vinegar. His case was very remarkable indeed; the great discharge of blood did not come either from the lungs, the windpipe or its termination the bronchia, but from the tongue from which it oozed and from the place where, several years ago, a molar tooth had been extracted. He only complained of a weariness in his limbs and great chillness and the day after the bleeding he felt quite giddy. He observed that he had suffered in a similar way when his face had swelled considerably. I thought it best to send him back to his settlement the second day after the bleeding commenced, as I did not consider him strong enough to undertake the journey overland.

July 9th & 10th. We reached the mouth of the small river Urana[1] at nine O'clock. We had now to continue our journey overland, and accordingly abandoned our bark canoes. I was disappointed in not finding the Maopityans whom I expected to have met there. Our baggage was soon unloaded and put into a temporary hut, and as the Maopityans did not arrive in the course of the day, I started next morning, leaving the greater part of the baggage behind to be brought after us. We proceeded however only a few miles this morning (the 10th) when the barking of dogs announced the approach of strangers, and our Taruma messengers accompamied by fourteen Maopityans stood before us. The latter differed in figure and dress from the tribes I had hitherto seen. Although they were only of middle stature they were thinner and more bony than the Tarumas, their heads laterally flatter and the eyes brilliant, but what most astonished me was the peculiar way in which they wore their hair; it was plaited in a long queue which hung down the back, and in lieu of being wrapped round with becoming black ribbon as we still occasionally see it among gentlemen of the old regime in Europe, the Maopityan had inserted it into a tube from ten to twelve inches in length, made of palm leaves and ornamented with numerous strings to which feathers of all colours were attached. There were four females among their number who, if they were a fair specimen of the Maopityan ladies, gave us rather a good opinion of their looks. We admired

[1] The Urana River is a right-bank tributary of the Essequibo.

the ingenious if not pretty way in which they wore the cincture which attaches the only piece of dress the females of the uncivilized Indians wear in the interior. A piece of round bone very neatly worked and from which long strings of beads were suspended was worn as an ornament in their ears. The men used for a similar purpose pieces of bamboo about two inches in circumference which were passed through the lobes of the ear; and below the angles of the mouth were holes for the reception of small sticks ornamented with feathers which very much resembled the antena of a May-bug.

They greeted us with much cordiality and as a severe shower of rain detained us for some time at the spot where we met, many a curious glance was thrown at us and at our baggage. The rain descended in torrents. I resolved to proceed only a short distance in order that the Maopityans might fetch up the baggage we had left at our last night's camp. Their own burdens which they had brought with them were therefore set down and they started off in all haste. Scarcely, however, were they out of sight when the Tarumas in our company fell, like locusts, upon the Maopityans' baggage, and commenced an examination of it, which for minuteness might shame the overhauling of our strictest custom house officials. Nor did they confine themselves to a mere examination but began appropriating to themselves not only the eatables but everything else they took a fancy to. Perceiving my astonishment they endeavoured to bribe my approval of their proceeding by the offer of one of the combs which the Maopityans execute very skillfully; but to their great surprize I expressed to their Chief my detestation of their conduct and insisted that he should order his people to restore everything that could be restored. The smoked Maipure (*Tapirus Americanus*) and Apuya (*Dycotales torquatus*)[1] which they had swallowed in a ravenous manner, could not, of course be restored, but I saw that every other article was packed up again, as they had found it, and keeping watch over it, I did not leave the baggage until I saw every person on his march. Our Macusis and Wapisianas had naturally no hand in this attempted plunder.

July 11th and 12th. Our course during the first of these days was E.S.E. The path led us over hills from 100 to 150 feet high which alternated with low swampy ground, overgrown with the Manica Palm (a species of Euterpe) the graceful *Mauritia aculeata*[2] (here, however, of such a height, from 60 to 70 feet, that I have some doubts whether it be really the M. aculeata of H & B.[3]) numerous Scitamineae, the Turu (*Oenocarpus Bataua* and *Bacaba*),[4] and Muru-muru palms (*Astrocaryum Murumuru*), and that remarkable palm the Iriartea exorrhiza, which raises its trunk by means of a number of roots similar to the Banyan tree six to eight feet above ground and which may be compared to scaffolding from the middle of which the trunk rises.*

* "Radices 8, 10, 20 et plures altitudine 6, 8 pedum e terra emergentes atque in conum vastissimum dispositae" – (Mart.)[5]

[1] Tapir and peccary respectively.

[2] *Mauritia martiana*.

[3] Humboldt and Bonpland.

[4] According to Mennega et al., Turu is the Arawak for *Jessenia bataua*. The *Oenocarpus bacaba* in the same language is Lu. According to the NMNH checklist, *J. bataua* = *O. bacaba*.

[5] Von Martius, *Historia naturalis palmarum*, 3 vols, Leipzig, Frid. Fleischer, 1823–50, II, p. 36. Schomburgk has not cited the passage entirely accurately. It should read 'Radices octo, duodecim, viginti et plures, altitudine sex ad octo pedum e terra emergentes atque conum vastissimum peripheriae nonnunquam viginti pedalis dispositae.'

Of equal interest was a palm which our Macusis pointed out to me with great delight and which furnishes the outer cases of their blowpipes. I consider it an Iriartea though it has not that peculiar growth of its roots.

We pitched our camp near numerous blocks of granite and starting next morning at ½ past 6, crossed soon afterwards the small river Onoro[1] which falls into the Essequibo. Our path was now more level as it followed the valley of the river. After noon we reached the foot of a mountain which our limbs fatigued with a march of six hours, refused to climb. I therefore gave orders to pitch our camp and with M$^{r.}$ Goodall and two guides went in the direction whence the sound of a cataract seemed to promise one of those sublime scenes of nature so frequent in the interior of Guayana. We had advanced about a mile when we came in sight of a magnificent cascade formed by the Onoro which precipitates itself from a height of about 100 feet into the valley. Our Maopityan guides remained at a respectful distance not venturing near for fear of spirits. M$^{r.}$ Goodall and myself got as near as we could and then scrambled to the summit.

The black masses of rock which rose above the foaming waters were clothed with verdure and chiefly overspread with a Lycopodium interspersed here and there with a scarlet justitia. I found there a plant of great interest to me it was a Solaneae of humble growth but whose flower resembled an Auricula in colour. I was astonished at the total absence of Orchideae which generally delight in those situations where the spray of a cataract preserves a constant moisture. The high trees prevented our enjoying a prospect from our elevated situation, but we could see to the South West, the mountains we would have to ascend the next morning.

The Thermometer stood under the tent at 3 o'clock in the afternoon 78° F., at 6 o'clock 74°.2 F., the difference between the wet and dry bulb, was during the former hour 1°.6, during the latter 0°.8 or $^{2}/_{5}$ of a degree.[2] The Barometer varied between 29^{i}.148 and 29^{i}.114.

July 13$^{th.}$ Started Soon after six, ascended and halted forty minutes after seven on the summit of Mount Zibinia atzacko.[3] Bunten's Barometer was here consulted*

*Barometer 28.749. Attached Thermometer 70°.43 F. Detached Thermometer 71°.35 F. Wet bulb 70° F.

and we found that our height was approximately not more than 370 feet. We again crossed the Onoro rushing turbulently towards the large Cataract. We thus ascended & descended hill after hill and although the intermediate valleys were always considerably higher than our camp last night, we descended nevertheless, in several instances from two to three hundred feet, in order to ascend a similar height at the distance of a few hundred yards. These mountain valleys between hill & hill were swampy and overgrown with Manica palms and with the beautiful *Mauritia flexuosa* or Ita palm. As splendid as this latter tree appears in the Savannahs which seems to be its favorite place of growth, it cannot vie with the specimens I saw here. Some of the trunks attained a height of more than a hundred feet before the beautiful fan-shaped leaves spread out in tropical grandeur. Their luxuriant growth was really surprizing, the more so as I had hitherto seen them only on plains and arid savannahs, while here at an elevation of not

[1] The Onoro is a right-bank tributary of the Essequibo.

[2] It is $^{4}/_{5}$ths. This has been corrected in the published version (p. 51).

[3] Zibingatzako Mountain.

less than twelve hundred feet above the sea their summits stood one hundred and twenty feet above the ground.*

* "Crescit raro in elevationem octigentorum pedum supra oceanum adscendit regiones maritimas potius quam interioris terrae constinentem amans." (Kunth).[1] I observed them on Savannahs as high as 3300 feet above the sea near Roraima.

The other species with a prickly trunk (*M. aculeata*) grew to a height of from fifty to sixty feet, and was much more robust than the specimens I had seen at the Rio Negro, nor did it grow here in groups. The Maopityans call the Ita, Kibi; the Tarumas Yuro-i.

We traversed the summits Honicuri, Yiatzo,*[2] and Kabai okitza and after descending into the valley between the sixth and seventh summit, we found the first rivulet which flows towards the Amazon. It was then about ten o'clock and our Barometer indicated a height of eleven hundred and thirty feet above the sea. The ridge which causes the division of the basins of the Essequibo and the Amazon, in these regions is one hundred and twenty feet higher. The small rill is the Caphiwuin or Apiniau[3] which receives the Wanamu[4] and forms at their junction the Caphu, the river Trombetas of the Portuguese.

We now followed more the direction of the ridges of the hills, having previously crossed them transversely. Our course continued E. by S. deviating scarcely half a point. The narrow valleys or glens which divided the hills descended towards the East. Having passed the summits Ketia-una and Kenuckawai,[5] the latter the twelfth hill in the course of the morning, we traversed three more and then descended towards the river Darura, one of the first tributaries of consequence which joins the Caphiwuin or Apiniau. It was just noon, the greater part of our people far behind and even Mr. Goodall, who otherwise proved himself such an excellent pedestrian had dropped among the reserve, and I was about to give orders for pitching our camp on the banks of the clear mountain stream, but our Maopityans anxious to reach their home described the distance of their settlement as only a cople of miles and after a rest of half an hour we commenced our march anew. I found it however, a good five miles before we entered the provision grounds and heard the barking of the dogs announcing our arrival.

We saw before us two huts, the one of uncommon large size, the other smaller, but both overtopped by that peculiar small roof which gives to them an asiatic appearance, the more since a number of pieces of wood, cut in different shape hung up along the eaves which the wind moved to & fro.[6] We directed our steps to the smaller of the two, where we were welcomed by the Chieftain (a Taruma by birth) and all the other inhabitants who had not come to meet us at the Essequibo. It was ridiculous to see with what fear depicted in their faces, the females gave us their hands; they had, no doubt, been schooled to tender their welcome in that way; but if they had expected an electric shock, the hand could not have been offered under greater apprehension. Our Macusis and Tarumas, who considered themselves so much better acquainted with European politeness laughed most

[1] Karl Kunth, *Enumeratio plantarum omnium hucusque cognitarum, secundum familias naturales disposita, adjectis characteribus, differentiis et synonymis*, 4 vols, Stuttgart and Tubingen, J. G. Cottae, 1833–43, III, p. 219.

[2] Honikuri-Yiatzo Mountain. No footnote accompanies this marker.

[3] Rio Cafuini.

[4] Rio Anamu.

[5] Kenukawai Mountain.

[6] See Goodall, *Sketches of Amerindian Tribes*, p. 76, for an illustration of one of these houses.

heartily at their awkwardness. I noticed among them a young woman with a frightful tumour on the left side of the abdomen above the hip, and what rendered her appearance still more shocking, she was enceinte. She possessed much more courage and vivacity than the others and with a smile gave me a hearty shake when I offered her my hand.

After the first burst of welcome was over I went to the large hut, a dwelling of all that remained of the tribe of Maopityans or Frog Indians. The interior of the hut was similar to those of the Tarumas, but surpassed in size any I had seen among the Indians. It measured eighty six feet in diameter and was of a proportionate height. That peculiar ornament, the painted trunk of a tree which the Maopityans call Aiyukuba was more adorned with Indian figures than I had seen it among the Tarumas.

The flatness of the head and consequently the long face and short circumference is peculiar to the tribe. I have not been able to learn upon the most minute inquiries, that the form is given to the head by artificial means. The occiput of the men is high and almost perpendicular above the front; the frontal bone is small with regard to extent, and in no comparison to the face below the eyes; the cheekbones are harsh and prominent; but the most remarkable part of the head is the great extent between ear and ear, if measured from the upper part of that organ and the line continued above the eyebrows, to the commencement of the other ear.*

*I add a comparative measurement to substantiate the foregoing statement.

	Maopityan Indians				Taruma	Wapisiana	Atorai
	ft in	ft in	ft in	ft in	ft in	ft in	ft in
Circumference of the head along the frontal sinusses and temporal ridge	1″ 9.3	1″10.	1″9.8	1″ 9.1	1″10.6	1″10.2	1″10.3
Long diameter of face	0″ 9.5	0″9.1	0″ 9.8	0″10.1	0″10.1	0″10.2	0″10.1
Diameter between the processes of the helix	0″11.8	0″12.5	0″13.1	0″12.0	0″11.2	0″11.1	0″11.0.

It surpasses the measurement of other Indians generally by an inch or two. The lower jaw-bone is of great depth, a formation which is generally considered as a sign of animal propensities, which however their high front seems to counterbalance. Their noses are good. The features of the females are regular, the brow delicately arched; the eyes are large and black and like those of other Indian females possess fire only when animated, at other times their usual expression is that of diffidence and bashfulness, and are seldom raised to the speaker. The average stature of the men is five feet six inches, that of the females four feet ten inches. The tallest of those who form the remnant of the tribe was only four feet eleven and a half inches; but their form is good, and the hands & feet very small.

The bows of the Maopityans are larger than those of the Macusis and Wapisianas being generally from six feet ten inches to seven feet in length. The lower or convex side runs more in an edge than in the bows of the former tribes and they are likewise differently strung. Iron is still scarce among them, and the greater number of arrow points are made of bone. They possess a kind of arrow poison which they prepare from a root, which was the only part of the plant I saw and from which I judge it to be herbaceous. The poison is neither so strong nor does it retain its quality so long as the Urari of the Macusis. The remarkable tube, or blowpipe, the Cura of the Macusis, is known to them only from description, nor did I see any fire-arms among them; but when I recollected that on my first visit to the Tarumas in 1837 the whole tribe possessed only two old fashioned fowling-pieces which they were sadly afraid to fire off and that at my present visit I saw scarcely a man without one, the Maopityans may as easily procure themselves these weapons so superior to their bows the more so as they are considered good trainers of hunting dogs, and among the Indians of the interior a good hunting dog like a marriageable girl commands a gun.

One of the men dressed himself in the way they use for great occasions and he felt quite proud when the artist of the expedition Mr. Goodall, took his portrait.[1] The upper arm is ornamented with bracelets which are four inches broad and which are made by winding the young leaflets of a palm round a form of wood of the same thickness as the arm: they are afterwards ornamented with Indian figures. They are worn round the upper arm, and a bunch of the large tail feathers of the red and blue Maccaw (Macrocercus aracanoa)[2] are fixed between the arm and the bracelet overtopping the head of the wearer by five or six inches; add to this the two parcels of parrots' feathers attached at right angles to two pieces of wood fixed in holes that have been bored, when young, below the angles of the mouth, and which, as already observed may be likened to the feelers or antennae of a May bug, & besides all this that peculiar head ornament, the queue encased in a tube and adorned with numerous strings of red cotton, to which feathers of divers colours are attached, and it will be confessed that the Maopityan in his holyday dress presented a very peculiar appearance. He had in his hand a thick piece of bamboo which in consequence of its being hollow, makes a loud noise when stamped upon the ground; but which for the sake of more noise is surrounded with hard shells of a seed*

*They appeared to me to resemble a Thevetia.[3]

which adds a peculiar shrill sound.

These people are very ingenious; the combs which they manufacture are really handsome. The teeth are made of hard palm wood and fastened into a piece of bone. At the distance of an inch and a half below this bone are fixed two pieces of palm wood, one on each side of the teeth and the space between the two pieces and the bone is plaited with red and white cotton which serves both as ornament and for fixing the teeth firmly. Like the Tarumas, they do not cultivate much cotton, and prepare their hammocks of the fibres made of the young leaves of the Mauritia palm. Their waist cloths are of spun cotton but they barter them from their next neighbours the Woyawais, who, it appears, are great cultivators of the useful cotton-plant.

These Maopityans are the last of their tribe; their whole number amounts to only thirty nine individuals, namely fourteen men, eleven women, eight boys and six girls. They were formerly divided into two small settlements, but latterly they united as if a sentiment of their approaching extinction had brought them together.

[1] See Goodall, *Sketches of Amerindian Tribes*, p. 74.

[2] It is difficult to identify this bird. The *aracanoa* could easily be a copying mistake for *ararauna*, but *Ara ararauna*, the Blue and Yellow Macaw has no red plumage. On the other hand, both the Scarlet Macaw (*Ara macao*) and the Red and Green Macaw (*Ara chloroptera*) have some blue plumage, so the reference may be to one of these last two.

[3] This is likely as the shells of the seeds of *Thevetia nereifolia* are widely used in the interior of Guiana for such a purpose.

In their vicinity was a small settlement of Taruma Indians, the Chieftain of which they invited to preside over them as Captain and he and his people moved over to the Maopityans, who thus connected themselves closer with the Tarumas. They are now living together in the great circular hut forming a community of sixty souls, isolated from other Indians by thick forests and high mountains; their nearest neighbours being, to the south, the Woyawais, to the west, the Tarumas at the Essequibo, both about four or five days'journey from them. They call themselves Mawakwa; the Wapisianas however call them Maopityans, from "Mao" a frog, and "pityan" people or tribe.

Their provision fields were very extensive, but they had been improvident, and the Cassada plant was not yet ripe for use; and the information of Yamoriko was quite correct that they mixed rotten wood with their Cassada flour to make it last the longer. Our glass beads and knives were, however too seducing and the women (to whose department the barter of the produce of the provision grounds belongs) readily parted with what they could spare.

I was naturally anxious for information as to the direction we should take in order to fall in with the head waters of the Corentyne. It appeared the intermediate space between this settlement and the river was quite uninhabited, and although the people knew of the existence of a river of the name of Curuni, the only practicable way, they said, to reach it would be to follow the Caphiwuin or Apiniau until it is joined from the North by the Wanamu then to ascend the latter to the settlement of the Pianoghotto and Drio Indians.[1] The Frog Indians were however no great navigators and their fleet was most miserably inefficient consisting merely of a few bark canoes in a rotten state. I therefore gave the necessary orders to construct woodskins or bark canoes for our own navigation and as in expectation of such straights the coxswain had been ordered to provide himself with the necessary tools and principally a pair of good American axes,*

*A traveller through the forests of the Interior should consider an American axe, an indispensable tool. A dozen other axes of the best european manufacture, will not prove equal to a good American felling axe. Two of these axes have been used during our expedition, which lasted four years and were at the end of that time as serviceable as at the commencement.

on our starting from Pirara. All hands were soon engaged in felling trees for the construction of our craft. I engaged at the same time two Maopityans to proceed next morning to the Pianoghottos at the Wanamu to inform them of our intended visit and to request them to meet us with a supply of provisions, as the scarcity among the Maopityans did not promise a large supply.

The Indians between the fourth and fifth parallels of latitude generally make use of the Bully-tree (*Mimusops Spec.*?)[2] for the purpose, but as this useful and magnificent tree does not grow here we had to use two other trees, one apparently allied to the Locust (*Hymenaea Courbaril*) and the other the white Maran or Balsam Capaiva tree. I was quite astonished when I became aware of the fact that the latter tree is likewise used for such a purpose.

[1] The Drio Indians are the Trio Indians, a Carib-speaking group that lives either side of the Brazil–Surinam border. The identity of the Pianoghotto (Eagle people) is more problematic. They no longer exist but were almost certainly a subgroup of the Trio. For a discussion of this, see Protasio Frikel, 'Das Problem der Pianakotó-Tiriyó', in *Beiträge zur Völkerkunde Südamerikas*, ed. Hans Becher, Hanover, Kommissionsverlag Münstermann-Druck, 1964, pp. 97–104.

[2] *Manilkara* spp.

The Coxswain who had been present when they felled the trees, told me that when they came near the heart the Balsam gushed out in such quantity that several gallons might have been collected. In the absence of bottles they brought several joints of the bamboo filled with it which being ultimately put into quart bottles filled five of them. It was perfectly white and transparent. The Maopityans like all other tribes call this plant Maran, but its medicinal qualities are unknown. Indeed I have never seen an instance of that peculiar disease for which it is used in Europe, among the tribes of the interior.[1] The Indians use the balsam like palm oil to anoint their bodies with.

The layers of the liber of a species of Amyris were found by Cailliaud[2] to be used by the Nubian Mohammedans as paper on which they write their legends;*

*Lindley's natural System of Botany 2d Edit p. 165.[3]

we may therefore conclude that the bark parts easily from the wood and this quality seems to belong to the whole order of Amyridaceae, to which Lindley has joined the genus Copaifera,**

**I do not follow Endlicher in his 'Genera Plantarum'[4] who considers Copaifera as belonging to the Leguminoseae, though it possesses a leguminous fruit; in every other respect it seems to belong to Amyridaceae.[5]

and the Indian selects the bark of these trees for the preparation of his slight skiffs. Some of these trees however must reach to an enormous size, as a single one gave us two bark canoes each thirty five feet long, and four feet five inches wide. The bark of the tree which I considered to be a species of Hymenaea is much thicker than the former and nearly half an inch in thickness. The wood is white and appears to be rather soft, but the heart is heavy and close-grained and apparently of great durability. One of these trees which was cut down near our camp measured one hundred and twenty five feet from the base to the top of the branches. The Maopityans call this tree Yaru-yaru.

July 16th. After suffering the pains of labour for nearly three days the young woman with the large tumour, this morning brought a child into the world. Previous to her delivery she had been kept in a small out house, but had afterwards walked to her accustomed place in the large hut. The Indians invited me to see the infant, and accordingly, provided with some suitable presents, I went. I have before observed that this women was more lively than the others and when she saw me approaching her hammock, she held the child up and appeared particularly proud that her first-born was a boy. Nature had here done all. If her deformed body and protracted labour be considered it is surprising in how short a time after, she could sit up, and smile at her baby and at the presents it received from me. She is a Maopityan and the second wife of a Taruma the father of the child. Polygamy appears very common, but in the present instance at least,

[1] Venereal disease.

[2] Frederic Cailliaud (1787–1869), French traveller and writer. He published extensive accounts of his journeys in Egypt and up the Nile between 1815 and 1822.

[3] John Lindley, *A Natural System of Botany: or, a Systematic View of the Organisation, Natural Affinities, and Geographical Distribution, of the Whole Vegetable Kingdom*, London, James Ridgway & Sons, 1838.

[4] Stephen Ladislaus Endlicher (1805–49), Austrian botanist and author of *Genera plantarum secundum ordines naturales disposita*, 1 vol. and 4 supplements, Vindobonae [Vienna], Apud Fr. Beck Universitatis Bibliopolam, 1836–40.

[5] The genus *Copaifera* belongs to the family Leguminosae (Caesalpiniacaea).

it is clear that beauty was no motive of choice in the second marriage – the first wife who has had two children was a fine looking Indian.

The newborn child had all the characteristics of the mothers tribe. It was not quite an hour old when I saw it, and the flatness of its head as compared with the heads of other tribes was very remarkable. Its size was 14¾ inches, the circumference of its head 13¾ inches. A female child of the same tribe not quite two months old measured 17 inches, the circumference of the head 13¾ inches, the circumference round the abdomen 14¾ inches. A third one four (lunar) months old measured 23 inches and was 17½ inches in circumference around the abdomen. In all three cases the profuse hair on the head of such young children appeared to me remarkable.

We had most unfavourable weather during the commencement of our stay at the Maopityan settlement. George Town on the coast is famed for the severe showers of rain to which it is subject; but the torrents which we experienced on the night of the 15$^{th.}$ July surpassed in violence anything I had ever witnessed; the noise of the descending rain overpowered the thunder which could only be heard when the thunder cloud passed over our head and clap followed clap. The succeeding morning was fair and cool, and the thermometer which during the previous rainy days had stood at 70°·F. now indicated only 69°·F. a degree of freshness which rather surprized me as we were not more than 880 feet above the level of the sea and near the equator.

The latitude of the settlement was to 1°25′18″N.*

*This is the mean of 43 circum meridian altitudes of * Centauri & γ Draconis.

and its distance by Chronometer (Frodsham's N°· 369) 14′37″ East of the Taruma place on the Essequibo.*

*58°6′14″ West of Greenwich.

The following table will give the necessary comparison; it rests upon forty observations –

Period & number of observations	Instruments	Mean	Maxima	Minima
From July 14 to July 18. 40 Observations	Barometer	$29^{i}.098$	15$^{h.}$ July 10 a.m. $29^{i}.173$	16$^{h.}$ July 5 p.m. $29^{i}.022$
	Attached Thermometer	72°.57	15$^{h.}$ July 3 p.m. 80°.42	17$^{h.}$ July 6 a.m. 66°.20
	Detached Thermometer	73.04	80.50	66°.00
	Wet bulb Thermometer	71.72	75.70	65.90

18$^{th.}$ July. Our bark canoes being ready, and the women having prepared us as much Cassada flour as they could spare in their straightened circumstances, we engaged six men from the village to accompany us to the first Pianoghotto settlement, and started on the following morning.

The Caphiwuin was scarcely fifteen yards broad where we embarked and full of rocks and otherwise obstructed by trees that had fallen across, through which we had in many instances to cut a passage for our boats; this together with the want of skill of the Maopityans frequently endangered our frail canoes. Indeed the awkwardness with which the Maopityans paddled drew down upon them the ridicule of our Macusis, who considered themselves very superior boatmen.

Several small rivers joined the Caphiwuin after which it increased in size and forced its way turbulently through the mountain chain. We passed Mount Yucawari on the 20th. I estimated its summit at 1500 feet above the [river] which was about the average height of the hills we had hitherto passed. The direction of these ridges is by no means uniform and deviates generally between North and East and North and West.

The flora of the banks of the Caphiwuin, exhibited all the luxuriance of the tropics. The scarlet Isertia,[1] the blue Petrea, the long pendulous flowers of the Posoqueria[2] with its rounded fruits of citron yellow and large green shining leaves[,] the flowing risps of Aublet's *Carapa Guayanensis*; the Ant tree (*Tachigatia pubiflora*)[3] with its upright panicles of yellow blossoms; the splendid *Clusia insignis*[4] with its large rose-shaped flowers and aspect as if formed of virgin wax, each petal tinged with rose, Aublets crimson Eperua, remarkable for its falcate pods pending from the trees on thread-like peduncles[;][5] the banks of the river adorned with graceful bamboo bushes and gigantic Musaceae, numerous palms and the *Theobroma cacao* growing spontaneously, the majestic Bertholletia, all united to render the picture highly interesting to the botanist and delightful to the eye. The falls and rapids became now numerous; we halted in the evening of the 20th. at the head of Uwiya the first fall which was of any consequence. The dykes which caused it were slaty greenstone and according to the chronometer we were then ten miles east of the Maopityan Settlement.

We passed, soon after noon of the 21. Mount Caramuzin about 2000 feet above the river or 2850 above the sea. It is on the river's left bank and is in its form more pointed than the other mountains we have passed. Our camp was pitched this night near another fall. The rock was here of greenstone passing into clay-slate and stretching right across, they caused the river to fall perpendicularly as over a mill dam into the basin below. Some cavites in the hard rock were very remarkable. They were shaped like soup plates or saucers, about eight inches in diameter and two inches in depth and quite smooth. I was unable to account for their formation. Observations of Northern stars gave for the latitude of this place 1°23′23″ and its meridian distance by chronometer was 13′11″ East of the Maopityan settlement.

The following day, we lost two of our bark canoes. One ran against a rock and split, and the second sunk; two of the Maopityans went through the heavy surge at the foot of one of the falls which caused it to be swamped.

We had proceeded only a few miles on the morning of the 23d July when our progress was stopped by the large Cataract Wamaru Serrika. Here we had to carry both the boats and the baggage over a distance of 700 yards which occupied the whole day. A barometrical measurement gave me 45 feet for the total height of the fall.

July 24th. The difficulties which the navigation of the river presented were still further enhanced by our not understanding the language of our guides. It appeared that only one of them had been down this river before, and then, as far as we could understand him, he was very young. This man was moreover sullen & not favoured by nature with much discernment. I gave him a seat in my own bark-canoe, which, by the Indians, is

[1] A genus of the family Rubiaceae.

[2] A genus of the family Rubiaceae.

[3] *Tachigali pubiflora*. The Guyanese Creole's Ant Bush is *Siparuna guianensis*, and Ants Tree, *Cordia nodosa*.

[4] *Clusia grandiflora*.

[5] Probably *Eperua falcata*.

Figure 10: Descending falls on the River Caphiwuin. Watercolour by Edward Goodall, reproduced by permission of the British Library (Add. MS 16936/32).

considered an honour, but it did not change his manners. We had scarcely advanced a mile on the morning of this day when another large fall was before us. We had again to unload and carry the baggage about 600 yards overland. The boats were lowered down the fall, and here I met with a painful accident. Anxious to witness the passage of the corials over the impediment, I crossed a small island from whence I could observe the manoeuvres of the crew. Deeply intent upon what they were doing I paid no attention to the objects immediately around me and had placed myself unwittingly and without hat, under the large nest of a brownish species of wasp. I must have touched it and roused their ire. The first intimation I had of the dangerous proximity was the violent pain I experienced from their stings in fourteen different parts of my head, and not being able to get the wasps out of my hair, they continued to inflict their wounds. Had I not been ashamed of showing any weakness before the savages, I might have cried out with the pain; which, strange to say communicated itself to the right arm chiefly under the arm pit. I passed a feverish night and felt the consequence of the stings for some days after. Having passed some more cataracts and rapids the river became comparatively smooth. It had scarcely increased in breadth for the last fifty miles, and in lieu of mountains groups of small hills scarcely 150 feet in height approached the banks while the stream itself was studded with numerous blocks of granite of the usual spherical form.

July 26th. The trees near the banks of the river were this morning enlightened by large flocks of small monkeys; they consisted chiefly of Pisas and Yarkis (*Pithecia Spec.*? and *Cebus Capuchinus*)[1] and some squirrel-monkeys (*Callithrix Sciureus*),[2] and astonished at seeing human beings passing their abode, they exhibited every attitude and grimace, which their agility and curiosity at the strange sight of us intruders called forth. We had hitherto been accompanied to a general silence, interrupted only by the noise of falling waters, our own voices and the splash of the paddles. This day it was otherwise, several feathered songsters raised their voices, discordantly broken in upon by the noise of an assembly of hawks (*Ibycter leucogaster* Vieill.) known in the Colony by the name of bull-dogs, and which when perched on the summits of the highest trees near the banks never fail to greet the passing boat with their stunning cry.

A large Coaita monkey (*Ateles paniscus*) resting with its long legs on a curved branch, while it kept its erect position by grasping with its spider-like arms some branch above it, reminded us forcibly, (at the distance we were from the animal) of the human structure the more so as she was a mother and had a young one clinging to her breast. We had not proceeded a great distance when the howling in unison of half a dozen Araguatos or howlers (*Stentor seniculus* Geoff.) resounded like the roar of lions through the surrounding forest. They did not observe us and we were able to come quite close to the trees where they were sitting. I counted six. Amidst the deep bass of several old ones, was distinctly heard a shrill fine voice, proceeding from what is called by the creoles "the Singman" or precentor; no community or group of howlers is without him and he is naturally distinguished by his diminutive size as compared with the others. So say the Creoles and although the fact is uncontested that whenever they commence their dreadful howling the shrill voice is heard amongst them, we as naturalists cannot agree with the woodman that nature should select one from the rest and give him a smaller size that he may act as leader in the discordant concert. It is most probably a female or a young aspirant who has not yet acquired the sonorous bass voice of the aged.

We had commenced shortening our allowance, and the Indians could not resist the temptation of replenishing the larder. The crack of the percussion cap,* the report of the gun, the cry of the startled singers and the heavy fall of one of them was the act of a few moments.

* The Macusi Indians call the percussion cap tacki-pang, the sound of the words intimating the explosion of the cap, and the heavier report of the gun.

The hollow sound of rushing waters informed us that we were approaching another cataract. We had passed in the course of the morning numerous small hills on our right and left without any serious obstacle interrupting our navigation, but on approaching a hill which extended north & South we saw another fall before us. The Maopityans call it Karamatahura: it afforded a very remarkable appearance for the greater volume of water flowed about 150 yards along a raised ledge of granite, like a natural aqueduct, in a South east by east direction, about 20 feet above a smaller branch of the river, which

[1] Schomburgk seems to have these names the wrong way round. There is a species of *Pithecia* (*P. pithecia*) called Yarké in Guyana, but this would then suggest that he is calling *Cebus capucinus*, rather than *Chiropetes satanas*, Pisa.

[2] The squirrel monkey is now placed in the genus *Saimiri* of which there is a species *scireus*.

precipated at the commencement of the ledge, into the basin below, and then flowed gently along the foot of the natural aqueduct, while the waters above were rushing turbulently towards the abrupt termination of the granite ledge, where it again united with the former stream by forming a large cataract. These shelves of granite astonished me by their vastness; the action of flowing water having denuded whole tracts of their earthly covering, they now appeared like gigantic terraces.

The rocks were covered with that strange water plant the Wirinye, belonging to the genera Mourera, Podostemon, Lacis, &c. and the river having fallen in the course of the last two days, they now raised their branches above the water, and were in full bloom. Here I secured a specimen of that kind of fish of a dark-blue colour, which on the one hand resembles closely the Pacu (*Myletes Pacu*) on the other Pirai (*Serra-salmo niger*)*

*Vide Fishes of Guayana in Naturalist's Library Vol.1, p.225 & 236.

and of which I saw the first in the river Padamo, a tributary of the Orinoco.**

**Journal of the Royal Geographical Society vol.X p.241.

The Maopityans call it Umursu, but with the exception of Sororeng, who had already accompanied me as interpreter on my former journey to the Orinoco, none of our Indians had ever seen one before.

While our boats were being lowered down the Cataract (which nearly occasioned the destruction of my own skiff) M$^{r.}$ Goodall occupied himself with making a sketch of this remarkable fish of which it is my intention to give elsewhere a full description.

Scarcely had we overcome this obstruction and proceeded half-a-mile further when we halted at another fall, where the river shooting over an inclined plain, for the distance of about 400 or 500 yards formed a precipitous cataract at its extremity. We had again to unload and as the day was already so far advanced I gave orders at once to encamp at the foot of it.

The black Pacu abounded here, our crew waged war against them with bow & arrow & cutlasses, indeed a wholesale slaughter took place and upwards of twenty were secured. It was quite an enlivening scene, every person wished to carry a prize away and many a tumble took place, when, from over eagerness, they paid no attention to the slippery rocks, or when too venturesome the force of the current carried them partly downwards to the foot of the fall.

We made out from our guides that this was the last fall of consequence we should meet with in descending the Caphiwuin. The Barometer showed a descent of two hundred and twenty two feet from the place of embarkation at the Maopityans, which in a direct line is merely 50 geographical miles from Karamatahura, which makes about 4.4 feet pr. mile. I formerly considered the Padamo (a tributary of the Upper Orinoco) as one of the rivers the most obstructed by falls and rapids, but the Caphiwuin far surpasses it. The latitude of the foot of the lower fall was 1°21′50″N. and its longitude 57°16′50″ West of Greenwich. The Mean of Barometrical observations from three o'clock in the Afternoon to six in the evening 29^{i}.349 Attached thermometer 79°.7 F. detached 79°.9 F. from which date the descent mentioned above has been reduced.

July 27$^{th.}$ This morning we chased a jaguar which was swimming across the river, but we were too far off to intercept its course. He turned himself leisurely round, throwing a glance at us, and measured the distance our boat was from him, and no doubt satisfied that he was safe, he continued his course, bearing his tail curled high above the water.*

*They always swim across the river with their tail cocked up, and we recognized by this circumstance whether it was a jaguar or any other animal that swam before us.

He ascended the bank by means of a tree that had fallen into the river, and shaking the water like a poodle-dog from his spotted skin walked leisurely into the thicket without giving us even a second glance. My rifle missed fire twice to my great annoyance as I felt most anxious to send a ball after him, if it had only been to make him trot a little faster out of our sight.

The jaguars appear to be very numerous here about; it was only yesterday that we had seen the marks of their feet near our camp, and a few days previously we saw one crouching on the trunk of a tree; when the boat got quite near, it jumped down and scampered into the bush. But our black cook[1] tells the best story, and still trembles at the recollection of his adventure. One night, he says, while lying awake in his hammock, he saw a huge animal who came up close to him smelling him all over. He feared to move and seeing a pair of eyes like burning coals, shut his own, but when the beast's snout came near his face, he could retain himself no longer and giving a tremendous shriek which awoke the whole camp, he jumped out of his hammock and nothing in the world could induce him to return to it that night. He preferred sleeping on the ground under the tent of the coxswain.

We passed about noon, the river Camu (Sun river)[2] which joins the Caphiwuin from the N.N.E. and is nearly of the same size as the recipient. It is now inhabited[3] and has its source, as we were given to understand, in some high mountains. After this junction the Caphiwuin turns a point more to the South and some of its reaches trend even to the West of South. Its banks are low, and it no longer meanders through small hills as it did yesterday. I estimated its average breadth at about 300 feet.

July 28th. An hour before noon we passed one of those temporary huts which the Indians when on fishing expeditions or while travelling erect on the banks of the river, and miserable as they are, they attest that human beings have constructed them for their abode. We had not seen one since we had left the cataract Uwiya, a sure token that ages had elapsed since human beings had travelled up or down this river. Uwiya appears to be the "Ultima Thule" of the Maopityans down the river & the hut we had just past "the farthest" of the Pianoghotto Indians upwards, while the intermediate 70 miles was the haunt of the Jaguar and Tapir, and as our Indians no doubt believe, the abode of all kinds of hobgobblins and other spirits mischievously inclined towards the human race.

A small fall which however occasioned us some difficulty induced me to encamp.While they were pitching our tents I observed some smoke at a short distance before us, a sight hailed with delight in the wilderness as a sure token of the presence of man. One of the bark canoes was quickly brought over the fall, and with our precious guide, the coxswain, and Sororeng we pulled towards the place and eagerly climbed the

[1] It is not clear who this is. The black cook, Hamlet Clenan, had been dismissed for theft. See p. 106 above.

[2] Rio Camú.

[3] This is a secure reading of the manuscript, but in the published version (p. 63) it reads 'uninhabited'. This seems more likely as otherwise Schomburgk would almost certainly have mentioned who lived there.

steep banks in expectation of finding people. The furious barking of half-a-dozen dogs soon proved that we had not been mistaken.

We found a man, a young woman, no doubt, his wife, and a young girl arrived at puberty, but who, like the poet, seemed to think that "beauty is when unadorned, adorned the most,"[1] and accordingly despised even the fig-leaf. A boy about thirteen years of age and in the same naked state as the young girl came out of the wood curious to see what caused the barking of the dogs and appeared awfully frightened at sight of my white face. It was a Zurumata family, a branch of the Pianoghotto tribe who were clearing a new provision field. Two small panaps or temporary huts served them as shelter. The man had all the appearance of a Maopityan, nor was the pig-tail wanting, sheathed in a tube of palm leaves & ornamented with coloured feathers. He wore below the knee a band of cotton strings, from which hung a large tassel, coloured red. Round the ankles were tied some strips of palm leaves. The young woman did not differ in her attire from the generality of the Indians; but her qué-yu,[2] which is usually made of glass beads, when this so much coveted article is to be procured, appeared to be made of seeds. The other two individuals, as already stated were naked. This family appeared to be short of provisions and we understood it to be their intention to return the next morning to their settlement, which they hoped to reach in five days.

The junction of the Wanamu and Caphiwuin appeared to be a few miles to the Eastward, but as I could not ascertain whether we should find people I was uncertain as to our route.

We returned to the Cataract where the ledge of rocks promised me an opportunity of taking astronomical observations for latitude; but with the exception of some circum-meridian altitudes of Mars and *γ Draconis, my object was frustrated by thick black clouds which rose in succession to the North East and spread over both hemispheres. Numerous shooting stars, some apparently taking their course upwards, became visible when the opaque clouds did not interfere and proved that the upper atmosphere was clear. The Barometer stood at half-past nine o'clock 29.382. The attached thermometer 73°.4 F. Our latitude was 1°7′N. and our meridian distance from the Maopityan settlement according to the Chronometer 1°13¾′ East.

July 29th. We reached, after three hours paddling, the junction of the Wanamu and Caphiwuin. The former joins from N. by E. while the last reach of the Caphiwuin before the junction is N.81E, the united streams after their confluence continue their course towards the East by South, taking ultimately, if the information of the Indians be correct a more southern coast. The Pianoghotto & Maopityans call the two rivers from their confluence, Kaffu or Caphu. I estimated their united breadth at from five to six hundred yards; the Wanamu is, at its mouth, about three hundred and fifty yards wide, and the Caphiwuin of a similar breadth. According to my observations of last night the latitude of the junction was 1°2½′N. and its longitude 56°48′43″ West from Greenwich. The Barometrical observations gave me a height of 540 feet above the Sea. From the

[1] Probably a slightly misquoted version from the Scottish poet James Thomson (1700–48), *The Seasons*, 'Autumn', l. 204.

> 'For loveliness
> Needs not the foreign aid of ornament,
> But is when unadorned adorned the most.'

[2] The apron worn by Amerindian women of the region.

information that I afterwards procured I have no doubt that the Caphu is the river Trombetas Oziximina or Acunhas[1] Cunuriz which falls in 1°57′,[2] near Obydos, into the Amazon.[3] The junction is, according to Von Martius 451 Parisian feet (about 480 English feet) above the sea.*[4]

*I consider Von Martius's numbers too high; he gives for Barra de Rio Negro 522 Parisian feet, (556 English), a height which Fort San Joachim does not possess though it is nearly 300 miles higher up the Rio Branco which has several rapids and a large fall.

The river Trombetas is remarkable as being one of the last passes where the fable of the existence of the Amazons has placed the republic of warlike women, who only once in the course of the year, namely in April received men into their society. It was at the mouth of this river where according to Father d'Acugna, Orellana found, in 1542, women fighting among the men;[5] and on my inquiries while travelling at the Rio Negro, that river was always pointed out to me as the one at whose sources the Amazons resided. The upper branches of the river Trombetas were perfectly unknown; large cataracts and the fear of savage Indians had prevented the inhabitants of the lower Amazon from ascending that river for any distance, and for want of better information it was the subject of the strangest stories.*

* M. Montravel[6] in his late report to the French Government, respecting the expedition up the Amazons, for the purpose of surveying that river, observes that the inhabitants near the mouth of the Trombetas pretend that it possesses rich gold mines near its sources. The regions through which I travelled did not exhibit the appearance of being rich in precious metals.

The Caribs of the Corentyne pretended that these women without husbands, inhabited the regions near the sources of the Corentyne, which we now know to be at no great distance from the northern branches of the river Trombetas. We have therefore as well from the South as from the North the same traditions that the Amazons of the New World inhabited a central district from whence the rivers flow northward towards the Atlantic and southward towards the Amazons. The route of the present journey which traversed these very regions, has only added to the conviction that the existence of Amazons was one of those inventions calculated to captivate the attention and add to

[1] Acunha's Cunuris. Christoval de Acuña (1597–c.1676), Spanish Jesuit, who accompanied Pedro Texeira on his second exploration of the Amazon in 1639. He wrote an account of the journey, *Nuevo descubrimiento del gran rio de las Amazones*, Madrid, en la Imprenta del Reyno, 1641, of which an English translation by Sir Clements Markham was published by the Hakluyt Society in 1859 (see 'New Discovery of the Great River of the Amazons' in Clements R. Markham, *Expeditions into the Valley of the Amazons*, London, Hakluyt Society, 1st ser. 24, 1859, pp. 41–142).

[2] South latitude.

[3] Oziximina is a misspelling for Orixímina, the name of the river on many earlier maps. Today it is the name of the town at the mouth of the Trombetas. Obidos is located further downstream.

[4] See Spix and Martius, *Reise in Brasilien*, III, 'Geographischer Anhang', p. 40.

[5] De Acuña, *Nuevo descubrimiento*, Chaps LXXI–LXXII. Francisco de Orellana was on Gonzalo Pizarro's 1541 expedition to cross the South American continent, and he, with a small band of men, became separated from the main party. They made their way down the whole length of the Amazon and back to the Caribbean.

[6] Louis Marie François Tardy de Montravel (1811–64), French naval officer and governor of French Guiana (1859–64). He commanded the French naval boat *La Boulonnaise* on a journey up the Amazon between 14 November 1843 and 12 February 1844. See Tardy de Montravel, 'Exploration du fleuve des Amazones, depuis Para jusqu'a Obidos', *Revue coloniale*, 3, 1844, pp. 457–512.

the wonders of which the new world was considered the seat. The inhabitants of the Central district were perfectly unacquainted with even the tradition of the existence of such a republic which I trust will henceforward be regarded as a mere groundless fable of the dark ages.

We now entered the Wanamu, or as the Indians at its upper course, call it, the Yauuh; it has, like the Caphiwuin yellow muddy water. Indeed, since when we left the Essequibo, I have not seen a single river with black water. Its current was about one knot and a half per hour; it was studded with blocks of granite and foaming rapids at a short distance from the junction and we anticipated many impediments in its ascent.

As our provisions were low we hoped the information we had received of the existence of Indian settlements near the mouth of the Wanamu would speedily be realized though we began to have our doubts, and noon passed without our meeting with any sign of an inhabited place, and on close inquiry by means of the few words I had collected of the language of my guides I ascertained to our great disappointment that the nearest village was a journey of eight days from the mouth of the Wanamu. We had been toiling eleven days without interruption, and eight days' journey were yet before us. Reduced to less than a single basket of bread and eighty pounds of Cassada flour we had now to adopt the most rigid economy.

The river as we advanced flowed between hills which approached its banks; their height did not exceed two or three hundred feet and as they presented the same outlines to which we had for some time been accustomed the scenery was monotonous. Rapids were frequent but by no means of such a height as those of the Caphiwuin. The banks of the latter river abounded in Carapa trees,*

* The Indians press out of the fruit an oil which they use chiefly for anointing their hair.

(*Carapa Guayanensis* Aublet) Brazil nut (*Bertholletia excelsa* H & K), and the wild Plantain tree (*Phanakaspermum Guayanensis*, Endlicher),[1] not one of which were to be seen during the first days of our ascent up the Wanamu. The nights were foggy and clouded, and a heavy mist generally obscured the sun at his rising, untill he reached from 20° to 30° above the horizon when it cleared and a strong breeze set in from the North East. The heat between one and two o'clock was frequently intense and the Thermometer freely exposed to the Sun rose to 128 F. In the morning at six o'clock it varied from 68° to 71°; otherwise the weather had been fair for several days past and it was evident that the dry season had set in. The river Curiau[2] joined the Wanamu in about 1°16′N. lat. I estimated its breadth, near the junction, at 200 feet, and according to the direction as pointed out by our guide, it comes from the North-West. Fifteen miles above the junction I discovered the first block of granite since we left the Essequibo, covered with Indian picture writing. I had in vain looked for such near the large falls of the Caphiwuin; but as the river was much swollen it is possible if there were any that they might have been covered by the water. At the present locality there was on the left bank of the river a large block of granite about forty feet in diameter, covered from the water's edge with figures, many of them much worn and the connection between the adjacent figures effaced; but Mr. Goodall and myself measured and figured as many as time would permit.

[1] *Phenacospermum guianensis*.

[2] Rio Curiaú.

August 3$^{d.}$ During the preceding night, when all was hushed, I had caught the sound of falling waters, but as after proceeding about two miles this morning we saw no fall before us I concluded I must have been mistaken. About half a mile above our last night's camp (in lat. 1°30′N) a river of the size of the Curiau joined the Wanamu, on its right bank. Our guide, who appeared the most stupid of all the Indians with whom our travels had made us acquainted gave me merely a broad ha! for answer when I enquired the river's name, and as I could not obtain anything further from him, I introduced this river under that name in the map.[1] We soon afterwards again heard the sound of rushing waters, and still no fall in sight. At last, after nine o'clock we found we had not been mistaken, and fall above fall presented itself to our eyes, while the river betokened the difficulties we should here have to surmount.

Little hills consisting of heaped up blocks and clothed in verdure encompass the river which has forced its way over the granite ledges. We had to unload the Corials several times, while passing over the impediments which the Cataract Zibi[2] threw in our way, and we did not pass it altogether without accident. I had a favorite hunting dog of the Taruma breed, which attempted to swim across the river a little above one of the Cataracts, but the current was too strong for him, he was carried down and rose no more. The name of this Cataract, Zibi, appears to have fixed itself in the memory of our guide in consequence of the tradition that the spirit which dwells there demands from every party that passes, a victim, as toll. He had spoken of this previous to our reaching it, and our Indian crew greatly dreaded the passage. The loss of the dog did not appear to have satisfied the spirit of the waters for one of the bark canoes was swamped while ascending the last fall of the series. I saw the accident from a distance, and a pang shot through my heart as I recollected that it contained our last basket of Cassada flour: it was however, most fortunately saved.

We halted a short distance above the great fall which according to my observations last night was in lat. 1°33′30″N. and 88 miles, east of Maopityan Settlement.

August 5. The crew of one of the boats came in after nightfall without their craft, which having split in two they had abandoned, and made their way through the forest. The loss was not great; the only difficulty was how to divide the crew, as our canoes were all overloaded, and the one which I was in, was now threatened with a similar fate.

Eighteen days had now elapsed since we left the Maopityans and we commenced to look with suspicion on our guide whose conduct had by no means inspired us with much confidence. This morning, however, we found some broken branches on the river, which appeared to have been broken only that morning or, at the farthest, the evening before. Full of expectation I encouraged the Indians who were in my craft to double the quickness of the strokes of their paddles, and the canoe in which I was, being in advance of the others, we were just turning round a point when I saw a bark canoe with two men and several dogs coming from the opposite direction towards us. Resembling in their attire the Maopityans, I conjectured they were the two men, I had despatched two days previous to our departure from the Maopityan settlement to inform the Pianoghotto Indians of our intended visit. Scarcely, however had the Indians got sight of us, when they turned the head of their boat and paddling with a swiftness

[1] This is probably the Rio Mahá.
[2] Cachoeira Zibi.

that fear alone accomplish, all our calling to them that we were friends proved of no avail. Our guide remained deaf to our entreaties to speak to them, though we knew that he spoke their language, and as the other Maopityans who might have been better inclined were far behind us, we could only follow the fugitives in the hope of overtaking them, and prevent their spreading any unnecessary alarm. We now saw a second canoe coming down the river in which there were likewise two men; but suspecting us as the others had done they also turned their boats and fled. We were close upon the hindermost boat, when they turned into one of the inlets, where supposing them to have landed at their settlement, we followed them; but to our mortification we found only their canoes with their hammocks and other things; and the dogs still tied up; the men had fled into the wood. I now forced our guide to follow them with one of our own Indians, and in order to pacify their apprehensions I ordered the other boats to proceed onwards, and took the lead. After a progress of about fifteen minutes, I heard the barking of dogs and turned into another inlet where I soon discovered the landing-place of the settlement. I hastened with the greatest eagerness up the high banks and saw a few huts before me, tenanted only by barking dogs, and our guide and the Macusi Indian, who not being able to overtake the strangers had arrived a little before us and found that all the inhabitants had fled leaving everything behind them. We found the Cassada upon the baking pans, the matappa[1] filled with grated Cassada roots, a pepperpot and some fresh Cassada on the spot where they had been eating, when the news must have been brought to them, all showing that they must have fled in the greatest consternation; they had not even taken their hammocks with them, which I never before knew the Indians to leave behind them.

We again despatched some of the Maopityans after the fugitives in the hope that they might yet overtake them, and had now leisure to look round us and brood over our ill luck. Our position was anything but cheering: there we were entirely unacquainted with our distance from the next settlement or the direction we should take to reach it while those from whom we expected information fled at our approach.

The place did not appear like a permanent abode; a small mud hut[2] not quite finished and several panappes or temporary huts constituted the settlement. We observed however eight or ten cutlasses, several new axes, knives & scissors, all of dutch manufacture, and an abundance of a coarse kind of beads, but I could not make out what they were made of. Indeed among all the contradictory information we had received from the Maopityans it appeared they had spoken truth when they had said that the Pianoghottos traded with the Indians and the Maroon negroes of Surinam.[3]

I gave strict orders to the Indians who had accompanied us not to touch anything of the different articles which had been left behind by the fugitives. Our Macusis and Wapisianas obeyed these orders, but not so the Maopityans who most ravenously fell on the eatables, alledging as an excuse their excessive hunger. Those whom I had sent after the fugitives returned towards evening without success; they had traced them for some distance but said they had lost their track. I think the provisions they had seen on

[1] This is the basketwork cylinder which is used for squeezing cassava.

[2] Schomburgk probably means a house with walls made of branches and covered with mud. It sounds less like a temporary encampment than a new village under construction.

[3] The Maroon or Bush Negro communities in the middle reaches of the Surinam rivers were, until the 1960s, the main suppliers of manufactured goods to the Trio Indians of the far interior.

arriving at the deserted village was most likely the cause of their early return. I now began to mistrust the Maopityans more than ever, and soon found I was justified in so doing. I had retired to my hammock, when a Wapisiana whom we called by the name of Moller and who was the only one who appeared to be on familiar terms with and understood our guides, came with the information that he had overheard them planning to pilfer the place during the night and then leave us to our fate. They had been occupied since nightfall in preparing Cassada bread and had likewise scraped together what eatables they could lay their hands on. My resolution was soon taken. Mr. Goodall and the Coxswain having received their orders, I put an end to the preparations of the Maopityans, and placed them all six in the round hut, and loading our guns before their eyes with double balls, assured them they would be used if they made any attempt at escape. Long after midnight I threw myself into my hammock for an hour, and was up before dawn, the coxswain having the watch from four in the morning. I interrogated him as to whether all was right, and he thought it was; but on inspection I found that three of the prisoners had escaped; fortunately the other three, and among them our precious guide and the chief man of their number, had not been equally successful.

With day light I was enabled to look about and found that previous to my stopping their proceedings last night, they had succeeded in pillaging the huts of everything valuable. On our arrival we had counted eight axes and a similar number of cutlasses, two only of the former remained and hammocks, pakals[1] with their little treasures, glass beads &c. were all gone, indeed they had almost succeeded in making a perfect clearance, and their design as communicated by Moller was placed beyond doubt.

I gave the three who remained to understand that the guide and the other principal man were to remain as hostages, and I threatened to fire upon them should they attempt to escape, while the third was desired to restore the stolen things and that the two would remain as prisoners until he and his accomplices whom we knew could not be far off had succeeded in finding the Pianoghottos and induced them to return to the village.

The one whom we had allowed to go free now went several times the adjacent forest and back no doubt carrying messages and advice from the prisoners to the three men whom the coxswain had allowed to escape. In an hour's time we saw nearly all the stolen articles returned and we were told that their three comrades were gone after the fugitive Pianoghottos to induce them to return. One of them arrived about noon in breathless haste communicating the information that some of the Pianoghottos were coming in a canoe, and he advised that we should allow the guide & the other prisoner to go to meet them and speak with them. This ruse however, did not succeed, and taught by our morning's experience we kept the closer watch over our prisoners.

The adjacent fields abounded in Cassada plants, and I employed our Macusis in preparing bread, keeping an account of the number of roots which we used for our most pressing necessity, in order to pay for it as soon as we should meet the owner, or if this could not be, to leave the payment behind.

Our situation was by no means pleasant and had it not been for the discovery of the plot, we should have been far worse off. Our hope was built on a friendly intercourse with the Pianoghottos, as sad experience had proved to us that the Maopityans did not

[1] Woven baskets with lids in which personal possessions are kept.

intend to act honestly by us from the commencement. All their information with regard to distances proved incorrect, and had they succeeded in escaping during the night with their plunder, which under existing circumstances it would have been impossible to replace out of our own stores, they would no doubt have embroiled us with the Pianoghottos even if they had returned, and who doubtless would have ascribed the theft to us. It was equally evident that the two messengers whom we had despatched from the Maopityan settlement for the purpose of informing the Pianoghottos of our intended visit, had never proceeded on their journey, and most likely only awaited our departure to return to the village.

The following days passed between hope and anxiety, sometimes buoyed up by the information that traces of the people had been discovered; then again disappointed. The few hours rest which our close watch over the prisoners allowed us to take in turn was disturbed by swarms of musquitoes, and to make matters still worse M^r. Goodall's health gave way, and I myself became unwell. Determined however to prevent the escape of our prisoners I ordered our tents to be set up right across the entrance, so that any person going in or out was obliged to pass under our hammocks. Our own Indians who equally depended upon our success for their safe return, and whom I knew I could depend upon, were quartered in the hut with the two prisoners, the Coxswain and one of the Canoemen, in a small hut behind to prevent their breaking out of the hut in that direction; and for keeping us all on the alert, those upon whom the watch fell as well as the rest, we trusted to the swarms of musquitoes which rendered it next to impossible to enjoy sound rest.

One of the hostages whom we had allowed to start with one of our own Indians in search of the fugitives returned on the 9th. Inst^t. with the information that while stealing softly along they had heard the cry of a child and soon after observed a woman squatting under a hut of palm leaves trying to quiet her infant. As soon as she became aware of the presence of the two men she cried out to spare her life. Her fears being quieted she related that the men in the canoes who had first seen us when we approached had thought we had come to murder the men and take the women into slavery, and had accordingly fled with the greatest precipitation taking nothing with them, and as the next settlement was five days' journey off, she being sick was afraid to undertake it, and with her husband and child had remained in the hope that not finding any person we would have left the place ere this. The cries of the child for bread had induced her to steal near the settlement and she had approached it close enough to see me sitting under the tent. She had taken some yams out of the field and returned to the temporary hut. Her husband was absent at the time the Maopityans found her and she would not return alone but begged the Maopityans to bring her some bread for the child, which was crying for hunger. I hastened them back with bread & presents; but they soon returned; the fears of the women must have been re-awakened for she did not wait for our men and had again fled.

In order to satisfy myself of the correctness of this account, I examined the provision ground along with Sororeng and he soon pointed out to me the marks of recent foot-steps. The impression of the childs feet, where the soil had been moistened by the dew were very distinct and we followed to the spot where the mother's curiosity had urged her forwards towards my tent while the child had remained behind. At least Sororeng drew my attention to the deeper marks of the child's feet which he ascribed to its remaining standing at the same spot for some time.

August 11th. The three Maopityans who, in the first instance had escaped during the watch of the Coxswain returned voluntarily this evening. They pretended having been in search of the Pianoghottos and said that they had fallen in with the camp of one of the parties on the third day (August 8th) but unfortunately the dogs of the strangers had given the alarm, and they had all fled leaving luckily for the starving Maopityans some smoked meat behind.

Our situation was really critical. We had sufficient evidence that several parties, one of which could not have numbered less than twelve persons, having passed our camp in different directions, no doubt spreading the alarm of the arrival of enemies; and that we might easily be overpowered will be evident when it is considered that our whole party consisted of only thirteen persons and two dogs, of which number the six Maopityans may be perhaps considered as so many enemies; At least I could not trust men who by their former actions had shown themselves deficient in a virtue possessed by all other Indian tribes, common industry. Indeed I had no doubt but they would gladly join any attacking party in order to share in the spoil & prevent our informing the Pianoghottos of their dishonesty.

Mr. Goodall's indisposition under these circumstances was the more unfortunate. I resolved, however, that as soon as he should be sufficiently recovered to push on towards the North leaving the baggage behind and taking only the most valuable instruments of the expedition. According to my observations Demerara was in the direction N.N.W. about four hundred miles; and the Corentyne, where it becomes navigable, most likely not more than sixty.

August 13th. One of the party who was despatched in search, returned this afternoon and brought the joyful news that they had fallen in with people near the small river Iriau[1] and that two of the Maopityans were coming with them by water. Our joy at this news was great. We fancied our difficulties were now at an end, and our Indians collected round the narrator to hear more of the particulars. The moments appeared hours before the strangers arrived; they consisted of a man, two women and two children, and we recognized in the Indian and one of the two women the Yarumata[2] whom we had met near the mouth of the Wanamu. As far as I could learn by means of imperfect interpretation, they had proceeded from the place where we had met them near the Wanamu to some Pianoghotto villages that lie to the northward, to convey the wonderful news of our arrival, and while there some of the fugitives from the present place had brought the news of our arrival taking us for others accompanied by hostile Tchikianas.[3] He had told them of their mistake and volunteered to proceed to this place for the purpose of proving that he had no fear. He said that eight Pianoghottos were to follow him next day to assist us as guides. We were all now most thankful that our affairs had taken such a favorable turn, and in our joy heaped presents upon our informants; a looking glass seemed to cause them great delight.

We waited until the 16th. for the arrival of the Pianoghottos, but in vain. And as Mr. Goodall was by this time sufficiently recovered, I considered it better, now that we knew where to find the people, not to delay our departure.

[1] Rio Iliau.

[2] Zurumata, a subgroup of the Pianoghotto (see p. 170 above).

[3] These are probably the Chikenas, three of whom Farabee met in the second decade of the twentieth century (*The Central Caribs*, Philadelphia, University Museum, 1967 [1924], p. 194). There has been no mention of them since then.

We had been at this place ten days, which were, I must confess passed in great anxiety of mind. My Astronomical & meteorological observations were however not neglected. The latitude of the settlement deduced from eighty four circum meridian altitudes of North & South stars was 1°40′5″N.; the longitude by chronometer 1°35′55″ in Arc, to the East of the Maopityan settlement.

The mean of Meteorological observations gave the following results –

Period & N° of observations	Instruments	Mean	Maximum	Minimum	Remarks
August 7th to			7th. 1h. a.m.	8th. 6h. p.m.	
August 14th	Barometer	29i.258	29i.341	29i.216	
inclusive			8th. 2 p.m.	14th. 6h. a.m.	
	Attached thermometer	81°.39	91°.56	66°.92	
	Detached thermometer	81.47	92.0	67.0	
	Wet bulb thermometer	77.05	82.9	66.5	

On the 26 August I boiled the Barometer thermometer; the mean of the boiling point was 211°.093; the Barometer stood then 29i.278. The attached thermometer 77°.54 F.; detached thermometer 78°.3; the wet bulb thermometer 76°; the wind blew from the South East with the Estimated force of 4 according to Capt. Beaufort's table.

The height of the village according to the Barometer, taking the mean of its observations as stated in the foregoing page and comparing it with the mean of observations in Georgetown would be 753 feet.

The weather was generally fine, the morning foggy and the evaporation amounted on the 13th. August, during the twelve hours of the day (from 6 a.m., to 6 p.m.) to 266 grains out of 1000 grains exposed to the air. The wind blew with considerable force on the 12th. of August as already indicated; but otherwise it was generally calm.

The magnetic intensity was indicated
Pr. needle L (a) by 100 vibrations at 82°F. in 2m 53s.75.
" " L (b) " " " " 87 " " 3. 42.60.

[****]

The third part of this manuscript, JMS 6/42 (c), consists of 41 pages and is a direct continuation from the previous part.

JOURNEY FROM THE PIANOGHOTTO VILLAGE TO THE RIVER CUTARI AND THENCE BY THE CORENTYN TO GEORGETOWN IN DEMERARA.

August 16. About the time that we intended to start, the Maopityans, with the exception of one, were no where to be found. The Zurumata and his family and his family had likewise vanished and we considered ourselves in the same predicament as on the 5th. Inst. I had, however, gathered sufficient information of our road, and as the Maopityan who had accompanied us was willing to accompany us and said the others would follow as soon as they had set a new bark canoe afloat, we started on our journey, and this the more readily as I expected to meet the Pianoghottos who had promised to come to our assistance.

We descended the Wanamu for about ten miles and then turned into the Iriau one of its tributaries which joins it on the right. The river was much impeded by trees that had fallen across and by small rapids; it was scarcely sixty feet abroad, and its current strong our progress was therefore slow. We heard on the morning of the 18th. the barking of a dog, and found on the river's left bank the Maopityans who had just arrived having walked overland. They told us we had now to abandon our canoes and continue our journey overland and that it would take us five days to reach the first settlement of Pianoghottos; but as a kind of consolation they told us the Pianoghottos whom they had expected the previous day would no doubt arrive in the evening. The Zurumata was with them and confirmed their account; but here again we were doomed to disappointment; fortunately for our larder a fine forest deer weighing about seventy pounds was brought in by our huntsman.

August 19th. The Zurumata and our villainous guide had taken their departure during night and only three Maopityans were left, who naturally pretended they knew nothing about the others. The day looked as gloomy as our situation, and we were after all reduced to the necessity of abandoning our baggage and proceeding on our journey. Indeed were it not for our faithful Macusis and Wapisianas who accompanied us from Pirara and Watu Ticaba, we should now be quite alone, in lieu of which we were at least able to save the instruments and documents of the expedition and some other articles. It was however with a heavy heart that the selection of the most indispensable objects was made. There lay the collections I had made since leaving Pirara; we had conveyed them many hundred miles and in spite of cataracts and miserable boats they had reached this spot in safety, only to be left behind with very little hope of seeing them again. How frequently did I reconsider our disposable force to consider whether certain objects of peculiar interest to science or to myself personally could not be stowed somewhere; but the few Indians we had at our service were already heavily enough loaded with the instruments and such articles of barter as were necessary to secure our subsistence and return to the coast.

Towards evening it began to thunder and so vivid and rapid was the lightning flash succeeding flash that the whole vault of heaven seemed on fire while the peals of thunder followed in such quick succession that the greatest interval between them only allowed me to count four while eight times out of ten I could only count two or three; and this lasted for two hours. There was little rain during the time the thunder continued; but after the electric clouds had passed which was about midnight, it came down in torrents and continued the whole night and following morning until 10 O'clock.

Augst 20th. We left the least valuable man in charge of our goods and chattels. Clothes were at a discount since necessity forced us to limit the quantity of our baggage, and Mr. Goodall accordingly rigged out the guardian of our baggage with some old trowsers, a jacket a pair of boots & a hat giving him with artistic skill the carricature resemblance of a gruffy old gentleman. A roof of palm leaves being constructed over our baggage we left it and departed.[1]

We had to cross the left bank of the Iriau, which from the heavy rain was rising rapidly. A tree that was lying across the stream was partly under water and moreover in

[1] The published version (p. 75) includes after the words 'the guardian of our baggage', the explanation 'a figure made of straw'.

an inclined position, so that I feared for the safe passage of our Indians with their burden; but they all reached the bank in safety.

Our path led us over hills and through swamps abounding in the graceful Uassa palm, (a species of Euterpe or Oenocarpus) the upper column of which contains the rudiments of the leaves, and affords, like the *Euterpe oleracea* an excellent vegetable. We may probably be accused of barbarism for destroying thirty or forty of these graceful palms to provide ourselves with a dish of Cabbage, but hunger has few scruples and must plead our excuse.

August 21st. Shortly before noon we reached a small stream running to the N.N. West, and consequently in a contrary direction to those we had hitherto crossed. We had now left the basin of the Amazon. Though the hills we passed over in the course of the morning were somewhat higher, none of them exceeded four hundred feet and those which formed the "divortia aquarum"[1] were scarcely one hundred and fifty feet high. A walk of a few minutes brought me to an abandoned house which my observations placed in latitude 1°49′ North and 3′14″ in arc, east of the Pianoghotto settlement at the Wanamu; we therefore crossed the division of the two fluvial systems in lat. 1°48′30″ North & in longitude 56°30′ West of Greenwich. The mean of the Barometrical observations compared with the means in Georgetown gave for the height of the abandoned settlement 794 feet.*

*The data were	Barometer	29i.195
	Attached Thermometer	74°.04
	Detached do	74.33.

The Wet bulb Thermometer showed from six in the evening to half-past eight next morning no difference with the detached Thermometer.

We found evident marks that people had been here very lately. A parcel of beads were hanging up in the hut and a fine large cutlass of English make with the old mark **G.R.** and the crown over the letters, was hid between the leaves of the hut. The adjacent forest abounded with the Majestic Bertholletia which bears the delicious Brazil nuts. They were bearing young fruit, but many of the last season were lying on the ground and afforded our Indians and ourselves a great treat. I found while returning from my astronomical observations that the Coxswain had erected our tents beneath one of those giants of the forest; which could not be less than 100 feet high before it branched out. I arrived too late to prevent him from doing so, but I really feared an accident might happen, if a heavy wind should shake one of the fruits off, which are, when full grown, as large as a child's head. And so it happened, for while Mr. Goodall and myself sat at dinner, the thunderstorm accompanied by a whirlwind swept over the place, and four of the fruits with a shell as hard as that of the Cocoa-nut, came down one of which fell near where I was sitting, with such force that it partly burried itself in the ground.[2]

The next morning we had to ford a river about eighty feet wide which the Maopityans called Arimattau.[3] Our route was more hilly; the valleys were as swampy as

[1] The watershed.

[2] The Brazil nut which one buys is merely a segment, like that of an orange, of the full nut which, as Schomburgk describes, is of considerable size and weight.

[3] Aramatau River.

they had hitherto been, with here and there large blocks of granite half buried in the ground. The three Maopityans who, contrary to expectation, had remained with us, had told Sororeng that we should this day (August 23rd.) reach a settlement of Pianoghottos. They were therefore directed to start an hour before us, to prevent any apprehensions and further flight of the Indians. Indeed we found sufficient evidence that the fugitives from the Wanamu had taken their road in this direction and no doubt spread the alarm of the arrival of the Tshikianas. We followed at our usual time of starting and had to cross several hills from five to six hundred feet high.

Various were our conjectures as to whether we should find any people to day and how they would receive us. A cry of astonishment from some of our Indians who had kept up with me, attracted my attention; a recent encampment was before us, the fires still burning – an Indian pot which evidently had been emptied in haste as part of its contents were lying on the ground – a bunch of plantains in a corner, and what might greatly have alarmed us, a pool of blood near one of the huts, all combined to make us believe that the three Maopityans on coming up with the encamped Pianoghottos had again put them to flight. The blood on closer inspection proved to be that from a bush hog of which we found some remains. Nevertheless these signs were by no means calculated to sooth our apprehensions and we redoubled our pace, anxious, if there were a settlement before us to ascertain whether it contained human beings or not. After crossing a hill about five hundred feet high we ascended another of about half its height and saw before us a cleared space, several acres in extent, being the provision fields of the settlement. All was silent, not the barking of a dog nor the sound of a human voice. I feared our worst apprehensions were again realized. Three huts now came in sight and "thanks to God here at least are people" burst from my lips as I saw a number of athletic men before me. I hastened towards them and tendering my hands received their ready grasp and was really overjoyed at the hearty welcome.

It appeared our Maopityans had met a hunting party this morning and after having partaken with them of their morning meal, during which they found time to explain our objects, they all proceeded without delay to the settlement to give information of our coming. I had seldom seen a finer set of men than those who now stood before me; some appeared to be 5 feet 6 and 5 feet 8 inches in height, their limbs strong and muscular; in their attire they resembled the Maopityans. That important piece of head dress, the queue, was attended to with such neatness, that it would have done honour to a Parisian coiffeur of the old regime. The hair of the hinder part of the head was all gathered up into the queue, that on the forehead was cut rather short, with the exception of two tufts sweeping from the ears towards the face much in the fashion of our gallants who are not graced with the head of an Apollo. Neither male nor female were painted in lines, but their whole body, with the exception of the face was covered with Roucou or red paint. The men wore a profusion of beads round their wrists, and across their shoulders, and, like the Zurumata, a band of cotton below the knee, with a long tassel hanging from it. The ankles were tied with strips of palm leaves ornamented with red and black paint. Their waste was girded with a broad piece of bark from which their waistlap was suspended. Almost every man wore one of those combs which we first saw among the Maopityans; they were tied to a string and hung round the neck so that they fell upon the breast. The females did not wear this peculiar ornament. The bows and arrows of these people were long, the former strung like those of the Maopityans; but that formidable weapon for close fight, the War-club, as among the Maopityans was not used. The females were less favorably gifted by nature than the men; and of ornaments, if I except their own manufactured beads, they wore but few. As if fashion were here reversed the females had shorn their hair short, which did not tend to improve their looks; two, however, had long hair which they wore in queue like the men.

They brought us presents of sugar cane, pine apples, cashews (*Anacardium occidentale*) and some new made bread for which I gladly gave them some glass beads & fish hooks. The settlement consisted of three huts, a round one resembling in its structure those of the Macusis, and two open sheds. With the exception of dogs and some fowls of a pure white colour, they had but few domestic animals.

My enquiries were now directed towards the continuation of our route, and from what I learnt, I have no doubt that the Curuni or Curuwuini of the Pianoghottos and the Curitani of the Maopityans is one and the same river.[1] Their next neighbours to the East are the Cocoipityans or Harpy-eagle Indians,[2] and five days' journey to the eastward are the settlements of the Mekurus or Maroon Negroes of Surinam.[3] The Pianoghottos trade with the Maroon negroes and give in return for axes, knives, & cutlasses which we saw they possessed in abundance, hunting dogs, waistlaps, hammocks and Cassada graters.

The small stream which runs at the foot of the mountains flows into the Cutari, which we crossed near its source on the 22^{d} of August. The Cutari joins the Curuni; both are said to be of equal size.

August 25. With the exception of some young puppies and the fowls, we are again the sole occupants of the place. I had yesterday a long palaver with the Chieftain in which I desired him to send his people for as much of our baggage as they could convey, and as I accompanied my request with a handsome present to himself and a promise to pay his people, he consented. At daybreak this morning I found the whole village in motion; men, women, children, & dogs all wandered out, proving that they had not acquired sufficient confidence in us to leave their women and children under our guardianship.

The three Maopityans whom I had paid for their services returned with them. I likewise sent payment to a fourth, who, I knew had failed to accompany us further only in consequence of a bad foot; but I thought I punished the other two very leniently by merely withholding their payment. Seven days were likely to elapse before the people return, and as I resolved on descending the Cutari, the Coxswain received orders to prepare the necessary number of bark Canoes for our navigation to the Coast.

As far as we could understand there is a tradition among the Pianoghottos that once upon a time, but long since, Caribs came up the Cutari, and one of the Indians who knows a few words of the Dutch Creole language from his intercourse with the Maroon Negroes, said, that after descending the Cutari for ten days, people are again met with,

[1] At this point Schomburgk was close to the Cutari or Kutari River. On the Guyanese map the Cutari and Aramatau meet, and shortly north of their junction they join the Sipaliwini to form the Corentyne. On the Surinamese map the stretch of the river between the junction of Aramatau/Cutari and the Sipaliwini and the junction with the New River is marked as the Coeroeni (i.e., Curuni or Kuruni), and only after that does the river become the Corentyne.

[2] This appears to be the only reference in the literature to a group of this name.

[3] The current Trio word for Maroons is Mekaro.

though he did not know to what tribe they belong. His imperfect knowledge of the Creole language did not give me sufficient confidence to rely implici[t]ly upon his information.

This settlement is pleasantly situated; it stands on the suummit of a hill which by a barometrical measurement rises 216 feet above the small brook which higher up forms some fine cascades. The summit is a circular plain, about 800 yards in diameter, and has been cleared by the Indians for their provision grounds, in which they grow Cassada, Plantains and Bananas, Yams, Sugar-cane, Pineapples &c. all of which appear to thrive equally well in the yellow soil.

There is another settlement about half a day's journey farther the West which lies on the Cutari where we intended to embark, and the Coxswain was sent to search for trees in the vicinity for the construction of our craft, while the remainder of our crew were busily employed in preparing bread for our Journey, as I had bought a whole field of Cassada from one of the Indians before we left.

August 28h. Mr. Goodall & myself were taking our breakfast this morning when I saw a strange Indian stealing round the large hut wistfully looking at us. When he became aware that I had observed him, he approached us uttering what I took to be a salutation. All at once he squatted on the ground, covering his face with his hands, and broke out into such a strain of lamentations that we were quite surprized. Our Macusi Indians collected round him and appeared equally astonished; but their curiosity had no effect upon him for he continued his lamentations his whole body appearing convulsed from their excessive violence. Each of his words was twice repeated in a chanting tone, first slow and secondly with a strong intonation and as they generally ended with a vowel the effect was by no means inharmonious.[1] Thus he continued for about twenty minutes without interruption, when he rose, wiped his eyes and addressed me again in his language of which I did not understand a single word. I had now more time to look at the individual; he was young and not so tall as the generality of the Pianoghottos I had lately seen, and of a slight but well proportioned figure. He was entirely daubed over, except his face, with red paint, and on the breast he wore a comb. His waist was girded with bark, to which the waistlap was fixed, the red colour of which was relieved by strips of palm leaves. He had neither bow nor arrows, merely a short dutch knife in his hand. He directed his looks solely towards me and did not deign to throw even a glance at the Macusis round us. His agitation had not yet subsided and he continued to tremble, either in consequence of his previous lamentations or through fear. A present of some trifles re-assured him and he now gave me to understand that he was hungry. Our huntsman had been fortunate enough to shoot a wild hog the previous day and with a large piece of meat and a cake of Cassada bread; he returned to the wood and disappeared, I suppose, to carry the news of what he had seen to his companions, as he pointed eastward and showed the fingers of his hands making at the same time a motion as if that number were coming.

August 29th. I was rather astonished this morning at seeing Mr. Goodall out of his hammock earlier than he was accustomed to be when not *en route* and his loud speaking

[1] Although not often nor widely reported from the Guiana region, a common form of greeting throughout Amazonia involves tears and lamentations, expressing sympathy for everything the traveller has undergone. This would seem to be an example of it rather than any personal expression of feeling on the part of the individual. This conclusion is reinforced by the similar behaviour by a party of visitors the next day.

to the Indians led me to suppose that he had been disturbed. I was soon made acquainted with his grievance; legions of ants had directed their march through his hut, and a detachment has assailed him in his hammock and driven him fairly out of it. The Chief column of the marching army of ants was about six inches broad & until nine o'clock they marched on without intermission through the tent, besides which there were several branches of minor extent.

I now examined my own hut more minutely and observed there several other columns; but of less breadth, that continued marching uninterruptedly until the heat of the sun caused them to retire into their caverns. Indeed the whole open space in front of the huts was traversed by numerous columns, carrying away towards their burrows, crickets, spiders, cockroaches, and other noxious insects, which they must have surprized during their nocturnal rambles, or hunted out of their holes. The ant was small and without those prickles which distinguish the genus *Atta* or Cushi ants of the colonists. They were probably a small species of that kind which is called the hunter or Yagerman and of which M^rs. Carmichael has given such a lively and interesting description.[1]

A party of Indians arrived about 10 O'clock and we recognized the one who visited us the previous day among them. An elderly and good looking Indian who appeared to be the chieftain, closed the party which consisted of ten individuals. He came towards me and uttered a similar salutation to the one yesterday and then pointed to the hut where he took his seat & accompanied by the other Indians they all broke out in lamentations in the same manner as the young man. The wailings over, the Chieftain rose and began to converse. In the absence of the inhabitants we did the honours of the village and placing some meat and bread before them they were soon engaged upon it. But now came the question, which they soon rendered intelligible by signs – What is in those boxes we see before us? and nothing less would satisfy their curiosity than opening the chests and displaying their contents. The lot fell first on my own canister, that being of larger size. My military appointment, as Colonial Aide-de-Camp raised great exclamations of wonder no doubt, in consequence of the silver lace upon it. I have remarked that among all the Indian tribes I have visited, silver has more more attraction for them than gold. Most of the Indian languages have a name for the latter, but for silver, they have adopted the Portuguese word "*prata*". In the Macusi language gold is called Carucuri.

My large telescope proved of great interest to them; but a still greater sensation was caused by the sight of our fowling-pieces, Arquebusa Arquebusa, escaped from almost every lip and women & children ran away crying when we took one of them up, fearing it might be fired off; but I refrained from causing such a shock to their nerves.

The dress of our visitors did not differ from that of the other Pianoghottos. There were three females among the party; but none of them more favoured with good looks that the ladies we had previously seen. A kind of stays (at least I can find no other name for that peculiar article of dress) which, of smaller dimensions, we had already seen

[1] Mrs A. C. Carmichael, *Domestic Manners and Social Conditions of the White, Coloured, and Negro Population of the West Indies*, 2 vols, London, Whittaker, Treacher & Co., 1833. Mrs Carmichael devotes several pages (II, pp. 326–33) to the description of both parasol ants and what she calls 'chasseur-ants'; the latter is obviously the French name for the 'hunter or yagerman-ant'. She describes how a column of chasseur-ants invaded her house in Trinidad and cleared out all the pests and vermin, including the mice and rats.

among the Maopityans, reached from mid-way of the back, for about eighteen inches downwards and kept their figure upright. It was made of seeds of the same kind as we had previously seen.[1] Large strings of the same description of seeds were fixed around the upper arms and wrist, and where their apron is attached to the stays, there hung a large tassel also made of these seeds; each string terminating with hard shell of a nut. Two of the females wore their hair in a queue, the third had it cut short.

I had now an opportunity of learning the mode by which they manufacture their beads, which is so ingenious that I must describe it. These seeds are of a shining black and almost twice as large as hemp seed. They are first perforated by means of a piece of hard wood, and while fixed to the wood the lower end is rubbed briskly over a rock (decomposing gneiss) which takes off the pointed part of the seed; it is then reversed and the same process repeated, after which they are strung upon a thread. If a considerable quantity is thus prepared and strung together, they take another kind of rock (decomposing mica slate) which they pound coarsely, and then having fixed firmly one end of the string of beads and holding the other end in the left hand, they take, in the right, a quantity of the pounded stone which they rub up and down the seeds till these have acquired a uniformly cylindrical shape. The tree which furnishes the seeds is of a large size, and when young, has its trunk and branches as also the mid-rib of the leaves covered with prickles; the leaves are without stipules, la[n]ceolate, abruptly pinnate, with pellucid dots, and of an aromatic pungent taste, which is likewise peculiar to the wood and branches; the seeds are shining and have a testaceous integument. Though I have not seen the inflorescence of the tree, nor the manner in which it bears its fruit, I have little doubt that it belongs to the Xanthoxylaceae.[2] The seeds have a very remarkable taste almost resembling spermaceti. The Pianoghottos call them Were,[3] and, as already observed, use them as a substitute for beads. The females make their short aprons of them, which constitute their full dress, and when finished, fringe them with the horny seeds of another tree (*Vantanea Guayanensis*)[4] which are much larger and make such a rattling noise at every movement of the wearer that the approach of a female is heard long before she comes near. Indeed when they go to fetch water from the brook which they always do in company the noise they make is both great & peculiar.

The rock which they use for grinding and which is decomposed gneiss is called Were Kitto, the other for polishing and rounding the seeds, called Tzai, is a decomposing mica-slate.

We soon made great progress in becoming acquainted with our visitors, and their curiosity at our doings became almost troublesome; besides which they had a custom which did not agree with our reduced stock of articles for barter being uncommonly covetous of whatever their eyes rested upon. To keep them in good humour I was frequently

[1] This item of dress had been abandoned by the Trio by the mid-twentieth century, but a drawing of one, made at the very beginning of the century, is to be found in De Goeje, *Bijdrage tot de ethnographie der Surinaamsche Indianen*, Leiden, E. J. Brill, 1906, Plate 1, 26.

[2] Modern spelling is *Zanthoxylum*; it is a genus of the family Rutaceae.

[3] The Trio still recognize a plant called 'weere' from whose seeds they used to make beads. De Goeje (*Bijdrage*, Plate I, 10 and Plate II, 7) illustrates the seeds (wĭrĭ) and a necklace made from them and a contemporary photograph appears to show such a necklace being worn (see De Goeje, *Beiträge zur völkerkund von Surinam*, Leiden, E. J. Brill, 1908, Plate XV, 1).

[4] *Vantanea guianensis*.

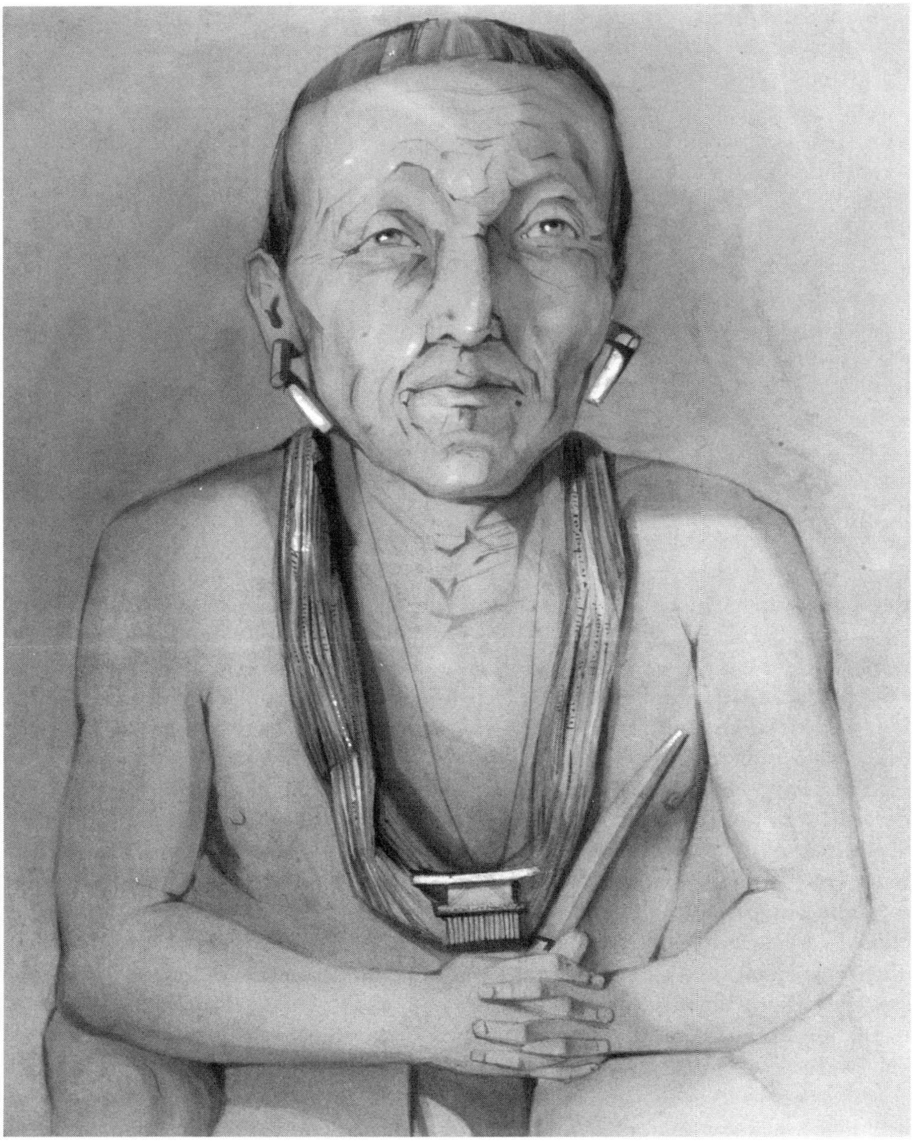

Figure 11: A chieftain of the Pianoghottos. Watercolour by Edward Goodall, reproduced by kind permission of the Pitt Rivers Museum, Oxford (1884.117.79). See p. 187, n. 1, for the identification of the subject.

obliged to encroach upon our store of glass beads. Indeed I have seldom seen Indians among whom the men have shown such a decided love for beads, even knives appear to have less attraction for them, and as for combs that article in so much demand among the Macusis, they do not care the least for them, well knowing that they are capable of making a substitute for them which answers all their purposes. By promising them some beads I easily induced the Chieftain and some others to sit or stand to M[r.] Goodall for their portrait.[1] He considered them, however, the most fidgety of all the Indians he has depicted, they could scarcely remain in the same position for a moment. The Old Chieftain has certainly the most characteristic face among them. Although he is of no great stature, being only five feet two and a half inches, he is portly and well made. His face has a thorough Indian expression, the forehead receding, and the eyes so oblique that he can almost look upwards without bending his head back. Another peculiarity, in which almost all the Pianoghottos share, is the great depression on the side of the head, below the parietal bone and between the outer angle of the eye and the ear (*os sphenoides*). The ear is uncommonly large (three inches in this individual) which I might, in this instance have ascribed to his wearing pieces of bamboo in the ear, if it were the *lobuli* which determined the size of it. The ears of a boy who was four feet seven inches in height were two inches and seven tenths in length and one inch and five tenths in breadth. The waist was small, the young men being seldom more than two feet three inches round; but this may be ascribed to their wearing from their earliest youth, tight girdles about six inches broad, made of the bark of a tree, a custom which must be injurious to their health. It appears they are as vain of a small waist as any fair lady of European birth and boarding-school education. It was ridiculous to see how they contrived to make their waste appear smaller, by drawing in their breath when I came to measure that part of their body.[2]

Their language has many words similar to the Macusi, Ina, yes; seni, this; amoré, you; urupa, bow; purau, arrow; weh, sun; &c are the same in Macusi and Pianoghotto. They call the sun Weh, like the Macusis, but the moon nuna, like the Caribs, and a paddle pura, like the Wapisianas; indeed it appears to be a language much intermixed with other words.

The custom of each tribe of Indians having their own names for the adjacent tribes, as also for their chief rivers and mountains renders it very difficult to identify the said the said tribes & rivers &c. Thus the Maopityans, whose name is a Wapisiana word are called Mawakwas by the Tarumas, and Tziraus by the Pianoghottos, who call themselves Pianahutto, the Tarumas Zarumas; the Atorais, Orais, &c. The river Wanamu is called Pianaghotto by the Tarumas, and Punama by the Pianoghottos.

Towards the East, from our present abode, it appears, dwell the Orokoyanna (Parrot Indians), the Hackuyanas[3] and Mekuros or Maroon Negroes inhabiting the affluents of the Marowini; towards the S.S E. the Tshikianas who inhabit the banks of the Kaphu and who are described as warlike; near the sources of the Wanamu, at a distance of five days' journey the Drios, a sister tribe of the Pianaghotto, a small number of whom live likewise in a village only a few miles further north from here. Far to the SE reside the formidable Maipurishiannas or Tapir Indians, who are described as Canibals, and who use the skulls of their enemies as drinking vessels; towards the S.W. reside the Tunayanna (Water Indians) and, farther westward, the Carawayanna, the Barakutyo or

[1] The sketch held by the Pitt Rivers Museum, Oxford (1884.117.79) and reproduced here as Figure 11, is captioned 'S. America – A Wapisiana from Watu Ticaba', whereas the virtually identical portrait in the British Library (Add. MS 16937.74), which is also reproduced in Goodall, *Sketches of Amerindian Tribes*, p. 78, is captioned 'A Chieftain of the Pianoghottos'. Linda Mowat has discussed this discrepancy and rightly, in my opinion, concluded that in this case the British Library version has the correct caption (see 'Four Paintings by Edward A. Goodall at the Pitt Rivers Museum, Oxford', *Journal of the History of Collections*, 5, 1993, pp. 223–9).

[2] See Goodall, *Sketches of Amerindian Tribes*, pp. 78–9 for portraits of Pianoghotto.

[3] It has not been possible to identify either the Orokoyanna or the Hackuyana.

Barecuto and the Woyawais. The West appeared uninhabited. On my mentioning the names of the Arawaak and Waraus, they pointed to the northward.[1]

The mode of singing among the Pianaghottos is very remarkable. One begins who in a voice and manner that we would call reciting gives utterance to his words in short phrases and in a plaintive tone, while the others, however many they may be, accompany these words by humming a plaintive melody. They are the instruments that accompany the recital as in some of our melodramas. I greatly regretted not knowing their language so that I might have understood the purport of their peculiar song.

August 31$^{st.}$ A great number of the Indians returned this day with some of our baggage and we subsequently purchased some of their curiosities. The greatest admiration is due to the Waist laps of the men, which are so firmly woven of indigenous cotton, that they would do honour to a European manufactory. They are the handiwork of the Pianaghotto and Drio dames. Their Spindles, although simple, and the circular piece at the end which sets it in motion, rather coarsely executed of bone, are nevertheless very neat and terminate in a piece of bone very neatly cut out, through which the thread is conducted. A pair of Pianaghotto ear-rings would prove rather too ponderous for our European ladies to have a chance of becoming fashionable. They are made of the large teeth of the Water haas (*Hydrochoerus Capybara*, Desm.) and provided with an ingenious spring which keeps them fast in the ear. Their baskets or Zumpas on which the bread is set before the men, when taking their meals, are very neatly made; exhibiting the grecian pattern. Their arrows are well made and richly ornamented where the feathers are tied to it and some of their bows made of letter-wood were from six feet eight inches to six feet ten inches in length.

September 1$^{st.}$ We broke up our encampment and prepared to leave this settlement in order to proceed to the next place, where they were preparing the bark canoes. We had been here nine days and though all the baggage had not yet reached nor was likely to reach us, I did not wish to delay our departure. The prolonged stay had enabled me to make a series of meteorological and astronomical observations, of which the results are as follow,

Results of 100 meteorological observations

Period & number of observations	Instruments	Mean	Maxima	Minima	Remarks
August 21$^{st.}$ to			30$^{th.}$ August noon		Weather
September 1$^{st.}$	Barometer	29^{i}.074	29.139.	29.028	generally
100			29$^{th.}$ August 2.30 p.m.		fine with
observations.	Attached Thermometer	80°.62	90.44	65.12	slight airs
	Detached Thermometer	80°.96	91.40	65.60	from the E.
	Wet bulb Thermometer	74.42	83.	65.1	& E.N.E.
					On the 27th August thunder in the West

[1] The Maipurishianna and Carawayanna remain unidentified. Maipurishianna translates as the Tapir People and the Carawayanna are possibly the Brazilians; the modern Trio word for Brazilians being Karaiwa. The Tunayana is a group that seems to come and go in the region. In the late 1970s–80s there were Amerindians who so designated themselves, living in the Trio village of Kwamalasamoetoe. For historical references to the Tunayana, see Bos, *Some Recoveries in Guiana Indian Ethnohistory*, Amsterdam, VU Uitgeverij, 1998, *passim*. The Barukutyo or Barecuto are presumably the Barokoto.

Height of the village above the sea 940 feet.

I found the latitude of the place to be 2°1′40″ North and the longitude 56°28′20″ West of Greenwich. The evaporation amounted during the twelve hours of day to 371 grains out of 1000 exposed in the morning. During the twelve hours of night (the vessel placed under the tent which was open at its sides) it amounted to 18 grains.

I have already observed that the settlement was 216 feet above the river Cutari, here about 24 feet wide. This small river forms a number of fine cascades and the rock over which it precipitated itself was a very coarse granite.

The path to the neighbouring settlement, which is inhabited by Drios led over a continuation of small hills, the highest of which was perhaps not more than 250 feet. The Brazil nut tree or Toka as the Pianaghotto call it is very abundant. Like many tropical trees, it was shedding its leaves and the path was thickly covered with them.

We crossed the Cutari after a march of three hours; a farther march of half an hour brought us to the settlement. Our canoes were not in so advanced a state as I could have wished for the sake of dispatch; the bark does not always part as easily at one period as another, and several trees which had been cut down proved useless for our purpose.

The following day we had a visit from some Drios who reside at the head of the Wanamu. The false alarm had reached them and they proceeded *en masse* to the place from whence their countrymen had fled at our arrival, here the mistake was explained to them and some of them anxious to see white people did not mind the journey of five days out of their way to satisfy their curiosity. They were tall and well made men, from five feet five inches to five feet seven inches in height and had ornamented their bodies by incisions like the South Sea Islanders. They were the first whom I met in Guayana who ornamented their body in this manner.[1] Otherwise they were dressed like the Pianaghottos, and following their custom of painting the whole body red, and allowing the face to remain of its natural colour, it gives them, like the former, rather a ghastly appearance. In some instances however they paint half the face red. I presented them with combs, fish hooks, glass beads &c and they departed shortly after greatly satisfied.

As far as I could understand from these people who corroborated, the account I received previously from the Pianghotto Chieftain, the Maroon Negroes on the Marowini, and its western tributaries are very numerous. They live in large villages in huts like those of the Indians and each village has a Chief who presides over it.

These Maroons[2] are the descendants of the fugitive negroes who forced, by the cruelties to which they were subjected, fled as early as the commencement of the last century into the forests & settled first near the rivers Surinam Saramaca, and Copename. Under the command of chiefs whom they selected from among themselves, they cultivated sufficient land for their subsistance and made frequent incursions into the neighbouring plantations. The colony was obliged to send several expeditions against

[1] It should be noted that Schomburgk reported 'body incisions'. His brother, Richard, turned this into 'tattoos' (*Travels*, II, p. 384). Although this is not impossible, the more likely thing is that the incisions were the result of scarification, which was widely practised in the area as a means of blood-letting to induce health, strength and skill.

[2] For more about Maroons, see Richard Price, *The Guiana Maroons: A Historical and Bibliographical Introduction*, Baltimore, Johns Hopkins University Press, 1976.

them which ill-conducted, failed, and their number increased considerably when in 1757 the slaves of the plantations near the river Tempati[1] revolted and joined the rebels.

The colony was under the necessity of entering into a treaty in October 1760 according to which they were acknowledged as free and bound themselves to assist in defending the Colony and that they should not only deliver up the slaves who had deserted to them since, the 14 October 1759 when a preliminary treaty was signed, but likewise every other slave, who should subsequently make his escape to their territory. The Colony bound itself to pay an annual tribute, or as in consideration for the feelings of the colonists it was called, annual presents.*

*These articles were signed at the plantation Rama opposite the negro camp; & sixteen of the chief Captains of the Maroons were present. It is related that they swore to keep the treaty, the obligation being solemnized, according to their manner for this purpose a small quantity of earth was mixed with water in a calabash, and each of the contractors having made an incision in his arm for the purpose of procuring some drops of blood, it was mixed up with the earth, of which the white deputies as well as the Negro Chiefs were obliged to drink in pledge of their faithful adherence to the contract. The name of the principal chief was Araby[2] and his authority was generally acknowledged by the Bush or Maroon Negroes all over Surinam. They have since remained quiet, and as far as I know have fulfilled their engagements.

Their number was estimated in 1760 at 20,000 which was no doubt an exaggeration. From Anka where their principal camp was formerly situated and where the Dutch Colony maintained a Resident, they have in time advanced further to the South-east, and chiefly to the South West where they are settled near the river Tapanoni, the south-western branch of the Marowini. Our Drios and Pianaghottos are their Mattees or friends; nevertheless the Indians and especially their better halves complain sadly that they drive such hard bargains, and are uncommonly sparing in their dealings in glass beads, and our liberality in this article won their hearts.

It was difficult to ascertain the number of Maroons near the Tapanoni, five of the Drio Indians pointed to their fingers & toes and then carrying, with violent gestures their hands to the head, pointed to their hair, by which they wished to impress twenty times more than they had hairs on their head.[3]

On the 5th September, the Coxswain announced that our bark canoes were ready. While felling the Copaivi and locust trees, he had been assisted by some of the Pianaghottos who, as Jason[4] related, broke out several times while working with him, in lamentations similar to those of the first Pianoghottos who came to visit us. Sororeng understood that a tradition prevailed among them that the arrival of the first white man betokened the extinction of their race. If that be the case we have to admire their hospitality which was no way diminshed by any angry feelings against those who brought this dread tradition so forcibly to their minds and rendered the period of their extinction definite.

One of the greatest ornaments of the forest round us is a lofty tree covered with a multitude of pink blossoms. The Drios call it Maipuremu; it proved to be Aublet's

[1] An upper tributary of the Commewijne River.
[2] For more about Alabi, see Richard Price, *Alabi's World*, Baltimore, Johns Hopkins University Press, 1990.
[3] I have never seen a Trio make a gesture like this, but it may well have been a practice that fell into disuse.
[4] This is the first reference to anyone called Jason on the expedition. Nothing more is known about him.

Vantanea Guianensis (Aub. p. 572 t.229) which, in consequence of the imperfect account given of it, has been shifted from place in the natural system. The peculiar formation of the anthers and the drupaceous fruits show it to belong to *Humiriaceae*, and it is according to Mr. Bentham very nearly allied to *Helleria* of v. Martius. The Drupe is furrowed like the stony seeds of our peaches, but they are twice as large. The Indians cut them in halves and wear them as ornaments particularly the Children.

Not more than half of our baggage had arrived and as misfortune had decreed it, the salt and our tent-covers and a great part of my collections remained behind, nor could we induce any of the strange Indians to undertake the long journey a second time. I therefore resolved on departing the next morning. Thirty circum-meridian altitudes of northern stars gave me for the latitude of the place 2°3′30″ North, and the Chronometer indicated that it was 6′19″ in arc to the West of the Pianaghotto village consequently in longitude 56°22′ West of Greenwich.[1]

I adopted this place of our embarkation as a new starting-point for the Chronometrical distances in our journey to the sea-coast.

September 6th. One of our new bark canoes had split during the night and the necessary repair delayed in some measure our departure. The Drios had brought us what provisions they could spare, which unfortunately consisted of yams a very bulky article. One of the Indians, however, consoled us with the assurance that after ten days' journey we should again find inhabitants.

Before we started I walked once more to the settlement our tents having been erected at some distance from it. Not a soul was to be seen; they had departed in the morning. Their apprehension that we might perhaps carry some of them off with us, no doubt induced them to retreat before our departure took place.

We embarked at noon in five bark Canoes. The Cutari was scarcely more than thirty feet in breadth; quite shallow and so impeded by trees fallen across it, that parties had been occupied the two previous days in cutting a passage for our craft. Nevertheless, we made but slow progress and after three hours paddling were only a mile and a half in a direct distance from the place of embarkation.

September 10th. The history of the preceeding five days may be comprized in a few words, – our advance slow and connected with unceasing labour. It appears the Cutari is only navigated when its bed is full to overflowing, and none of the trees thrown down by the fury of the winds or torn off by the torrent, were now lying across the river had been cut through to admit even the passage of such small craft as ours. Every morning regularly we despatched half of our party in advance to cut such trees as offered the greatest obstacle, while the rest of us followed about noon. The river is monotonous; its banks offer scarcely any plants of interest to a Botanist. I observed a few cocoa or chocolate nut trees, with ripe fruit on it and numerous Bambusaceae. But the river itself abounds in Haimaras and we, this evening caught seven in the short space of an hour. One of the Indians was very severely wounded in the thigh by one of these fishes.

September 11th. The Cutari took a more Western course deviating in its direction perhaps one or two points on each side of West. It was increased to a breadth of about 80 feet, frequently crossed by ledges of granite and impeded by rapids. We arrived by 11

[1] This was a mistake. Schomburgk had subtracted rather than added the difference in arc to the longitude of the Pianoghotto village. Accordingly the longitude was 56°34′39″W, which appears correctly in the published version (p. 87).

o'clock at the junction of the Aramatau with the Cutari. We crossed the former river on the morning of the 22 August while walking from the Iriau to the Drio settlement. It is near its mouth about 20 feet wide and the united rivers are about 500 feet wide and take a North North West course. About two miles from the junction commence a series of falls and rapids of which the largest is from 12 to 15 feet high. They extend North North East for about 2 miles; after which the river again flows North North West and at the distance of four miles from where it received the Aramatau joins the Curuni[1] forming at the junction two large falls.

I have related in my Journey to Esmeralda *

*See our Journal Vol. X. p.240.

that the river Kundanama joins the Padamo in a similar manner and what is the more remarkable, the recipient is there free of falls. In the present instance both rivers are impeded at their junction by extensive Dykes which consist of granite composed almost entirely of red felspar[.] These rocks were more or less covered with Podostemaceae, and the pretty Mourera with its lilac blossoms presented a very pleasing appearance among the foaming water.

We observed numerous Pecus[2] apparently of a different species, or perhaps merely a variety of the common Pecu (*Miletes pecu**).

*Fishes of Guayana in Naturalist's Library Vol.1 p.236.

Their back appeared black & their sides reddish, but as we did not succeed in procuring one I could only judge of them from seeing them swimming in the limpid stream. We were highly amused at the Macusi Carutshi. A large sting-ray (*Trygon Garrapa**)

*Fishes of British Guayana in Naturalist's Library Vol.2 p.182.

was lying partly concealed in the sandy bottom at a spot where the river was shallow and anxious to procure it for his dinner, he took his paddle and standing up in the boat, struck it on the head; this rough treatment appeared to rouse it from its lethargy and with one leap it sprang out of the water directly into the boat where Carutshi was standing, who, nearly as quickly as the fish had jumped in, jumped out into the water, and it was some time before he could muster sufficient courage to attack the sting-ray, who, so unceremoniously had taken possession of the boat, much to the merriment of the Indians who witnessed the affair. I could scarcely have thought it possible that an animal so unwieldy in appearance possessed sufficient agility to leap from its hiding place under the water at least eighteen inches above the surface, into the canoe.

The Curuni, as far as we could see up its last reach before the junction appeared, like the Cutari to be studded with rocks it comes apparently from the W.N.W.[3] and from what we heard at the Drio settlement, it has its source near those of the Tapanoni, the western branch of the Marowini. Its upper course is inhabited by the Eagle Indians.

[1] As mentioned above, the names of these rivers vary on different maps. However, most agree that the river here, which Schomburgk calls the Curuni, is the Sipaliwini.

[2] Pacu.

[3] Once again Schomburgk has his directions back to front. West-north-west is the direction in which it flows.

According to my observations of North & South Stars, the junction of these two rivers of equal breadth at their confluence takes place in latitude 2°20′50″*

*This is the result of 65 circum-meridian altitudes of 34 & 38 Sagittarii, δ Cygni, α Pavonis, α Cygni and α Cephei.

and the Chronometer showed that we were 7′52″ in arc West of our embarkation, the direct distance of which was South 25° east nineteen miles and we had required six days to accomplish it. The Barometer as compared with its height at the coast showed we were 612 feet above the sea.

It smelt round our camp like an onion fair. That liana which smells so strongly of garlic grew in abundance round us and a great many of the branches having been cut to clear the spot, the smell was really oppressive. I have met with it frequently, but have not been fortunate enough to find it either in blossom or bearing fruit in order to determine its character. I have no doubt that the Powis (*Crax alector*) feed upon the berries or the leaves of this ligneous twiner, as the flesh of that bird, at certain periods, tastes strongly of onions.

September 12th. The united rivers presented a continuous course of rapids; they were not dangerous but they delayed our progress. After ten o'clock in the morning we came however to a fall where the water rushed over the ledges, the first forming a Cataract from eight to ten feet in height, and the second from fifteen to sixteen feet. Near the foot of the latter rose from amidst the foam of the Cataract a solitary rock in the form of a huge pile, studded with laminar mica. The morning sun threw his rays upon the northern side of this singular rock and rendered [it] of dazzling brightness realizing the picture which Raleigh drew of Guayana, where every rock was described as argentiferous. While we were occupied in carrying our baggage overland and drawing the boats after us Mr. Goodall took a sketch of this cataract which for want of another name I have inserted as Sir Walter Raleigh's Cataract in my map.[1]

The heat at two o'clock in the afternoon was very oppressive. The thermometer rose to 142° F. in the sun. The barometer had risen; at least it stood higher at four o'clock than the previous day and our descent that day could not have been less than from thirty five to forty feet which ought to have depressed the mercurial column. I have frequently observed that this predicts rain, while a low barometer, in rainy weather, frequently predicts a change for the better. The effect is therefore the reverse of what takes place in Europe.[2]

September 13th. My surmises were correct, heavy thunder and rain rendered our night quarters very uncomfortable and we greatly felt the absence of our tents. We started in the morning during rain. The river continued its monotonous course towards the North by West. It was now about three hundred yards broad and small hills lined with low trees and thick bushes approach its banks. Here and there are seen wide branching silk-cotton trees (*Bombax Spec.*) or the stately purpleheart which rose above the lower vegetation, consisting of the wild Arnatto (a species of *Bixa*), the Water Guava (*Psidium aquaticum*), an Inga and numerous Jacana trees (*Triplaris Americana*) and

[1] It still has the name today on the Guyanese map.

[2] Compare this with the published version (p. 90): 'The barometer had fallen; at least it stood lower at 4 o'clock than the previous day; and our descent that day could not have been less than from 35 to 40 feet, which ought to have caused the mercurial column to rise.'

Cecropias (*Ceprop ia peltatae* and others of the same genus). There was almost a total absence of Orchideae, a circumstance which seldom occurs in the tropical and moist regions of Guayana. The current is very trifling amounting to about a mile in an hour, where it is not accelerated by rapids or falls. The latter are caused by ledges of rocks of Hornblende Gneiss here and there traversed by a vein of quartz. We halted for the night near the mouth of a small river, which joined the Curuni on its left bank; a few yards lower down, another stream came in from the right. Our latitude this night was 2°34′ North. Here I suffered a loss which I felt very severely. It will be recollected that I lost a very fine Taruma dog at one of the Cataracts and of all the dogs I had purchased, one only now remained, an animal the more remarkable from its being a cross breed between the Taruma dog and the *Canis cancrivorus*, Desm. A Jaguar during the night carried him off from the midst of our camp, and thus was I not only deprived of what I considered a treasure, but what was worse his place as a good purveyor of our larder, could not be supplied in the wilderness.

September 14th. We passed the river Siparawuini[1] at half past seven; the rapids continued and the heat was oppressive in the afternoon. The Thermometer rose, in the sun, to 135½° F. a heat to which we were greatly exposed, as the small size of our woodskins or bark canoes did not permit the erection of any thing to shelter us from the burny rays. I regretted this the more for the sake of Mr. Goodall who continued indisposed and I myself was then suffering under acute rheumatism.

We were drenched with rain. During the night, a severe thunderstorm passed over our camp, against which the temporary tents made of wild plaintain leaves could afford no protection. I shall not easily forget one of the thunder Claps. I was just on the point of rising out of my hammock to see whether I could not make the roof a little more water tight when I saw the tent clearly illuminated with a rose-coloured light and the thunder following instantly after it, might have been compared to the firing off close by of the heaviest pieces of ordnance.

"Twas listening fear, and dumb amazement all."[2]

I stood transfixed anxiously listening whether I should hear cries of distress but, as if nature had been for a while appeased by the effort, the heavy rain ceased for a while and all was calm. The voices of our Indians relieved the suspense which during that awful silence they had no doubt equally shared with me. On inquiry I found that all was well, and the electric fluid had no doubt struck the river close by. The following day we met with an accident; in descending one of the falls, the small boat which contained our kitchen utensils swamped and sunk; our loss, however, was but trifling, and as the boat fortunately did not sink in deep water, we were able to recover it again. It was to be ascribed to the imprudence of our cook, a most ill fated being, who, when I travelled without a european companion from Roraima to the Cuyuni on a former occasion likewise sunk a boat, and we were not then so lucky but lost every article of kitchen utensils, without saving so much as a plate, knife, fork or spoon, or a grain of salt; and thus was I obliged to do without these comforts for nearly a fortnight. The Rock which caused the cataract, in the present instance, was mica slate which occurs much more frequently in the Curuni than in the other rivers we had passed. It is, here and there, traversed by veins of quartz.

[1] This is the Kamani River of both the Guyanese and Surinamese maps.

[2] James Thomson, *The Seasons*, 'Summer', l. 1128: ''Tis listening fear, and dumb amazement all.'

The heavy rain had caused the river to swell. In the evening we halted near a small stream which was pouring out its whitish waters with such impetuosity that they rushed for several hundred feet into those of the Curuni, forming a white band, narrow near the mouth of the river and from thence spreading like the tail of a comet. We found several bark canoes tied to the banks of the river, and a narrow foot path. We little thought that they would prove the last vestiges of man that we should see till we reached the Caribs at the lower Corentyne. We had been informed that a path of three days journey led to the settlements of the Indians located at the upper Sisparawuini.

The cloudy state of the weather precluded my taking more than three altitudes of the *α Lyra when near the meridian; they were, however, sufficient to prove that our progress had been slow,*

*The latitude was, according to these observations, 2°48′, and a set of hour angles gave for our distance West of our embarkation about 22 miles.

and I was therefore under the necessity of putting the expedition upon short allowance, and to declare war against all animated nature, not excepting the Iguanas, of which large numbers, roused by the sound of our paddles, jumped from the adjacent trees, with a loud splash into the river and were sure to cause us the heaviest disappointment by escaping from our pots. On the evening of the 17th. we were successful enough to hook a large Paruaruima (*Phractacephalus Hemilopterus*)[1] measuring four feet two inches in length and two feet eleven inches in girth. It is a pity that this fish is almost all head and tail. The mishapen head is covered, as it were with a cuirass, and in the present specimen it measured one foot five and a half inches, and thus very materially diminished its usefulness as food for fifteen individuals.

In the afternoon of the 18th. we found that the river took a western and frequently a southern course, consequently retrograde; its bed was so studded with iles and falls that it appeared almost impossible to find a spot where it flowed gently along. Small hills of from two to three hundred feet in height approach its banks, and it divides into so many channels that we scarcely ventured to lose sight of each other for fear of being seperated. Mr. Goodall, who was now in tolerably good health, had unfortunately, since we had been put upon small allowance been restored to his best appetite, and both of us encourage the rest of the crews by paddling like themselves from morning until we pitch our camp at night.

Numerous ledges of rocks or bars extend right across the river in a northwest and southeast direction and cause dangerous falls. These rocks were, for the most part, composed almost wholly of felspar. On arriving at one of the Cataracts I led the way with my own little bark and the others followed. Hitherto, thanks to a kind Providence, we had been successful, though we came from time to time in rough contact with some sharp rock which pierced our skiffs, none of them half an inch in thickness,*

*The one I was then in was only four lines or one-third of an inch. Mr. Goodall's, which is the thickest, five lines.

and cause the water to spout in like a fountain, when pieces of shirts and trowsers are resorted to for stopping the leak and which always proved sufficient for the purpose.

[1] The Red-tailed Catfish (*Phractocephalus hemioliopterus*).

With the exception of one bark canoe which we were obliged to abandon we have hitherto suffered no loss since the canoe with our Kitchen sunk to the bottom.

The labyrinth of Islets (indeed the river cannot be less than eight miles in breadth, at least this is the distance between hills and hills) the numerous cataracts and rapids induced us to call this region the falls and rapids of the thousand isles.

The flora of these islets is much diversified. The *Elizabetha Coccinea* the Jacaranda, the *Laurus Surinamensis* which margined the river's banks, a *Strychnos* with clusters of white flowers of delicious perfume, the *Clitoria Arborea*,[1] struck me as not hitherto observed. I ought not to forget an uncommonly pretty Asclepideae, it flowers of a rich velvety purplish brown colour hung down in large racemes; it appears to be a new genus of this difficult order. The rocks, where they above water were covered with an arborescent *Solanum*; Orchideae were still wanting at least I saw none near the banks nor in the adjacent forest where we encamped. Our latitude was this night (Sep^t 19th.) 3°21′30″ North. Our progress westward from our embarkation 41′47″ in arc. During the last two days we had made 13½ miles westing. This slow progress and the probability that we should not find human habitations for a week or ten days more obliged me to curtail our allowances still more so that we were reduced to four ounces of cassada bread or Farhinha de Manioc and such game and fish as luck procured us. The river being swollen it was very difficult to obtain fish.

Sep^t. 20th. Island, rapids, and falls continued; the deep and sonorous roar of falling waters which broke upon us from time to time as the woods carried the sound apprized us of some great obstacle which we would have to surmount, and after paddling for about two miles we found ourselves at the head of some falls the height of which was fifty two feet,*

*The whole height of these falls from their head to the place where we embarked again was 65 feet according to the Barometer.

and which for grandeur vied with any I had seen in my peregrinations.[2] We had here to unload and carry the baggage and boats more than a mile over low hills about one hundred and twenty feet high, to the spot where the river again becomes navegable; a labour which it took us more than a day to accomplish. The rocks at the head of the Cataract were gneiss and here and there coarse quartzose sandstone or conglomerate cemented by iron, but near the foot of the Cataract huge blocks were lying about in the greatest confusion, as if they had been shattered by some great convulsion of nature. These blocks consist of fine-grained gneiss, traversed by layers approaching in appearance decomposed epidote (?).[3] The gneiss is stratified and on some of the blocks it appears much contorted.

[1] Species not reported from Guiana region.

[2] As will be seen below, Schomburgk called these falls King Frederick William Falls, by which name they are still known today. These falls are downstream from the mouth of the New River which Schomburgk apparently failed if not to note, at least to report. Given the width of the river and the numerous islands in it creating many channels, his failure to observe this tributary would not be surprising. However, maps henceforth show the River Pani, about which he had been told when on the Upper Essequibo in 1838, flowing into the Corentyne at almost the correct place. Barrington Brown, hugging the left bank of the Corentyne, found it in 1871 and called it the New River (see *Canoe and Camp Life in British Guiana*, London, Edward Stanford, 1876, pp. 328–9). The size of the New River is such as to have raised the question as to which is the mainstream; this has had consequences for the defining of the Guyanese/Surinamese boundary (see Appendix 1).

[3] The question mark is carried over into the published version (p. 94). Epidote, according to the *SOED* is: 'A mineral common in many crystalline rocks, consisting largely of the silicate of iron and lime'.

This delay gave me an opportunity of noting the hourly observations of the Barometer and thermometer on 21st Sept. which were as follow.

Hourly Observations on September 21st. 1843. Height 330 feet above the sea.

Instruments	Hour of day	Highest	Hour of day	Lowest	Mean
Barometer	11h. a.m.	29i.712	7. p.m.	29i.640	29i.668
Attached Thermometer	1. p.m.	84°.56	5. a.m.	71°.60	75°.56
Detached Thermometer	1. p.m.	85.00	5. a.m.	71.50	75.70
Wet bulb Thermometer	1. p.m.	79.40	5. a.m.	71.00	74.24
Temperature of the river	2. p.m.	88.00	3. a.m.	81.70	83.23

It was fortunate that the Cataract obliged us to stop this day as our reduced circumstances would scarcely have allowed such a delay for the purpose of making observations.

Since we had entered this continued series of rapids and falls we lost all sight of game and fish; we had to preach philosophy to our murmuring stomachs, to satisfy themselves with four ounces of bread. It is a great misfortune at such a time to be blessed with a good appetite.

We started early in the morning of the 22d from the foot of Frederick Williams Cataract, under which name, in honour of my sovereign the King of Prussia this magnificent fall will appear in my maps; and which according to the Chronometer was 45′15″ west of our embarkation or 57°7 West of Greenwich.*

*I did not succeed in procuring any observations for latitude, but as our progress on the 20th. was trifling, and the latitude having been found the night previous, to be 3°21′30″N. and on the night of the 23d. 3°38′38″ by assuming Frederick William's Cataract 3°30′N., we cannot be a mile from the truth.

The river widened considerable, extended, no doubt by the numerous islands. In the afternoon it presented, where the sight was not interrupted by islets a breadth of about four miles; ledges of gneiss traversed the river in a N.N.E and S.S.W direction, one following the other, like furrows in a ploughed field. Large spherical blocks of a coarse grained granite, or, at other times, shaped like an obelisk, were lying upon these ledges. A small chain of hills, their highest summit about 400 feet above the river extended from the left or western bank of the river for about two miles inland, and closed the back-ground of the picture on approaching from the North as we did. Our latitude determined from some very good observations was 3°38′38″ North, our distance West 54' 43″ in arc from our embarkation; our course, since the 17th. had been N.W. by W.

We nearly lost one of our canoes the next morning on descending a fall. Some of the baggage got wet; but the sun shone bright and the mishap was soon remedied. We paddled on unceasingly for nine hours, and stopt at night on a small island, our larder enriched with some Iguanas and a crane; the latter promised a delicious soup; but unfortunately we spoiled it by using gunpowder as a substitute for salt of which we had been deprived for some days past. The taste of the Nitre and Sulphur perfectly spoiled an entertainment so eagerly anticipated by our hungry stomachs.

September 24th. While hauling one of the boats over a rapid, the poor Arecuna Indian, who, the reader of the geographical journal will perhaps recollect was severely wounded by a Sting-ray during our ascent of the Takutu met with a similar accident this day. He suffered most excruciating pains; the wound had been inflicted near the ankle.

The juice of the moco-moco (*Caladium arborescens*)[1] having been recommended to be taken inwardly mixed with a little spirit and outwardly squeezed into the wound was tried, with the exception of the spirit of which we had none and for which water was substituted; but it did not appear to diminish the pain.

The rocks which formed the rapids were compact red felspar stratified in an East and West direction and traversed from North to South so as to form oblong squares. Large blocks of the same rock were lying upon it.

We halted at eleven o'clock at the head of a large Cataract and it became evident that we should have to unload the canoes and carry the baggage overland. While I was looking out for a fit place to transport it to the foot of the Cataract, I observed on the banks of the river the remains of some old huts and, on closer inspection found, that we were at the path which leads from the Corentyne to the Essequibo; and, as a further proof that my surmizes were correct, we now discovered two corials hauled up on the land, one above forty feet long, the other about half that size. While ascending the Rupununi last March, I saw at one of the settlements a Carib who told me he had lately come with others, from the lower Marowini and entered the Corentyne to join their relations at the Rupununi and that they had left their craft at the place where the path leads from the Corentyne to the Essequibo. We were now persuaded that after so many toils we had, at least, reached a part of the river which we knew had been visited by human beings, while I am almost convinced that for ages man has not traversed that part of the river which we had navigated since we left the spot where we found the old rotten bark canoes and since which period (Sep^t.^ 15th.) we had been struggling with cataracts and had suffered half starvation. Though we were well aware that our trials we[re] not yet over, it was consoling to know we were navigating the Corentyne and that men, before us, had passed from the spot up and down the river.

The antiquity (as far as this word can be used in things relating to America since its re-discovery) of this path rendered it very interesting to me. It must have been already known in the 17th. Century; at least D'Acunha found in 1639 among some of the Indians near the Rio Negro, iron tools, such as hatchets, bills, knives &c. and, on inquiry, was told that they had received them from the people who dwelt in the country nearest the sea on that side, and who had light hair; by which D'Acunha supposed they alluded to the Dutch.[2] Dr. Southey while speaking of Ajuricaba, one of the most powerful Caciques of the Manaos, and who lived about 1720, observes that he made an alliance with the Dutch of the Essequibo with whom he traded by the way of the Rio Branco. The trade on his part consisted in slaves. In order to

[1] Moco-moco, or Moka-moka, is *Montrichardia arborescens*, and its sap is used by the Palikur Indians of French Guyana as protection against the discharge of electric eels. A species of *Caladium* is used as protection against poisonous snakes. *C. bicolor* is of great ritual and shamanic importance. *Montrichardia* and *Caladium* are both genera of the Family Araceae. See Grenand et al., *Pharmacopées traditionelles en Guyane, Créoles, Palikur, Wayãpi*, Paris, ORSTOM, 1987, pp. 132–6, 141–2.

[2] De Acuña, *Nuevo descubrimiento*, Chap. LXIIII. The metal tools found among the Amerindians of the Rio Negro in the Seventeenth century could certainly have been of Dutch origin, traded either up the Essequibo and across the Pirara portage, or by the trail from the Corentyne to the Berbice which had been an Amerindian trade route for centuries. The first recorded journey by an European along this trail is in 1718–20 (see Bos, 'Atorai, Trio, Tunayana and Wai Wai in early Eighteenth Century Records', *Folk*, 27, 1985, pp. 5–15(6)).

obtain them, he hoisted the Dutch flag, scoured the Rio Negro and captured all the Indians on whom he could lay hands.*

*Southey's History of Brazil Vol. 3 p. 710–11. Ribiero reports that Ajuricaba was taken prisoner with 2000 (?) Indians, while Jao de Maya da Gama was Governor, and carried in chains to Para, and a mutiny on board the vessel having been discovered he jumped into the river to drown himself.[1]

Only a few of these slaves were carried down the Essequibo, the greater part being taken to Surinam by the path which from the inlet Primoss leads across the river Berbice to the Corentyne which river it meets at the Cataract where we were encamped, and this path is still called among the Macusi indians, the slave path or Juari Yemori, and this is likewise the way by which Don Francisco Jose Rodriguez Barata went twice from the Amazon to Surinam on affairs of his government in 1793.*[2]

*Humboldt's Personal Narrative Vol. 5 p. 480.

Though I had only walked along that path from the river Berbice to the Essequibo, my present journey made me likewise acquainted with the part where it strikes the Corentyne.

The Caribs those scourges to the less valiant tribes, settled from Surinam along the Rupununi, and by means of another much frequented path held intercourse with the Caribs of the Caroni and the Orinoco.

The Scenery is uncommonly picturesque near this Cataract; blocks of granite of a spherical form from fifty to sixty feet high, lie about in great confusion crowned with tropical vegetation; the wax-flowered Clusia the yellow Cyrtopodium, the thread-like Rhipsalis[3] with its pearly fruit, and upright or columnar Cerei give to these singular rocks a marked tropical appearance. We here enjoyed, after a long period the first emotions of satisfaction and I gladly availed myself of the privilege of discoverers, more particularly as I could not at the time learn whether the cataract possessed a name or not, to introduce it in my map as Lord Stanley's Cataract.*

*On our arrival at the Carib settlement I learnt that it was known to them under the name of Aritapu, I have nevertheless, preserved the name I gave it under our first impressions.[4]

Upwards of fifty circum-meridian altitudes of the stars δ Cygni, α Pavonis, α Indi, α Cygni, α Cyphei gave me as a mean 3°59′16″ North for the latitude.

Our Indian discovered with great joy some Kaitshara or Penguisa (*Bromelia karatas*)[5] with ripe fruit. This fruit resembles a Banana in miniature and grows in the middle of

[1] Robert Southey, *History of Brazil*, 3 vols, London, Longman, Hurst, Rees, Orme & Brown, 1810–19. Ribiero is Francisco Xavier Ribeiro de Sampaio who kept a diary of a tour of inspection he made up the Amazon, Rio Negro and Rio Branco in the years 1774–5. This was published as *Diario de uma viagem que em visita, e correição das povações da Cap de S. José do Rio Negro fez o Ouvidor e Intendente Geral da mesma, Fco. Xavier R. Ribeiro de Sampaio, nos annos de 1774 e 1775*, Lisbon, 1825.

[2] Barata travelled, both coming and going, via the Essequibo (see 'Diario da viagem que fez á Colonia Hollandeza de Surinam').

[3] A genus of the Family Cactaceae.

[4] It is marked on the Guyanese map as 'Lord Stanley's or Aritapu Falls'.

[5] The reading 'Penguisa' is not secure. *Bromelia karatas* is known as either 'anareke' or 'shiwishiwi' in Arawak, and as 'waurinit' in Warao.

the plant, perhaps from 80 to 100 nestling together. They contain a succulent flesh of an agreeable vinous taste when ripe, and although this was not quite the case in the present instance, they were highly acceptable, and by boiling them we counteracted their astringency. They were the more welcome as neither of the huntsmen we had sent out on our first arrival, had brought in anything and the attempts of our fishermen during the night had met with no better success.

The Indian whom I had met at the Rupununi, and who told me of having left his Corial at the path had given me permission to use it, if I should want it, and he directed me to deliver it to his brother, the Carib Chief William, at Tomatai.

According to the Barometer, we were still 230 feet above the sea, and we had yet to overcome many an impediment before [we] could get into smooth water. I therefore took the Corial and had it repaired as far as it was possible to do so in a few hours.

We started on the morning of the 25$^{th.}$ September at halfpast six o'clock, but we had scarcely proceeded a mile, when another Cataract, where we again had to unload, impeded our progress. The rocks which here consisted of quartz and felspar pebbles cemented by iron, were lying about in the greatest confusion imaginable, and had the appearance of having been vitrified in a furnace. On some spots where the iron cement was more prominent, they were so smooth, that we had to guard ourselves against slipping. The smaller rocks, where heaped together looked strikingly like the refuse from a smithy. Our course varied between North and N.N.E. the river still studded with Islets. While we turned round one of the islands, my boat being as usual in advance, we saw a large puma (*Felis Concolor*) drinking at the margin of the river; though he observed our boat, he scarcely took any notice of us, unfortunately the only gun we had in the boat, was wet from the severe rain, and it missed fire. We were not fifteen yards from him, and he leisurely put himself upon his haunches, like a dog, and appeared rather to have every inclination to attack our boat, than to be frightened at our approaching him. The Indian Sororeng was so exasperated at his audacity and the inefficient gun that he threw a large piece of wood at him which was lying at the bottom of the boat; this induced him to return leisurely into the wood.

Violent rain induced us to pitch our camp in the afternoon of the 26$^{th.}$ as early as half-past one o'clock, and we were so hungry that we thought success might perhaps crown our endeavours to procure some game or fish. The Indians brought in a Maam (*Tetrao Guayanensis* Gmel.)[1] of the size of a pullet, which to divide among fifteen would have required a pair of compasses and scales.

We saw on the morning of the 27$^{th.}$ a large amphitheatre of hills about two hundred to two hundred and fifty feet high, which surrounded the broad expanse of the river, here from East to West about three miles broad, the hills covered with a thick forest of umbrageous trees. The intermediate space between hills & hills was broken up by islets; blocks of granite heaped up in the most fantastic forms, met the eye in every direction and predicted the dangers we should have to encounter. In advance of the others I gave orders to the Coxswain to halt at one of the rocky islets, and climbing to the summit our route lay before us. The remembrance of what I felt at the sight before me will remain as long as memory lasts. Enormous masses of stone, black as Iron, extended as

[1] This is a synonym for the more usual *Tinamus major*. Gmel. is Johann Friedrich Gmelin (1748–1804), a German naturalist.

far as the eye could reach; against which the river dashed its waters with the greatest fury; the white foam forming the strongest contrast with the gloomy tint of the rocky masses; and when my looks fell upon the miserable skiffs to which our lives would be confided, skiffs whose construction could not boast of a woody fibre, when I recollected the privations we had all suffered and our consequently enfeebled state, our only hope of surmounting the dangers before us rested upon him who is Almighty, and whose aims and ends we poor mortals cannot fathom.

It was impossible to give orders, or maturely to consider and pre-arange what was to be done, we had to leave our actions to circumstances, and till the moment should arrive for averting the dangers we had to brave. As I have generally done, so I did in the present instance, I took the lead in my boat & ordered the others to follow it at the necessary interval, if it should pass successfully, or to profit by any disaster which might befall us by avoiding the danger. Far be it from me to seek to impress the reader with the belief that I felt no apprehension while my light skiff dancd upon the contending waves and passed between the black pointed rocks that beset its course – though I have passed hundreds of cataracts it would be sheer bravado to assert that I did not breath quicker or I feel my chest more contracted than usual when passing a fall, but I may with truth declare that those feelings are trifling when compared to what I experienced when, having arrived at the foot of the cataract, my boat bears round to await the descent of the others and I witness the imminent danger of every successive skiff as it is hurried down with the swiftness of an arrow and when the slightest mismanagement might entail destruction to its inmates. And how frequently had I not to endure this painful excitement during the course of this day. I now called to mind, that about the same time of the year in 1836 when at the foot of the great cataracts the Caribs had told me, it was impossible to pass the falls of the river above them, and that it was only when the river was full to overflowing that they could be surmounted.

Worn out with fatigue we halted at four o'clock in the afternoon. We had no guides, and circumstances, and our experience of former situations of the same nature, could alone direct us in selecting our path where, as far as the eye could reach, rocks and foaming masses of water were visible. We had to contend with legions of ants for the possession of our camp, and received many a bite while preparing our night quarters.

Hunger is a very disagreable feeling one's ideas get confused, & there is felt a vague sensation which ultimately engenders headache, and also an uneasy spasmodic contraction near the region of the stomach; such were my feelings this day; I did not dare enquire after those of the others, I was sure their answers would have aggravated my own suffering. The report of a gun roused me from my revery; what can the lucky hunter have killed was my first thought, and the moments which succeeded seemed as hours. At last an Indian arrived with heavy step, carrying on his shoulders a large female black spider monkey. I glanced at Mr. Goodall whose countenance depicted disappointment and disgust, but which sad necessity and the large vacuum that two ounces of farinha must have left in his stomach induced him to get the better of. He watched the preparations as the Indians proceeded step by step, first singeing off the hair from this human like form and then placing it in an upright position, with the arms crossed, when, the skin looking white now the hair was off, the sight proved too much for him, and I myself felt something like disgust at the meal before us. The sound of a heavy body falling on the ground drew my attention to a different direction, and to my great joy I beheld a fine young forest deer, over which young Hammon stood leaning on his gun with proud satisfaction. This was indeed a happy turn in our affairs. I have tasted the smaller kind of monkeys several times, but have never partaken of one which approached so nearly to the human form as this. The indians were less scrupulous.[1]

September 28th. We had to unload our craft twice in the course of the morning. In order to avoid the great cataracts, I sent off, at the Dawn of day, the interpreter Sororeng and another Indian to seek for the portage which I knew from my former journey up the Corentyne to exist, but the position of which the Caribs had kept secret from us in 1836. We had been fortunate enough to select the right direction for our descent and the leading canoes had already arrived at the embarkation by half-past six o'clock. We now unloaded for the third time, and took baggage and boats for about a mile overland, and arrived at the other end of the portage. I was here among objects familiar to me, the great Cataracts of the Corentyne. Barometrical observations gave me for the descent from the place where we disembarked to the deep and extensive bay at the foot of the great falls 90 feet; but from Lord Stanley's cataract 163 feet.

Recollecting how successful we had been in fishing and hunting, while here in 1836, we did not proceed any further and all hands that could hold a hook and line, or a rod or were able to shoot were sent in quest of game or fish; the hunters returned unsuccessful and the fishermen reported that they had not had a single nibble.

How shall I describe the three succeeding days! The river being much lower than we had found it in 1836 rapids and falls impeded its course, where, seven years before we had found no difficulties, and now with increasing labour our exhausted state was the more keenly felt. Several of the Indians suffered from fever and our little company was so divided that we could scarcely spare the physical force of a single individual; the greatest persuasion was therefore necessary to encourage our people and show them in perspective the plenty they would enjoy when [we] should once gain the lower Corentyne and the settlements where we had sojourned on a former occasion.

We started from the Great Cataracts with eight pounds of farina (a coarse kind of Tapioca)[2] and had a journey of from three to four days before us ere we could reach human habitations, supposing even that the Indians had not removed. I had therefore to reduce our allowances to three ounces a day. Since our provisions had been so short Mr. Goodall scrupulously divided the quantity of Farinha into fifteen portions, and we allowed every other person to take his share reserving the two last for ourselves, in order to prove to the Crew that we did not wish to fare better than they. I had been in similar straights before, and the small quantity which fell to my share was partaken of in small pinches and thus prolonged the enjoyment, but poor Mr. Goodall knew no economy in such case and dispatched his share instanter; the consequence was he suffered more severely. In spite of our debility we paddled from daylight until five and six o'clock in the afternoon and from the 29th. of September to the 1st. of October we failed in shooting any bird or catching any fish, to add to our meagre allowance.

On the morning of the 1st. of October we divided the last of our provisions amounting to something more than two ounces to each individual; but happily on the evening

[1] Monkey forms a major component in the diet of many Amerindian groups, and spider monkey, particularly when it is fat, is highly appreciated. It seems extraordinary that Schomburgk and his European companions had not eaten it frequently on their journeys, especially given how often he complains of a shortage of food.

[2] Made from manioc.

Figure 12: Carib settlement at Tomatai on the Corentyn. Picture taken from the copy of Richard Schomburgk, *Travels in British Guiana*, I, p. 128, in the editor's possession.

of that day, after having travelled twenty six days without meeting a human being we entered the Carib settlement of Tomatai.

We had started from our night camp very early, but a heavy easterly wind, and the broad expanse of the river caused waves for which our craft appeared but ill calculated, and our progress during the morning hours was not very rapid. About five o'clock I recognised familiar objects; Christian, the Carib settlement, was before us. The grass had overgrown the path which leads from the landing place to the houses and told us plainly that the place was deserted. This was a great disappointment, we however found some Papaw trees bearing fruit, in the abandoned provision fields; these we considered more than delicious, and somewhat refeshed we urged on towards Tomatai.

The Corials tied near its landing-place and a couple of dogs gamboling among the rocks in front of it, presented to our longing eyes a most delightful picture. Our skiff had scarcely touched the ground when I immediately jumped out and hastened towards the huts, unobserved by any one till I approached quite close. Old William at this time the Indian Chief of the place was at work upon a basket; he raised his head and though seven years had elapsed since he had seen me, and without having the slightest knowledge of my coming he immediately recognised me. When I told him of our sufferings and whence we had come, the good natured squaw hastened to bring some Cassiri,[1] which for the occasion she tendered in a fancy tumbler, and a plate of delf loaded with fish and fresh Cassada. Mr. Goodall who soon followed was equally served and the smoking pepper pot and some cakes of Cassada were placed before our starving crew.

Old William's curiosity about the Indians who inhabit the upper Corentyne was insatiable for all that the Caribs knew about that river extended no further than to the slave path, above that in their belief were regions inhabited only by the water-mama, sprites and hobgobblins.

The next morning I continued my journey. The previous day had convinced me of the impossibility of entering the sea-reach of the Corentyne with our bark canoes,[2] and I accordingly hastened towards the Post to make arrangements for proceeding without delay to the coast. Mr. Goodall had been desired to follow the day after, in order to allow our worn-out crew, a day of rest.

The hospitality so peculiar to the West Indian was heartily lavished upon us on our arrival at the cultivated parts of the Colony, and we ultimately reached New Amsterdam in Berbice on the 9th. of October and Georgetown, the capital of British Guayana, on the 12th. of October 1843.[3]

We had every reason to be grateful to a Kind Providence which had thus conducted us safely through so many dangers, had crowned with success the mission which had been our object and enabled us to perform our arduous duties without the loss of a single human life.

These satisfactory feelings are greatly increased by the circumstance that with the exception of a few thermometers and the mountain Barometer constructed according to Fortin's[4] plan, the instruments which were carried over several thousand miles of pathless regions were brought safely back to the Colony, where I thus had an opportunity of verifying them.

The General map of British Guayana which I have constructed from my exploratory expeditions is based upon the following data viz. the determination of the latitude of 174 different points, obtained by 4824 altitudes of heavenly bodies & the determination of hour angles for meridian distances and the rate of the Chronometers for 223 different stations, ascertained by 5801 Altitudes of the sun or stars.*

*If from the above observations, amounting to 10625 we subtract 230 made by Lieut^t Glascott R.N. then Assistant Surveyor of the Boundary Expedition, the remainder were taken by me individually, and that number does not include the lunar distances and altitudes connected with those observations, which amount probably to 1500 more.

[1] A manioc-based drink flavoured with sweet potato.

[2] It is not clear why Schomburgk was still using his barkskin canoes as he gave the impression that he had taken the canoes which he found above the falls (see p. 200 above).

[3] Elsewhere (RGS, RHS 1841–50, Letter of Schomburgk to Jackson, 20 October 1843) he reported that he arrived back in Georgetown on the 13 October.

[4] Jean Nicolas Fortin (1750–1835), Parisian instrument maker.

The Meteorology of the regions I have examined has received particular attention. The Barometer, with the Attached and detached thermometers were regularly observed wherever we remained stationary; besides which the temperature of the water, the evaporation, the quantity of rain which fell, the extreme of the heat of the sun and several other meteorological investigations were carried on wherever circumstances allowed it. The number of Barometrical & Thermometrical observations alone amount to 6692, not taking into account the other meteorological data which I procured during my journeys.

It is no doubt very difficult, under any circumstances, to convey with safety a collection of objects of Natural History from foreign parts to Europe; but the greatest difficulties are connected with conveying objects of that description over pathless regions such as those in the interior of Guayana where beasts of burthen are unknown and where on arriving at the rivers still greater hazzards are encountered in the shape of dangerous cataracts and rapids.

I have frequently lost by accidents which occurred to our boats the collections which I had brought together with the greatest exertions, or as in the instance related in the foregoing pages, have been obliged to abandon them. It gives me therefore great pleasure to be enabled to add something to the national collections in the British Museum, where I have deposited about 2500 specimens of dried plants, 100 specimens of woods, a quantity of dried fruits; a flower and young leaf of that wonderful plant the *Victoria Regia*, and several other botanical objects preserved in spirits; a collection of birds skins; upwards of 100 specimens of fishes in spirits from the rivers of the interior of Guayana; a geological collection extending from the Orinoco to the Equator, and an ethnological collection consisting chiefly of numerous weapons and the household utensils of the Indians of Guayana.

The Royal College of Surgeons has been presented with some skulls and with a perfect skeleton as well as a number of plaster casts of Indians.[1] The Model-room of the Admiralty has received 106 specimens of woods, many of them accompanied by dried twigs, some of them with the flower of the tree which furnished the specimen; and different Scientific Societies & Institutions have been presented with various curiosities from Guayana. The Royal Garden at Kew has received several living plants, chiefly of that interesting family the Orchideae.

I trust therefore that my exploring tours in a province so little known previous to my investigations and which combines tropical exuberance with fertility and extent may not only have contributed to the extension of Geographical Knowledge, but may also have proved beneficial to different branches of Natural History.

[1] These items appeared to have consisted of a skeleton of an African, the skulls of a male and female African, the skull of a female Amerindian, and twelve casts of different Amerindian tribes. Receipt of these from Schomburgk were acknowledged by the Royal College of Surgeons on 18 November 1844 and entered in its Museum Accession Register the following day. Schomburgk had earlier (1840) sold the skull of a Macushi Indian to the College, and again on this occasion it paid for the Amerindian remains, but those from Africans were donated by Schomburgk. After the Second World War, in which the Royal College of Surgeons was extensively damaged, such objects were transferred to the Natural History Museum. The latter's catalogue shows that the museum holds the two Amerindian skulls and twelve plaster casts. The African remains have apparently gone missing and it is possible that these were stored in the West African section which suffered badly from bomb damage. I am very grateful to Simon Chaplin of the Royal College of Surgeons and Rob Kruszynski of the Natural History Museum for this information.

I cannot conclude this memoir without drawing particular attention to Bunten's Syphon Barometer whose advantages are incalculable for those who traverse wild and pathless regions. The Barometer N°. 430 of his construction has accompanied me during my last expeditions, through forests and over mountains and in my boat navigation of impetuous rivers, and on my return to London I found on comparing it again with the Barometer at the Royal Society that in spite of all the severe trials to which it had been exposed it had not varied. On previous expeditions I have used Troughton's, Englefield's and Newman's[1] Barometers, and though every precaution was taken with them I never succeeded in bringing any one of these instruments safely back to the coast regions.

The all-interesting question in Physics, Terrestrial Magnetism, was not neglected during my journeys in Guayana, and I found opportunities for vibrating a pair of Hansteen's needles (the use of which Colonel Sabine R.A. had kindly procured me*)

*They are the needles L(a) L(b) mentioned in Table LII of the Magnetic Survey of Great Britain, and worth their weight in gold.

at 17 stations extending North and South from the 8th. to the 1st. Parallel North of the geographical equator, & from the 56th. to the 62nd. meridian West of Greenwich. The magnetic inclination and chiefly the declination were likewise ascertained in many instances.

I have now drawn attention to the general results of my exploring tours; many of the elements collected remain in their crude state; my active life not having permitted me as yet to turn them to account but I trust an opportunity will be afforded me for working up these materials in order to advance our knowledge of the Physical Geography and Natural History of Guayana, as also of *Man* – chiefly as he is exhibited in the few remains of the aboriginal races, thinly dispersed over many hundred thousand square miles, and fearfully hastening as by a divine decree to complete extinction.

R.S.

[****]

Richard Schomburgk and Fryer who had returned from Watu Ticaba on 21 May had arrived back in Georgetown in the second-half of June.[2] All the members of the expedition then remained there, completing the calculations and the maps, until 19 May 1844, on which date they, except for Richard Schomburgk who, because of the size of his collections, had to await a later merchant ship, took the mailboat for England.

Schomburgk never returned to British Guiana, nor to South America, but he left behind a legacy which continues to this day. This is the question of the boundaries of the country, the so-called 'Schomburgk line'. This matter is far too complex to deal with between the covers of this volume, but, at the same time, far too important to ignore. Accordingly a summary of it will be found in Appendix 1.

[1] John Frederick Newman (1790?–1860?), an instrument maker of London. He made both standard and portable barometers for James Ross's Antarctic expedition of 1839–43.

[2] Richard Schomburgk's dates at this point are unreliable. He states that the captains of the canoes arrived at Pirara on June 7th and that the baggage had been transported to the landing stage by the 18th. He gives the date of his final departure from Pirara as June 11th and his arrival in Georgetown as June 19th, but refers to it having taken thirteen days from Pirara to Bartica Grove (see *Travels*, II: pp. 315, 316, 323, 326).

EPILOGUE

Robert Schomburgk was back in London by 25 June 1844, when he reported his arrival to the Colonial Office.[1] There was a general feeling of a job well done. Murchison, in his 1844 Presidential address to the Royal Geographical Society, described Schomburgk in glowing terms:

> Few men are better formed than the Chevalier Schomburgk for the very arduous task of conducting exploratory expeditions in unknown regions. Conciliatory in his manners, yet firm, cool in judgement and prompt in action; enured to privation and fatigue, and undaunted by difficulty and danger, zealous and persevering – Such is his moral character; and when to this are added his various acquirements as an astronomical observer, as a botanist and naturalist, it will be conceded that he is one of the first travellers of the day – one of those, in fact, formed in the school of Humboldt, whose researches and observations extend alike over every subject of interest, and make us fully acquainted with the regions they explore.[2]

Light congratulated his Lordships at the Colonial Office for the successful termination of the survey, referred to the 'danger and fatigue' of those involved and of his own part wrote, 'but for the fortunate command I had, of funds independent of the colonists, [the survey] must have been long delayed, or perhaps have totally failed'. Goodall was recommended for his 'zeal and diligence' in the performance of his duties and Fryer for his 'exemplary conduct'.[3] It was, however, Schomburgk himself who rightly received the most praise. A Colonial Office minute on one of Schomburk's letters reads 'I believe that no man has undertaken more in the Public Services from the pure love of science and with a moderate pecuniary recompense'.[4] Lord Stanley, seeking public honour for Schomburgk, recommended him for a knighthood to which Queen Victoria assented in December 1844, and which he received in May 1845.[5]

Schomburgk's main concern on his return was to finish his map but this and other plans were disrupted by continuing ill health. For example, the day after his arrival back in London he wrote to George Bentham, saying that once his brother had arrived with the collections in about three weeks, he would like to be invited to stay at Pontrilas House, Bentham's home in Herefordshire.[6] Then, on 28 August, he stated an intention to attend the British Association for the Advancement of Science meetings in York and

[1] TNA, CO 111/218, Letter of Schomburgk to Colonial Office, 25 June 1844.

[2] Murchison, 'Address to the Royal Geographical Society of London', *Journal of the Royal Geographical Society*, 14, 1844, p. xcvii.

[3] TNA, CO 111/210, Letter of Light to Colonial Office, 16 May 1844.

[4] TNA, CO 111/218, Minute to Letter of Schomburgk to Colonial Office, 25 June 1844.

[5] TNA, CO 112/25, Letter of Stanley to Home Office, 14 December 1844; Letter of Colonial Office to Home Office, 6 May 1845.

[6] RBGK BC IX, Letter of Schomburgk to Bentham, 25 June 1844.

to return to London via Hereford, but also mentioned that he was suffering from severe headaches.[1] On 9 September he wrote to the Colonial Office requesting an extension of time for completing his map until the end of October.[2] Two weeks later he told Bentham that he had a severe cough that had been so troublesome since his return to Europe that he had been advised to give up all work in order to allow the doctor's prescription a better chance. However he still intended to go to the meetings at York that coming Monday, and if health permitted, to visit Hereford afterwards.[3] On 12 October, we learn that Schomburgk had returned to London direct from York because of his health, and that he was still not well.[4] The next month it is the same story. Schomburgk had planned a trip to see friends in York and Birmingham and then to come to Hereford, but a fever and a sore throat have caused him to cancel these arrangements.[5]

In early December Schomburgk left for the continent to meet Humboldt in Paris, then visited Berlin and Dresden[6] before returning to London during the second week of May 1845.[7] His health was still not good. He moved out of London to live in New Kingston (Surbiton) and his doctor advised him to winter outside Europe. This recommendation fortunately coincided with an invitation from the Barbados Government to advise on the construction of a railway line. As the British government had failed to come up with any offer of employment, Schomburgk decided to accept the offer, go there in November, and visit British Guiana before returning to England in May.[8]

As soon as the map of British Guiana was completed, Schomburgk had turned to the question of his future employment. On 1 November 1844 he requested that he might be involved in the British Guiana boundary negotiations. On the 30th of the same month, having heard he was being recommended for a knighthood, he asked that he might be made a consul somewhere. He repeated this request in late December, and also suggested that he might undertake the survey in Oregon that was being proposed. The Colonial Office replied that there was nothing available at the moment, but that Schomburgk would be kept in mind.[9]

In the meantime he busied himself with various publishing ventures including botanical descriptions.[10] He tried to persuade the Colonial Office to subsidize, at the cost of £500, an ethnographic study of the Guianese Amerindians, illustrated with Goodall's and his own drawings, but while the Colonial Office was sympathetic the Treasury turned the proposal down.[11] In June 1845, he attended the British Association for the Advancement of Science meetings in Cambridge and read two papers, one on the

[1] RBGK BC IX, Letter of Schomburgk to Bentham, 28 August 1844.

[2] TNA, CO 111/218, Letter of Schomburgk to Colonial Office, 7 September 1844.

[3] RBGK, BC, IX, Letter of Schomburgk to Bentham, 18 September 1844

[4] RBGK, BC IX, Letter of Schomburgk to Bentham, 12 October 1844.

[5] RBGK, BC IX, Letter of Schomburgk to Bentham, 13 November 1844.

[6] RBGK, BC, IX, Letter of Schomburgk to Bentham, 13 November 1844; TNA: PRO CO 112/218, Letter of Schomburgk to Colonial Office, 11 December 1844.

[7] RBGK, BC, IX, Letter of Schomburgk to Bentham, 12 May 1845.

[8] RBGK, BC, IX, Letter of Schomburgk to Bentham, 25 September 1845.

[9] TNA, CO 111/218, Letters of Schomburgk to the Colonial Office, 1 November 1844, 30 November 1844, 26 December 1844. Letter of Colonial Office to Schomburgk, 13 January 1845.

[10] See, for example, *Die Barbacenia Alexandrinae und Alexandra Imperatricis, entdeckt und beschrieben*, Braunschweig, Von Friedrich Vieweg und Sohn, 1845 and *Die Rapatea Friderici Augusti, und Saxo-Fridericia Regalis, entdeckt und beschrieben*, Braunschweig, Von Friedrich Vieweg und Sohn, 1845.

[11] TNA, CO 111/229, Letter of Schomburgk to Stanley, 8 August 1845.

Murichi or Ita palm and the uses made of it by the Amerindians,[1] and the other a fierce response to an American, Van Heuvel, who had dared to reintroduce a 250-mile long Lake Parima in the interior of British Guiana in his book *El Dorado*.[2]

The projected trip to Barbados came off that winter although it did not include an extension to British Guiana as originally planned, mainly it would appear because he was long delayed in Barbados. He left England in early November, was still in Barbados during the first week in August 1846 and was not expecting to leave for another two weeks. This suggests that he was not back in England until some date in September, although we know he was during the first week in October.[3] The journey gave rise to a number of more popular journal articles,[4] and a substantial volume entitled *The History of Barbados*.[5] There followed in the next two years other important works of which the best known is his edition of Ralegh's *Discovery of the Large, Rich, and Beautiful Empire of Guiana* for the Hakluyt Society of which he was an early member of Council.[6] He also busied himself translating the travels of Prince Adalbert of Prussia in Europe and Brazil,[7] and writing articles on Amerindian linguistics and ethnography.[8] Otherwise we have little information about Schomburgk's activities between his return from Barbados in October 1846 and August 1848, although we know that he lived in London, in Golden Square, during the winter and in Surbiton during the summer, and was 'on no pay'.[9]

[1] It was later published as 'A Description of the Murichi, or Ita Palm of Guiana (*Mauritia flexuosa, L.*) and its Uses by the Aborigines', *Simmonds's Colonial Magazine*, 6, 1845, pp. 43–9.

[2] Jacob A. Van Heuvel, *El Dorado; being a Narrative of the Circumstances which gave rise to Reports in the Sixteenth Century, of the Existence of a Rich and Splendid City in South America, to which that Name was given, and which led to many Enterprises in search of it; including a Defence of Sir Walter Raleigh, in regard to the Relations made by him respecting it, etc.*, New York, J. Winchester, New World Press, 1844. Schomburgk's response is 'On the Lake Parima (The "El Dorado" of Sir Walter Raleigh) and the Geography of Guiana', *Simmonds's Colonial Magazine*, 5, pp. 381–91.

[3] RBGK DC, LXX, Letters of Schomburgk to Hooker, 7 August 1846, 7 October 1846

[4] See 'A Visit to Turner's Hall Wood, Barbados, in 1846', *Simmonds's Colonial Magazine*, 8, 1846, pp. 482–8; 'Remarks on the Cultivation of the Cochineal Insect in Barbados', *Simmonds's Colonial Magazine*, 9, 1846, pp. 211–14; 'Steam-boat Voyage to Barbados', *Bentley's Miscellany*, 22, 1847, pp. 30–41; 'Pictures of Barbados', *Bentley's Miscellany*, 22, 1847, pp. 286–92; 'Rambles in Search of the Picturesque in Barbados', *Bentley's Miscellany*, 22, 1847, pp. 443–9.

[5] *The History of Barbados comprising a Geographical and Statistical Description of the Island; a Sketch of the Historical Events since the Settlement; and an Account of its Geology and Natural Productions*, London, Longman, Brown, Green & Longmans, 1848. This is the only work of Schomburgk's still in print. It has been republished a number of times, most recently by Frank Cass in 1998.

[6] It was the third volume published, and together with the first two was the subject of a review by James Anthony Froude ('England's Forgotten Worthies', *The Westminster Review*, new ser., 2, July–October 1852, pp. 32–67). Froude was highly critical of the first two volumes but praised Schomburgk's work. 'We cannot reconcile ourselves to the mortification of having found but one volume in the series to be even tolerably edited and that one to be edited by a gentleman to whom England is but an adopted country – Sir Robert Schomburgk ... in everything but its cost a very model of an excellent volume' (p. 35).

[7] See *Travels of His Royal Highness Prince Adalbert of Prussia, in the South of Europe and in Brazil, with a Voyage up the Amazon and the Xingu*, trans. by Sir Robert H. Schomburgk and J. E. Taylor, London, David Boyne, 1849.

[8] See 'On the Natives of Guiana', *Journal of the Ethnological Society of London*, 1, 1848, pp. 253–76; 'Remarks to accompany a Comparative Vocabulary of Eighteen Languages and Dialects of Indian Tribes inhabiting Guiana', *Simmonds's Colonial Magazine*, 15, 1848, pp. 46–64; 'A Vocabulary of the Maiongkong Language', *Proceedings of the Philological Society*, 4, 1850, pp. 217–22.

[9] See RBGK, DC, LXX, Letters of Schomburgk to Hooker, 20 November 1846, 8 January 1847, 22 May 1847, 24 August 1847, 25 August 1848.

On 14 August 1848, he presented a paper to the Ethnological Section of the British Association meeting at Swansea on Guianese Amerindian languages. Then, a week later, on 21 August his looked-for preferment materialized. He received the offer of appointment as British consul in Santo Domingo, in the newly independent Dominican Republic, a country in which there had not previously been any consular representation. Schomburgk immediately accepted the offer but requested that his departure for that country be delayed until December 1 as he had private affairs to attend to, above all to visit his aged father of whom he has 'no hope of meeting again in this world'. If necessary, he added, he would suppress this 'filial desire', but permission was given and he was away from mid-October for about four weeks.[1]

Schomburgk found himself almost immediately involved in Dominican affairs. A party of Dominican emissaries, who had been in Spain and France trying to arrange treaties of friendship and commerce with those countries, arrived in London in September 1848 to do likewise. Schomburgk was certainly present at some of the negotiations which proved inconclusive as no agreement could be reached over certain clauses. Accordingly Schomburgk was made plenipotentiary to complete the Treaty of Friendship and Commerce between the two countries once he was in post.[2]

He arrived in January 1849 in the Dominican Republic, a country sadly depopulated, in economic ruin and political anarchy. He left again in April 1857 and the period he was there saw a degree of political turbulence that is remarkable even for that country. The disturbance was not simply internal, although there was plenty of that, but also external, in which it took two forms. First, the Dominican Republic had only gained its independence from Haiti in 1844 and the latter country was determined to regain control. Accordingly there was a constant threat of incursions as well as full-scale invasions, one of which occurred four months after Schomburgk arrived. Second, and more insidiously, the attempts by various foreign powers, France, Spain and the United States of America, to exert their control over the country, often with the connivance of Dominican politicians who sought to avoid submission to Haiti through the protectorship of such a power. These were different interlocking aspects of a single political scenario and Schomburgk found himself fully engaged in them all. Their complexity is of a nature that defies summary description, and this is not the right place for a lengthy account. However, some brief word on Schomburgk's actions in all this are required.

The policy that Schomburgk had to follow was to maintain the independence of the Dominican Republic. He worked hard towards the signing of a peace treaty between the Republic and Haiti, often in conjunction with other foreign powers. The interests of these foreign powers was geopolitical more than anything else. The large Bay of

[1] TNA, FO 23/1, Letter of Palmerston to Schomburgk, 29 August 1848; Letter of Schomburgk to Palmerston, 28 August 1848; Letter of Schomburgk to Bidwell, 15 October 1848. Palmerston's letter of 28 August was the formal appointment; the original offer of the post was made on 21 August and it was gazetted on 25 August. Schomburgk later claimed (see below p. 213) that Prince Albert had been influential in his gaining this post. It is not clear when and how exactly Schomburgk and Prince Albert became acquainted but given their common German background and interest in science it is not too surprising. It was, however, clearly more than just a formal relationship as, in 1853, Schomburgk sent him two boxes of preserved Dominican fruits for the children of the Royal Household (TNA, FO 23/16, Letter of Schomburgk to Prince Albert, 3 July 1853).

[2] TNA, FO 23/1, Letter of Schomburgk to Palmerston, 11 September 1848; FO 23/1, pp. 51–88 consist of a series of memoranda and minutes, dated 6 September to 12 December 1848, relating to the treaty.

Samana in the north-east of the country would afford an ideal natural harbour for a naval fleet and there were said (wrongly) to be coal deposits in its neighbourhood. Accordingly Schomburgk found himself working with or against these foreign powers to prevent their designs on the future of the country. This, in turn, placed him in ever-shifting relationships with Dominican politicians who allied themselves in opportunistic ways with these same foreign powers. For the same reason, it was difficult to conclude the Treaty of Friendship and Commerce and it was not ratified until September 1850.[1]

Alongside these political and diplomatic matters, there were many run-of-the-mill consular affairs to keep him busy. These mainly consisted of sorting out the difficulties into which British merchants, seaman and others got themselves. On the whole Schomburgk was successful in what he did, and he had the undeniable advantage, when things got difficult, of being able to call for a British warship – something he did more than once.[2] On one occasion he managed to coerce the Dominican Government by closing down the consulate – an act tantamount to cutting off diplomatic relations, which would not have been to the Republic's advantage.[3]

When appointed, Schomburgk had assumed that his appointment would give him plenty of time for pursuing his scientific interests, and he agreed to make botanical collections for Lindley and others. He was quickly disillusioned and his plans in that direction were continuously blocked by the state of hostilities, accusations of his spying, the demands of his job and his health. Although Schomburgk complained about how little time he could afford for scientific matters, in practice he accomplished quite a lot. There was also a lot to be done for, despite it being the country of the New World where first European settlement took place, next to nothing had been undertaken in the way of scientific description. As usual Schomburgk collected everything; animals, plants, rocks, astronomical and geodesic observations, social and economic statistics, archaeological remains and ethnographic objects. He made various major journeys across the island.

The first of these tours was made during his first year in the Dominican Republic, in October and November 1849 and lasted some seven weeks. He visited the south-west of the country as far as the frontier with Haiti. All travel was on horseback as the only roads were bridle-paths and the country still showed the devastations of the war earlier in the year. It was probably on this trip that he saw and reported on the cave paintings at El Pommier near San Cristobal, some 20 miles west of Santo Domingo, and mapped a ceremonial plaza near San Juan de la Maguana.[4] He also began at this time his work on the currents and tides around the Dominican Republic and surveying the various

[1] TNA, FO 23/8, Letter of Schomburgk to Palmerston, 16 September 1850.

[2] See, for example, TNA, FO 23/19, Letter of Schomburgk to Clarendon, 7 October 1854. Schomburgk complained that the compliance of the Dominican Government was not available 'except it be done by the canon's mouth; I am sorry to say probably the only effective measure to bring them to reason' (TNA, FO 23/22, Letter of Schomburgk to Clarendon, 16 May 1855).

[3] TNA, FO 23/23, Letter of Schomburgk to Clarendon, 2 July 1855.

[4] Schomburgk wrote a letter to Prince Albert on 15 March 1851 describing these archaeological sites. Prince Albert had it forwarded to the Ethnological Society of London which published it in its journal as 'Ethnological Researches in Santo Domingo', *Journal of the Ethnological Society of London*, 3, 1854, pp. 115–22.

harbours and landing places along the coast, which gave rise to numerous articles in *The Nautical Magazine*.[1]

In the same months of 1850 he set off to explore the east of the country, visiting Hato Mayor, El Seibo, Higuey, Macao on the coast to the north-east, and Boca de Yuma to the south-east. However, after just over two weeks he was recalled to Santo Domingo to deal with problems that had arisen in his absence.[2]

The third major journey, which he undertook between 22 April and 15 August 1851, was through the rugged central districts to the northern region, visiting Puerto Plata, Santiago and the Peninsula of Samaná. The journey was extremely exhausting because of incessant rain and the terrible state of the roads. Even so he managed to visit all the important places in the north and to make botanical and geological collections. His official report is packed with information on a variety of matters. Perhaps politically the most important resulted from his inspection of the coal deposits at Samaná about which he concluded that it was not coal but lignite and of no use for driving the steam engines of battleships. However, he did not rule out the likelihood of true coal being found at great depth.[3] He returned to Samaná in 1854 when HMS *Calypso* conducted a further survey of the entrance to the bay, and once again optimistically predicted that coal, which would be of 'incalculable importance to Steam navigation with the West Indies and the Gulf of Darien', would be found in the lower strata.[4] Schomburgk returned to Samaná again that year, this time on board HMS *Devastation*, to investigate rumours that the United States of America had purchased part of the peninsula and were planning to build a fort there.[5] On both occasions he travelled in one direction overland.

In May–June 1852, he made another journey to the central district. No official report of this expedition appears to exist, and may not have been written although it was promised. There is, however, a brief account in *The Athenaeum* describing part of the journey.[6] One of the important outcomes of his various expeditions was a map of the Dominican Republic, published in Paris in 1858, largely based on the numerous observations he had made. There are also various publications on botanical matters, but on the whole his attempts at plant collecting were not very successful, and he was only able to make collections for the British Museum, Berlin and Paris. He complained to

[1] See ' On the Currents and Tides of the Mona Passage', *The Nautical Magazine*, 19, 1850, pp. 585–8; 'On the Currents and Tides of the Southern Coast of Santo Domingo', *The Nautical Magazine*, 21, 1852, pp. 289–92; 'On the Geographical Position of the City of Santo Domingo', *The Nautical Magazine*, 21, 1852, pp. 412–18; 'Harbour of San Domingo', *The Nautical Magazine*, 22, 1853, p. 276; 'Remarks on the Principal Ports and Anchoring Places along the Coast of the Dominican Republic', *The Nautical Magazine*, 22, 1853, pp. 202–7, 225–32, 304–13, 340–49.

[2] TNA, FO 23/8, Letter of Schomburgk to Palmerston, 14 November 1850.

[3] The official report of this expedition is in TNA, FO 23/11, Letter from Schomburgk to Palmerston, 25 August 1851. It was not until April 1853 that he sent a description and map of Samaná (TNA, FO 23/16, Letter of Schomburgk to Clarendon, 5 April 1853). These were forwarded by the Foreign Office to the Royal Geographical Society which published them in 1853 under the title 'The Peninsula and Bay of Samana, in the Dominican Republic' (*Journal of the Royal Geographical Society*, 23, 1853, pp. 264–83). The shelfmark of its manuscript is JMS 5/24. *The Athenaeum* had earlier published an account of another part of the journey as 'Visit to the Valley of Constanza in the Cibao Mountains of the Island of Santo Domingo, and to an Indian Burial-ground in its Vicinity', 1291, 1852, pp. 797–9.

[4] TNA, FO 23/18, Letter of Schomburgk to Clarendon, 18 February 1854.

[5] TNA, FO 23/19, Letters of Schomburgk to Clarendon, 28 August 1854 and 20 September 1854.

[6] Schomburgk, 'The Loadstone Mountain at Santo Domingo', *The Athenaeum*, 1338, 1853, pp. 738–40.

Bentham that conditions for collecting were almost worse than in Guiana.[1] There are also other published pieces from his stay in the Republic, such as his detailed description of the hurricane that struck Santo Domingo in August 1855.[2]

Schomburgk was not happy in Santo Domingo. After only three and a half years in post, he had written a private letter to Lord Stanley requesting a transfer to either Paraguay or St Thomas, saying that his life and work in Santo Domingo were very irksome, and shortly after he sent a more formal letter to the Foreign Office, asking to be moved.[3] He mentioned in that letter that he had also written on the subject to Prince Albert who had helped him get the post in the first place and to whom he was to write again soliciting help.[4] Requests to be transferred elsewhere occurred regularly after that,[5] although he also asked that if he had to stay put, whether his salary might be increased as the cost of living in Santo Domingo was very high. The responses to his request were not unsympathetic. Lord Malmesbury, in a minute dated 26 September 1852 attached to Schomburgk's request, directed that Schomburgk be informed that his presence in Santo Domingo was required while the troubles continued and that he must stay where he was 'a short time longer'. The minute also contained the information that Lord Malmesbury's intention was to recommend to the Queen that Schomburgk be offered the Paraguayan post if relations were established with that country. Nothing came of these requests for some years, but he had more luck with that for an increase in salary when this was raised by one-third (from £600 to £800 a year) in May 1855.[6] Finally, in March 1857, Schomburgk got the news that he has been appointed British Consul in Bangkok. He left Santo Domingo on 7 May after being delayed by an attack of rheumatism.[7]

There had been various reasons why he did not like being in Santo Domingo, one of which was ill health. The rheumatic fever which had so troubled him on his return journey from Esmeralda in 1839 recurred on several occasions,[8] he also suffered from malaria,[9] complained of failing eyesight,[10] and may have experienced an epileptic fit although he refers to it as a 'nervous attack'.[11] His other major problem was the lack of any social or cultural life and it would appear that his main pleasure was derived from experimenting with plants from temperate regions in his garden, which, he reported to William Hooker in 1854, contained fifty-one different types of rose.[12]

[1] RBGK, BC, IX, Letter of Schomburgk to Bentham, 20 October 1852.

[2] Schomburgk, 'Hurricane at Santo Domingo', *The Athenaeum*, 1462, 1855, pp. 1273–4.

[3] TNA, FO 23/14, Private letter of Schomburgk to Stanley, 6 August 1852; Letter of Schomburgk to Foreign Office, 23 August 1852.

[4] TNA, FO 23/16, Letter of Schomburgk to Prince Albert, 3 July 1853.

[5] See, for example, TNA, FO 23/16 Letter of Schomburgk to Clarendon, 3 July 1853; FO 23/18, Letter of Schomburgk to Clarendon, 6 April 1854; FO 23/22, Letter of Schomburgk to Clarendon, 7 February 1855.

[6] TNA, FO 23/21, Letter of Foreign Office to Schomburgk, 12 May 1855.

[7] TNA, FO 23/30, Letters of Schomburgk to Clarendon, 21 March 1857, 30 June 1857.

[8] As well as the attack in 1857, see TNA, FO 23/26, Letter of Schomburgk to Clarendon, 23 May 1856.

[9] RBGK, DC, LXXI, Letter of Schomburgk to Hooker, 18 February 1851.

[10] TNA, FO 23/4, Letter of Schomburgk to Bidwell, 23 December 1849; TNA, FO 23/8, Letter of Schomburgk to Palmerston, 1 September 1850. In the latter he wrote 'my eyes ... suffered to such a degree that it is now next to impossibility for me to read or write during night by candlelight'. This was probably the onset of glaucoma (see 'Introduction', Vol. I, p. 3, n. 3).

[11] RBGK, DC, LXXI, Letter of Schomburgk to Hooker, 7 March 1855.

[12] RBGK, DC, LXXI, Letter of Schomburgk to Hooker, 17 June 1854. See also, 'Effect of a Tropical Climate upon Plants of a Temperate Zone', *The Journal of the Horticultural Society of London*, 8, 1853, pp. 32–6.

It is a curious state of affairs but Schomburgk's name is possibly better known among the public in the Dominican Republic than in Great Britain, for he has been twice, at least, the subject of articles in Dominican newspaper 'colour supplements'.[1] His role both as consul and scientist is also appreciated by Dominican historians. In the first he is seen as the protector and saviour of Dominican independence,[2] and in the second as providing the first scientific accounts of the country.[3]

Schomburgk arrived back in London in the first week of June 1857 and was instructed to depart for Bangkok as soon as possible. He sailed on 4 September.[4] In the intervening months, Schomburgk was kept fairly busy. He requested that the Foreign Office purchase for him photographic apparatus in order for him to make contributions to natural history, geography and ethnology, and pay for instruction in its use, to which the Foreign Office consented.[5] Because the administration of justice was to be an important part of his future consular activities, for reasons explained below, he was told to attend police courts in London to familiarize himself with procedures.[6] He also spent some weeks of August on the continent, visiting relatives and friends (including Humboldt), in Paris, Leipzig and Berlin.[7]

The decision that Great Britain should have consular representation in Bangkok was the result of the Bowring Treaty[8] between the two countries, signed in 1855 and ratified a year later. Among other things, this treaty gave Great Britain extraterritorial rights over British subjects including those from neighbouring colonial dependencies.[9] A consular authority was required to handle these matters and the treaty in Article 2 specifically covered this point. It read:

> The interests of all British subjects coming to Siam shall be placed under the regulation and control of a consul, who will be appointed to reside at Bangkok. ... Any dispute arising

[1] See *Listín Diario Suplemento*, 18 November 1978 and 23 March 1985. Schomburgk, in 1849, also designed a flag for the Dominican Republic and a copy of his original design was available as a postcard from the Museo Nacional de Historia y Geografía in 1996.

[2] See, for example, Rodríguez Demorizi, Emilio, 'Schomburgk, primer cónsul de Inglaterra en Santo Domingo', *Boletin del Archivo General de la Nación*, 10, 50–51, 1947, pp. 3–8, and Perez, Carlos Federico, 'Sir Robert Schomburgk y la invasión haitiana de 1855–56', *Aula. Revista General de la Universidad Nacional Pedro Henriquez Ureña*, Año 2, Número 4, 1973, pp. 11–29. It was remarked to me by an historian when I was in Santo Domingo in 1996 that Avenida John F. Kennedy would more deservedly be called after Schomburgk.

[3] In the article about Schomburgk in the *Listín Diario* of 18 November 1978, he is referred to as 'el Humboldt de Santo Domingo'; a title that Schomburgk would have delighted in.

[4] TNA, FO 69/6, Letters of Foreign Office to Schomburgk, 6 June 1857, 7 July 1857; TNA: PRO FO 69/7, Letter of Schomburgk to Clarendon.

[5] TNA, FO, 69/6, Letter of Foreign Office to Schomburgk, 29 June 1857; TNA, FO, 69/7, Letter of Schomburgk to Clarendon, 23 June 1857. This is the last we hear of the photographic apparatus; there is no mention of Schomburgk ever taking photographs, and if he did none appears to have survived.

[6] TNA, FO 69/6, Letter of Clarendon to Schomburgk, 27 July 1857.

[7] TNA, FO 69/7, Letter of Schomburgk to Hammond, 27 July 1857.

[8] Sir John Bowring FRS (1792–1872) had a wide and varied career. A close friend of Jeremy Bentham he was one of the first editors of the *Westminster Review*, Member of Parliament for Clyde (1835–7) and Bolton (1841–7), and founding member of the Anti-Corn Law League in 1838. He was appointed consul at Canton in 1847, became Plenipotentiary to China in 1854 and Governor of Hong Kong. It was while holding these positions that he negotiated the Bowring Treaty.

[9] Not of importance here, but it might be noted that the economic clauses of the treaty which included the removal of all restrictions on trade and the setting of very low import and export duties, had great long-term consequences for the political, social and economic formation of the country.

> between British and Siamese subjects shall be heard and determined by the consul, in conjunction with the proper Siamese officers; ... But the consul shall not interfere in any matter referring solely to Siamese, neither will the Siamese authorities interfere in questions which only concern the subjects of her British Majesty.[1]

Although a treaty between Siam and the East India Company had been signed in 1827, recent attempts at renegotiating it between Great Britain and Siam had been unsuccessful. Success on this occasion was to some extent the result of the succession to the throne in 1851 of King Mongkut,[2] who, unlike his predecessor, was open to and interested in western science and technology. Despite the fact that the Emperor of China claimed sovereignty over the country, in practice this involved little more than triennial tribute, and the king was the absolute ruler of the country. As well as the First King, there was also a Second King whose role was rather ill defined, and although his status was very similar to that of the First King, he played little part in the running of the country. The day-to-day affairs of the country were in the hands of other officials which included a prime minister (Phra Kralahom) and a minister for foreign affairs (Phra Klang). It was with these two, together with a personal representative of the First King, his younger brother, Krom Hluang Wongsa, that Schomburgk had most dealings, although final decisions on anything had to be referred to the First King.[3]

Schomburgk was delayed in Singapore due to the lack of a Royal Navy ship to convey him to Bangkok, which he reached on 6 December 1857. He had insisted on a warship on the grounds that the French had lost of lot of face when one of their envoys arrived at the Siamese capital in a merchantman.[4] Life for Schomburgk as consul in Bangkok was very different from that in Santo Domingo. There were no external invasions, the diplomatic intrigues of other powers were limited to French designs on Cambodia, formally a dependency of Siam, and the naval bombardment of the small sultanate of Terengganu to ensure that a British-approved succession took place there.[5] An early and relatively easy success was to persuade the Siamese to permit the construction of

[1] TNA, FO 97/368, 'Treaty of Friendship and Commerce between Great Britain and Siam concluded and signed at Bangkok April 18 1855'. Schomburgk was not the first British consul to be appointed to Bangkok. That was William Hillier, who arrived in Bangkok on 5 June 1856 and died there on 18 October 1856 (TNA, FO 69/4, Letter of Hillier to Clarendon, 11 June 1856; Letter of Bell to Clarendon, 22 October 1856).

[2] King Mongkut, or to give him his full title Phra Bard Somdetch Phra Paramendr Maha Mongkut, is best known in the west through the representation of him by Yul Brynner in *The King and I*, the musical film version of Anna Leonowens's autobiographical account of her time as governess to the royal children (*The English Governess at the Siamese Court: being Recollections of Six Years in the Royal Palace at Bangkok*, London, Trübner & Co., 1870). Charles Keyes (*Thailand. Buddhist Kingdom as Modern Nation-State*, Boulder, Col., Westview Press, 1989, p. 50) has written: 'The popularity of the movie in the West has created an unfortunate impression there of a man who was unquestionably one of the most acute and learned of any of the people who have shaped the destiny of Thailand'. Although Leonowens's stay in Bangkok overlapped with Schomburgk's, she makes no direct reference to him even though she does threaten someone that he will be reported to the British consul-general. Schomburgk, for his part, did not think very highly of Mongkut. He wrote of him to Hooker that 'The nimbus which surrounds his head according to reports at a distance is found to be formed of tinsel on closer inspection', and on another occasion described him as 'a shabby monarch' (RBGK, DC, LVI, Letters of Schomburgk to Hooker, 4 September 1858, 28 January 1861).

[3] For an account of Siam in 1855, see John Bowring, *The Kingdom and People of Siam; with a Narrative of the Mission to that Country in 1855*, 2 vols, London, John W. Parker & Son, 1857.

[4] TNA, FO 69/8, Letters of Schomburgk to Clarendon, 20 October 1857, 13 December 1857.

[5] TNA, FO 69/32 is entirely devoted to correspondence and reports on this affair.

the India to China telegraph line across the country.[1] On the whole, however, he did not find dealing with the Siamese easy. The etiquette and scheming of an oriental court that were necessary in order to promote British interests proved unpalatable to him. To the Earl of Clarendon, he described the King as 'obstinate, vain and pompous' and his recommended solution to problems did not differ from that in Santo Domingo; the regular visits by naval ships as it 'produces a better effect than a note written in the strongest terms'.[2]

One of duties, however, which must have occupied a great deal of Schomburgk's time was the settling of civil and criminal disputes, for, as mentioned above, the Bowring Treaty gave the British consul sole jurisdiction over British subjects in Siam, and this included citizens of neighbouring British dependencies, and also involvement in the case of disputes between British and Siamese subjects. In 1860, for example, no fewer than 160 cases were heard.[3]

Schomburgk had a much larger consular establishment in Bangkok than he had had in Santo Domingo. In February 1858 he reported that the personnel attached to the consulate consisted of himself, an interpreter (this was later increased by the addition of two student interpreters), a first and second assistant, a doctor (a Royal Navy surgeon), two constables, one attendant, one Siamese writer, one linguist, one office coolie, one messenger, one porter, two watchmen, one scavenger, one head boatman and twelve boatmen.[4] However, ill health, and death, afflicted his European staff and on one occasion it was reduced to two through absences caused by sickness or sick leave. Nor was Schomburgk spared and it is rare that in his letters to Hooker he did not allude to his ill health. He reported dysentery, rheumatism that deprived him on occasion of the use of his arms, hands or toes, and poor and painful eyesight.[5]

Despite the combination of ill health and pressure of work consequent on staff shortages he still found time for scientific pursuits. He made plant collections for Kew, Berlin and Paris, but confessed his unfamiliarity with eastern flora and was uncertain whether the plants he was sending were new species.[6] He also sent back two boxes of geological specimens for the Museum of Practical Geology,[7] and published papers on other aspects of the country.[8] However, during his time in Siam, he made only three journeys away

[1] TNA, FO 69/12, Letter of Schomburgk to Malmesbury, 31 May 1858; TNA: PRO FO 69/12, Letter of Schomburgk to Malmesbury, 28 June 1858.

[2] TNA, FO 69/11, Letter of Schomburgk to Clarendon, 19 April 1858. Nor was Schomburgk willing to put up with what he regarded as disrespect on the King's part. In April 1859 when, two days running, the King had kept Schomburgk and the Captain of HMS *Niger* waiting for an hour and a half for audience, Schomburgk walked out (TNA, FO 69/17, Letter of Schomburgk to Russell, 18 April 1859).

[3] TNA, FO 69/24, Letter of Schomburgk to Russell, 10 January 1861.

[4] TNA, FO 69/11, Letter of Schomburgk to Clarendon, 9 February 1858.

[5] RBGK, DC, LVI, Letters of Schomburgk to Hooker, 4 September 1858, 16 June 1859, 28 January 1861; RBGK, DC, LVII, Letters of Schomburgk to Hooker, 26 December 1862, 24 January 1863.

[6] He wrote to Hooker on 10 November 1859, 'I shall feel diffident to collect any thing during my contemplated journey, as most probably it would only correspond with what has been collected previously in these regions' (RBGK, DC, LVI).

[7] TNA, FO 69/22, Letter of Foreign Office to Murchison, 3 October 1860.

[8] See 'The Vegetable Products of Siam', *The Technologist*, 1, 1861, pp. 355–62; 'Siamese Products', *The Technologist*, 2, 1862, pp. 444–50; 'The Vegetable Products of Siam', *Pharmaceutical Journal and Transactions*, 2nd ser., 3, 1862, pp. 123–8; 'The Paper-tree of Siam (Ton-khai), *Trophis aspera*, W.', *The Technologist*, 4, 1864, pp. 337–9.

from Bangkok. The first, in August 1858, was made, following an attack of dysentery, on the recommendation of the consulate doctor, to the sea at Anhin, possibly modern Ang Silas, on the eastern shore of the Bight of Bangkok, in order to recuperate.[1] He did a similar thing in May 1859 on which occasion he went on a ten-day visit to the coastal town of Phet Buri to the west of the mouth of Chao Phraya River.[2] The account he wrote of this journey was his last to be published in the *Journal of the Royal Geographical Society*.[3] As well as these two convalescing trips, he also undertook a more serious journey.

The question of cutting a canal or laying a railway across the Isthmus of Kra in order to save ships the voyage round the Malayan Peninsula on the passage from Calcutta to Bangkok and China had been raised by a Mr Henry Wise of Lloyd's and Schomburgk was instructed to look into it.[4] Schomburgk replied in May 1858 saying that he would undertake the survey of the Kra Isthmus himself and would get a lift to the mouth of the River Champon, modern Chumphon, on a Royal Navy ship expected to call in the near future. In the meantime he enclosed a memorandum on the subject. In fact nothing came of this, because he was absent from Bangkok on his convalescing trip to Anhin when the warship visited.[5]

Then, in November 1859 the plan to explore the Isthmus of Kra was resurrected. Schomburgk proposed to travel up the Chao Phraya and Mae Ping rivers to Chiang Mai from where he would head south-west across the mountains to Moulmein in Burma, on the Gulf of Martaban. From there he would proceed down the coast and recross to the Gulf of Siam via the Isthmus of Kra. He set out on this journey on 12 December but the journey was slow and he did not reach Chiang Mai until mid-February. He left by elephant on 27 February for Moulmein, which he reached on 22 March. Here he borrowed a steamer to take him south to Tavoy where he arrived on 3 April. At some point Schomburgk abandoned the idea of crossing to the Gulf of Siam over the Isthmus of Kra and instead, on 13 April, he headed east for Kanchanaburi by elephant, from where he continued by canoe to Bangkok, arriving back on 26 April. In reporting his journey, which, by his standards, he did very briefly, he mentioned that a full account was in preparation, but he does not appear ever to have completed more than a very small part of it. In the same breath, he also said that, now he was nearly sixty years old, it was unlikely he would ever undertake another expedition.[6]

[1] TNA, FO 69/12, Letters from Schomburgk to Malmesbury, 31 July 1858, 30 August 1858. The doctor, James Campbell, had doubts about his recovery (TNA, FO 69/14, Letter from Admiralty to Foreign Office, 21 October 1858).

[2] TNA, FO 69/18, Letter of Schomburgk to Russell, 21 November 1859.

[3] See 'Boat Excursion from Bangkok, in Siam, to the Pechaburri, on the Western Shore of the Gulf of Siam', *Journal of the Royal Geographical Society*, 31, 1861, pp. 302–21.

[4] TNA, FO 69/10, Letter of Foreign Office to Schomburgk, 6 February 1858.

[5] TNA, FO 69/11, Letters of Schomburgk to Malmesbury, 31 May 1858, 30 August 1858.

[6] TNA, FO 69/18, Letter of Schomburgk to Russell, 6 December 1859; TNA, FO 69/21, Letters of Schomburgk to Russell, 21 January 1860, 16 February 1860, 7 May 1860. RBGK DC, LVI, Letters of Schomburgk to Hooker, 10 November 1859, 28 January 1861. It should be noted that the letter to Hooker describing the journey was written nearly twelve months after the event. Furthermore a number of the dates in the two accounts vary. For example, he informed Russell that he left Tavoy on 13 April, and Hooker, on 8 April. He mentioned to Hooker that he was still writing up his account of the journey, but was impeded by his painful eyes and understaffed consulate. An account of the first part of the journey, that from Bangkok to Chiang Mai, appeared in the *Journal of the Asiatic Society of Bengal* ('A Visit to Xiengmai, the Principal City of the Laos or Shan States', 32, 1863, pp. 387–99). A detailed report of the very first part of the journey is also to be found in the archives of the Royal Geographical Society (MS 8/27/3).

Although Schomburgk rarely mentioned his well-being in official correspondence, there is evidence from his letters to Hooker; not only did he mention directly his declining health during the early 1860s, but also significantly remarked that a year after his expedition of 1859–60 that he had been too disheartened to sort out the plant specimens he had collected.[1] There is also a noticeable reduction in both length and number of his despatches to the Foreign Office that may be symptomatic of the same thing. In December 1863 Schomburgk wrote to the Foreign Office with a request for twelve months' leave. He pointed out that he has been at his post in Bangkok for six years, 'the insalubrity of which place is so strongly substantiated by the number of Officers of this consulate to whom the climate proved fatal'. The request was granted, and before Schomburgk left Bangkok on 30 May 1864, he informed his staff of his intention to seek permission to retire on grounds of ill health when he arrived back in England.[2]

Schomburgk arrived back in England on 11 August, but did not make his formal request to retire until October, when he wrote to Lord Russell 'Reluctant as I am to retire from active service I have to come to this resolution if I desire to prolong my life for a few years more'. He asks for a pension and Lord Russell wrote to the Treasury in support of this in very glowing terms. He pointed out that, whereas Schomburgk had had only sixteen and a half years of consular service, his previous work for the Colonial Office, his scientific attainments and his exertions in the cause of humanity made a strong case for an augmented pension. The Lords of the Treasury agreed and whereas he should have received only £352 a year it was increased to £500.[3] Schomburgk did not have long to enjoy his pension; he left for Germany in December and died at Schöneburgh, near Berlin, on 11 March 1865.

[1] See RBGK DC, LVI, Letter of Schomburgk to Hooker, 28 January 1861; RBGK, DC, LVII, Letter of Schomburgk to Hooker, 24 January 1863. He also wrote to Hooker, after his return to London in 1864, that 'I was such an invalid and had been so for some years that they had to send me from my couch on board the Steamer'. RBGK DC, XLII, Letter of Schomburgk to Hooker, 16 August 1864.

[2] TNA, FO 69/30, Letter of Schomburgk to Russell, 20 December 1863; TNA, FO 69/33, Letter of Foreign Office to Schomburgk, 17 February 1864; TNA, FO 69/34, Letters of Schomburgk to Russell, 21 April 1864, 29 May 1864; Letter of Knox to Russell, 18 June 1864.

[3] TNA, FO 69/34, Letters of Schomburgk to Russell, 12 August 1864, 5 October 1864; TNA, FO 69/35, Letter of Russell to Treasury, 1 November 1864; Letter of Treasury to Hammond, 30 December 1864. Schomburgk's salary in Bangkok had been £1,200 a year plus £120 housing allowance. The interpreter, the next most highly paid member of the consular staff, received £500 a year.

APPENDIX 1

The Boundary Dispute

In 1840 the Foreign Secretary, Viscount Palmerston, directed that the countries neighbouring British Guiana (Brazil, Venezuela and Surinam) should be informed of the boundaries claimed by Britain for its colony and if those countries had any objections to state their reasons for them. At the same time he instructed that the boundary be surveyed and permanent markers be erected along it. The boundaries presented to those neighbouring countries were those drawn up by Schomburgk and illustrated in a map of 1840 that accompanied his memorandum on the subject. Schomburgk, in drawing up the boundaries, had subscribed to the principle that they must, as far as possible, be permanent, natural physical features that were readily recognizable. There is nothing wrong with this principle, except that Schomburgk was prone to identify features that lay beyond what might reasonably be expected by the neighbouring countries. The consequences of this remain unresolved today. There is a vast and growing literature on the question of Guyana's boundaries, and there is no intention to give here more than the most summary account of the events of the past 160 years. This is most easily done by considering each country in turn.

The main difficulties arose with Venezuela. The Venezuelans claimed, and still claim, that their boundary ran from the mouth of the Moruka River south to strike the Essequibo at its junction with the Cuyuni, and then down the left or west bank of the Essequibo to its source. By the 1840 map, the British claimed that the boundary ran from the point where the Amacuro River meets the Orinoco, along the former river and then up its left-bank tributary, the Cuyupuni, to its source. From here the line ran south so that the basins of the Barima and Barama Rivers lay in British Guiana. The boundary crossed into the Cuyuni basin at the source of the Parawayauru, cut the Cuyuni at about 59°50′W, and continued up the right-bank tributary, the Aruarua, to its source. From here the line was drawn south-westward to Mount Roraima. The major problem with this boundary, as claimed in 1840, was that the area was unsurveyed, and thus no reliable maps of it existed. When Schomburgk undertook his survey, he found that the Barima and Barama rose much further west than had been represented on the 1840 map. Accordingly the boundary was now drawn further west, thus accrediting more territory to British Guiana at the expense of Venezuela.

The Venezuelans reacted quickly to the news of Schomburgk's erecting boundary markers at the mouth of the Barima and Amacuro Rivers in 1841, and sent that same year a delegation to Georgetown to object. There was a certain amount of backpedalling at this point: it was argued that the markers were erected for purely scientific purposes and it was agreed that they would be removed. In fact, they were not as it was pointed out that they were located in a remote place to which it was not worth making a journey solely for this purpose. The negotiations continued and in 1844 the British offered a boundary running due west from the mouth of the Moruka River on condition that Venezuela would never cede to a third power the territory between the Moruka and Amacuro. The point about this being that the British did not want a third party gaining control over the mouth of the Orinoco. Nothing came of this offer and in 1850 it was agreed to regard the disputed territory as neutral and that neither country would occupy it; a hope dashed in the same year by the finding of gold in the area. From then on there was a steady infiltration of the disputed zone by Guyanese subjects although there were counter-accusations of Venezuelan intrusion into the area. With no progress in the negotiations, in 1876 Venezuela recruited the support of the United States of America, which country explained its intervention by the need to uphold the principle of the Monroe Doctrine. Initially the presence of the USA failed to have any effect at all and Britain responded, during the 1880s, by extending its territorial claim to include the whole of the Cuyuni basin, thus pushing the colony's boundary nearly to the Caroni River. Then, in 1895, as a result of increasing pressure by President Cleveland, Great Britain agreed to go to arbitration. The commission appointed to look into the matter consisted of two British and two American judges with a Russian chairman, but notably lacked any Venezuelan member. A treaty was signed in 1899 laying down the boundary along its present course, which coincides with the boundary of British Guiana as represented by Schomburgk's map of 1844. The Venezuelans were extremely dissatisfied with this outcome, claiming that it was a result of collusion between the British and the Russian chairman, and that they had been forced to sign by the United States. Indeed the Venezuelans never accepted the treaty's legitimacy and in 1962, with Guyana's independence pending and with that country in political turmoil, they declared it null and void. On the independence of Guyana, Venezuela recognized the new state but reaffirmed that its boundary lay along the west bank of the Essequibo. In 1966 the Geneva Agreement set up a new attempt at reaching a settlement. It resulted in the Protocol of the Port-of-Spain in 1970 under which cooperation in the economic and environmental development of the north-west area was agreed to. There matters rest, with Venezuela still claiming a vast area of Guyana.

On the Brazilian front, diplomatic negotiations in the 1840s failed to settle the line of the frontier, partly because agreement on this matter got caught up in the negotiation of a commercial treaty which, in turn, could not be separated from the question of the Brazilian slave trade and slavery. After that the matter was more or less forgotten except for occasional complaints from both sides about breaches of the neutrality of the disputed territory but nothing serious occurred and the affair was not high on either country's agenda, partly because nothing valuable such as gold was found there. Then, spurred on by the apparent settlement of the British Guiana/Venezuelan boundary question, the two countries agreed in 1901 that the matter should go to arbitration. The arbitrator was the King of Italy who gave his decision on 6 June 1904. He was unable to accept that either country had established unquestioned right over the disputed territory and, finding that it could not easily be shared equally, he gave the major portion to Britain, fixing the Mahu and Takutu Rivers as the boundary. Brazil has never actively contested this ruling although there was (and is) a general feeling in the country that the arbitration was less than just. Perhaps the interesting thing is that today, nearly a hundred years on, as a result of events over the past decade, there are now probably more Brazilians than Guyanese in the region granted to British Guiana.

Guyana's third boundary, that with Surinam, the former Dutch Guiana, seemed totally unproblematic. In 1840 when Palmerston sent round that abrupt communication about British Guiana's boundaries, the Dutch replied that they were happy to accept the Corentyne River as the frontier. Indeed, in 1843, when only the survey of that boundary remained to be done, it was very nearly cancelled on the grounds that it was an unnecessary exercise, and thus entailed unnecessary expenditure. There was, however, a problem even if it did not emerge until much later. In 1838, when Schomburgk was exploring the headwaters of the Essequibo, a mountain range was pointed out to him, from which, he was told, one of the chief branches of the Corentyne River, the River Pani, flowed. This was almost certainly the New River. However, when descending the Corentyne in 1843, Schomburgk failed to describe or report it. It would not have been too surprising if he had missed it. The Corentyne here is about eight miles wide, studded with islands, and with the ration allowance down to four ounces of cassava flour a day there was no time for exploring. On the other hand there is evidence that he did observe it or learn about it as it appears on his 1844 map, joining the Corentyne in the more or less correct geographical position. Formally it was not discovered until October 1871, when the geological surveyor, Charles Barrington Brown, ascending the Corentyne and hugging its left bank, found and named it the New River. The size of the New River is such that it has given rise to a disagreement about which is the mainstream and which the tributary. It has been argued that Schomburgk failed to report it just because of this. If the New River were the mainstream the boundary would then lie much further to the west, and thus a large extent of territory would be lost to British Guiana. Because the boundary was considered so unproblematic, it was not until the 1930s that an attempt to formalize it was made. A draft agreement with the Netherlands about this boundary was reached, in which it was accepted that the Corentyne and Cutari would form the boundary. The war then intervened, the agreement was not signed and the question was not returned to until the 1960s. But when it was, the Netherlands went back on the earlier draft agreement and now claimed that the New River should form the boundary. Great Britain argued that it was too late to renegue on the earlier agreement, and negotiations got nowhere. Surinam, on independence, inherited this claim, and the ownership of the large triangle of territory between the Corentyne and the New River remains in dispute.

APPENDIX 2

Glossary of Vernacular Plant Names

Vernacular names for plants often refer to a number of different genera and species. It is normally not possible to tell to which particular species Schomburgk is referring. Accordingly all the plant species covered by the vernacular term have been listed, with the exception of the Passion flower of which there are twenty known species in the region so called. A further problem is that many of the scientific names have, since Schomburgk's day, been revised. In order to identify plants I have relied to a large extent on E. A. Mennega et al., *Check List of Woody Plants of Guyana*, although various other sources have also been used, including Allsopp's *Dictionary of Caribbean English Usage*, and Bennett's *Arawak–English Dictionary*.

The letters in parentheses after an entry refer to the origin of the term. Ak. for Akawaio; Ar. for Arawak; C. for Carib; Cr. for Creole; M. for Macushi; P. for Patamona; W. for Wapishiana; Wr. for Warao.

Ackayari, see Acourie.
Acourie or Acourie-Broot (Cr.): *Licania densiflora*.
Akoukoua: Unidentified, but see Ikanna.
Amapaima (M.): *Aniba canelilla*.
Anareke (Ar.): *Bromelia karatas*.
Ant Bush (Cr.): *Siparuna guianensis*.
Ants Tree (Cr.): *Cordia nodosa*.
Araho, possibly Reho.
Ariha, possibly Reho.
Ayawiye (M.) same as Tshiwipa (C): *Martiodendron excelsum*.
Balata (P.): *Manilkara bidentata*.
Balsam Tree (Cr.): *Clusia amazonia*.
Bannia, see Banya.
Banya (Ar.): *Swartzia bannia*.
Barata (Ar.): *Ecclinusa psilophylla*.
Baromalli (Ar.): *Catostemma altsonii*; *C. commune*; *C. fragans*; *Scleronema guianense*.
Bitter Ash (Cr.): *Quassia amara*.
Bitter Bush (Cr.): *Quassia simarouba*.
Boba (Ar.): *Iriartea exorrhiza*.
Bouba, see Boba.
Bourra-Courra, see Letterwood.
Bullet Tree, Bastard: Unidentified, perhaps *Chrysophyllum* spp.
Bullet Tree, Male (Cr.): *Chrysophyllum argenteum*.

Bulletwood (Cr.), *Manilkara bidentata*.
Bully Tree (Cr.): *Manilkara bidentata*.
Bulukush (W.): *Mauritia martiana*.
Burra-burraro, possibly Buruburulli.
Buruburulli (Ar.): *Licania divaricata*; *L. heteromorpha*.
Cackerally, see Kakaralli.
Camiau (W.), see Cedar, Red.
Canung (M.): Unidentified species of *Psidium*.
Carapa, see Crabwood.
Caraveru: *Bignonia chica*.
Caruman, probably Karamanni.
Casami (M.): Unidentified species of *Eugenia*, possibly *E. patrisii*.
Cashew, see Ubudi.
Cauabahiracahrie: Unidentified.
Cedar, Red (Cr.): *Cedrela odorata*.
Cedar, White (Cr.): *Tabebuia insignis*; *T. stenocalyx*.
Chocolate tree: *Theobroma bicolor*.
Cookarally, see Kakaralli.
Corró-va, possibly *Attalea sagoti*.
Coucourite, see Kokorite.
Courába: Unidentified.
Courida (Cr.): *Avicennia germinans*.
Crabwood (Cr.): *Carapa guianensis*.
Cumaka, see Kumaka.
Cumara, see Kumaru.
Curahara, see Kurahara.
Curare (Cr.): *Strychnos toxifera*.
Curida, see Courida.
Curua, see Kurua.
Curukaiya, possibly Kurokai.
Cuyama, see Kuyama.
Dehpoyeh (M.): Unidentified.
Determa (Cr.): *Nectandra rubra*.
Ducaly, see Dukali.
Dukali (Ar.): *Parahancornia fasciculata*.
Erra-warang: Unidentified.
Eta, see Ita.
Etaballia, see Iteballi.
Fig: Fifteen species of the genus *Ficus* are known as Fig in Creole.
Greenheart (Cr.): *Ocotea rodiaei*.
Haiahaia (Ar.): *Sapium aucuparium*; *S. cladogyne*; *S. hippomane*; *S. jenmanii*; *S. pauci-nervium*.
Haiari, Black (Cr.): *Lonchocarpus chrysophyllus*.
Haiari, White (Cr.): *Lonchocarpus martynii*.
Haiariballi (Ar.): *Derris pterocarpa*. *Alexa imperatricis*; *A.leiopetala*; *A. wachenheimii*.
Haiawa or Haiowa (Ar.): *Protium altsonii*; *P. aracouchini*; *P. plagiocarpum*; *P. guianense*; *P. heptaphyllum*; *P. hostmannii*.

Hakea, see Hakia.
Hakia (Ar.): *Tabebuia capitata*; *T. serratifolia*; *T. subtilis*.
Hennicarro, possibly Haiahaia.
Hog-plum (Cr.): *Spondias mombin*.
Houcouya, see Ironwood.
Hubu (Ar.): *Spondias mombin*.
Hupu, see Hubu.
Hya-hya, see Haiahaia.
Ikanna, possibly Itana.
Imba (P.): *Caryocar nuciferum*.
Impa, see Imba.
Incense Tree: *P. guianense*; *P. heptaphyllum*; *P. hostmannii*. See also Haiawa (Cr.)
Irimariye (M.): *Couratari guianensis*.
Ironwood (Cr.): *Dialium guianense*; *Mouriri huberi*; *M. sideroxylon*; *Swartzia leiocalycina*; *Tabebuia capitata*; *T. serratifolia*.
Ita or Ite (Ar.): *Mauritia flexuosa*.
Itana (Ar.): *Philodendron* sp.
Iteballi (Ar.): *Vochysia costata*; *V. crassifolia*; *V. glaberrima*; *V. schomburgkii*; *V. surinamensis*; *V. tetraphylla*.
Itikiboroballi (Ar.): *Swartzia benthamiana*; *S. davis*; *S. dipetala*; *S, grandifolia*; *S. guianensis*; *S. laevicarpa*; *S. leiogyne*; *S. sprucie*; *S. xanthopetala*.
Itsha: Unidentified.
Kaitshara: *Bromelia karatas*.
Kakaralli (Ar.): *Lecythis alutacea*; *L. chartacea*; *Eschweilera alata*; *E. coriacea*; *E. decolorans*; *E. grata*; *E. longpipes*; *E. parviflora*; *E. sagotiana*; *E. schomburgkii*; *E. subglandulosa*; *E. wachenheimii*.
Kakerabally: Unidentified.
Kamarakata (Ar.): *Acosmium nitens*.
Karaba (Ar., Ak., P.), see Crabwood.
Karamanni (M.): *Symphonia globulifera*.
Kirahagh-palm: Unidentified.
Karia (Ar.): *Astrocaryum sciophilum*.
Kokorite (Ar.): *Attalea regia*.
Kokorite, Sand (Cr.): *Attalea sagotii*.
Kokoritiballi (Ar.): *Ichthyophora elegans*; *Oxythece ambelaniifolia*; *O. guianensis*; *Planchonella elegans*; *Pouteria dura*; *P. egregia*; *P. reticulata*.
Kumaka (Ar.): *Ceiba pentandra*. See also Makau and Silk-cotton Tree.
Kumaru (Ar.): *Dipteryx odorata*.
Kurahara (Ar., C.): *Calophyllum lucidum*.
Kurua (Ar.): *Attalea sagotii*.
Kurokai (Ar.): *Protium crenatum*; *P. decandrum*; *P. sagotianum*.
Kuyama (Ar.): *Xylopia aromatica*; *X. frutescens*; *X. longifolia*; *X. nitida*; *X. pulcherrima*.
Lana or Lanah (Ar.): *Genipa americana*.
Letterwood (Cr.): *Brosimum guianensis*.
Locust (Cr.): *Hymenaea courbaril*; *H. oblongifolia*.
Lou, see Lu.
Lu (Ar.): *Oenocarpus bacaba*.

Maccau, see Makau.
Makariballi, see Makwariballi.
Makau (C.): *Ceiba pentandra*. See also Kumaka and Silk-cotton Tree.
Makwariballi (Ar.): *Forsteronia corymbifera*; *F. gracilis*; *F. guyanensis*.
Manaka (Ar.), see Manicole.
Manariballi (Ar.): *Elizabetha princeps*; *Pithecellobium corymbosum*; *P. gonggripjii*; *P. pedicellare*.
Manariballi, Black (Ar.): *Parkia nitida*.
Mangrove (Cr.): *Rhizophora harrisonii*; *R. mangle*; *R. racemosa*.
Mangrove, Black (Cr.): *Avicennia germinans*.
Mangrove, White (Cr.): *Laguncularia racemosa*.
Mangrove, Wild (Cr.): *Tovomita obovata*.
Manica, see Manicole.
Manicole (Cr.): *Euterpe oleracea*.
Manni (Ar.): *Symphonia globulifera*.
Maparakuni, probably Maporokon.
Maporokon (Ar.): *Inga alba*.
Maran (Cr.): *Copaifera guianensis*; *C. langsdorfii*; *C. publiflora*.
Maraniara, see Purpleheart.
Mararen, see Maran.
Marima, possibly Marimari.
Marimari (Wr.): *Couratari guianensis*.
Maripa (C., M.): *Attalea regia*.
Maropi (M.), see Hog-plum.
Massara, see Musara.
Milk Tree, see Haiahaia.
Mocco-mocco, see Moka-moka.
Moka-moka: *Montrichardia arborescens*.
Mora (Ak., Ar.): *Mora excelsa*.
Muc-mucu: Unidentified
Murre-marri, see Irimariye.
Murumuru (C.): *Astrocaryum vulgare*.
Murumutu, see Murumuru.
Musara or Mussara (C.): *Aspidosperma excelsum*.
Nastus: Unidentified bamboo.
Okheri-puima: Unidentified.
Orinok, see Oronoque.
Oronoque (Cr.): *Erythrina fusca*.
Ouboudi, see Ubudi.
Oubati: Unidentified.
Ourali, see Curare.
Paddlewood (Cr.): *Aspidosperma excelsum*; *A. oblongum*.
Paranguai, see Paranka.
Paranka (M.), see Cedar, Red.
Parata: Unidentified. Possibly Balata or Barata, 'b' and 'p', and 'l' and 'r' being very close to each other in the native languages of the region.
Parika (Cr.): *Anadenanthera peregrina*; *Mimosa schomburgkii*.

Paripi Palm, possibly *Bactris gasipaes*.
Paruacussana, unidentified.
Passion Flower (Cr.): There are twenty species of the genus *Passiflora* in Guyana known as Passion Flower.
Pendare, see Wild Star Apple.
Penguin, see Kaitshara.
Pentari, see Wild Star Apple.
Pijao Palm, see Paripe.
Piriko (M.): Unidentified species of *Psidium*.
Podai: Unidentified.
Popa, probably Boba.
Pourouck, see Bullet Tree.
Purpleheart (Cr.): *Peltogyne porphyrocardia*; *P. venosa*.
Purpleheart, Bastard (Cr.): *Peltogyne paniculata*.
Reho (Ar.): *Euterpe stenophylla*.
Sandpaper Tree (Cr.): *Curatella americana*.
Sawaray or Sauwarai (C.): *Astrocaryum jauari*.
Sawari or Sawary (Cr.): *Caryocar nuciferum*.
Sawari, Water (Cr.): *Caryocar glabrum*.
Sawari, Water and/or Bats (Cr.): *Caryocar microcarpum*.
Scimitu, see Simitu.
Shiruaballi (Ar.): *Aniba hypoglauca*; *Licaria canella*; *Ocotea canaliculata*.
Shiwishiwi (Ar.): *Bromelia karatas*.
Silk-cotton Tree (Cr.): *Bombax flaviforum*; *B. globosum*; *B. surinamensis*.
Silk Tree (Cr.): *Ceiba pentandra*. See also Kumaka and Makau.
Silk Grass: *Bromelia* spp.
Simitu (Ar.): *Passiflora laurifolia*.
Sirabally, Siruaballi, Sirwabally, see Shiruaballi.
Souari, see Sawari.
Suari, see Sawari.
Tá-au: Unidentified.
Table Tree (Cr.): *Cordia bicolor*; *C. exaltata*; *C. fallax*; *C. hirta*; *C. sericicalyx*; *C. tetrandra*.
Tasoka: Unidentified.
Tatabu (Ar.): *Diplotropis purpurea*.
Tederma, possibly Determa.
Tiger Wood: Unidentified.
Tonka bean (Cr.): *Dipteryx odorata*; *D. oppositifolia*; *D. punctata*.
Toucoushou-aheru, see Tukushuahiru.
Touro, probably Turu.
Trooly, see Truli.
Truli (Cr.): *Manicaria saccifera*.
Trumpet Tree (Cr.): *Cecropia angulata*; *C. sciadophylla*.
Tshiwipa (C.) same as Ayawiye (M.): *Martiodendron excelsum*.
Tukushuahiru (C.): *Brownea latifolia*.
Tunkara: Unidentified.
Turpentine Tree: *Bursera gummifera*; *B. simaruba*.
Turu (Ar.): *Jessenia bataua* (= *Oenocarpus bataua*).

Uassa, possibly Wasi.
Ubudi (Ar.): *Anacardium giganteum.*
Urukush, see Bulukush.
Wallaba, Clump (Cr.): *Dicymbe altsonii*; *D. corymbosa.*
Wallaba, Ituri (Cr.): *Campsiandra comosa*; *Eperua grandiflora*; *E. jenmanii.*
Wallaba, Soft (Cr.): *Eperua falcata.*
Wamara (Ar.): *Swartzia leiocalycina.*
Wanama: Unidentified bamboo.
Wanasoro (Ar.): *Cecropia angulata; C. sciadophylla.*
Wanussuri, see Wanasoro.
Warada, possibly Warapa (P.): *Macoubea guianensis.*
Warrakaro, possibly Warukuri (Ar.): *Tabebuia insignis.*
Wasi (C.): *Euterpe oleracea.*
Water Guava, *Campomanesia aromatica.*
Waurinit (Wr.): *Bromelia karatas.*
Wichabai (W.): *Manilkara bidentata.*
Wild Cashew, see Ubudi.
Wild Cherry (Cr.): *Eugenia patrisii.*
Wild Silk-cotton Tree (Cr.): *Bombax globosum*; *B. surinamense.*
Wild Star Apple (Cr.): *Couma guianensis*; *C. macrocarpa*; *C. rigida.*
Wina (Ar.): *Lecythis corrugata.*
Witchaway, see Wichabai.
Wourari, see Curare.
Yabirou or Yarourie or Yarula or Yarurá or Yaruri or Yaruru (Ar., Cr.), see Paddlewood.
Zuruaku: Unidentified.

BIBLIOGRAPHY

The bibliography is in two parts. The first part lists the publications of Robert Schomburgk, whether referred to in the text or not. The second part consists of the details of publications cited, by either Schomburgk or the editor, and is arranged according to the Hakluyt Society house style with the titles of any given author ordered alphabetically. In the former, Schomburgk's works are ordered chronologically, but alphabetically within any single year. The point of this is to give the reader a biographical overview of his literary production. It should be noted that, although the list contains a very high proportion of his publications, it is not intended to be exhaustive.

Part 1: The Works of Robert H. Schomburgk

1832, 'Remarks on Anegada', *Journal of the Royal Geographical Society*, 2, pp. 152–70.

1833, 'Berichte über die Kulturpflanzen Westindiens', *Linnaea*, 8, pp. 229–44, 257–92.

1834, 'Ueber Benutzung einiger westindischien Pflanzen', *Linnaea*, 9, pp. 511–13.

1835, 'A Refutation of Certain Strictures on the Geographical Situation and Localities of Tortola as a Harbour and a Roadstead', *The Nautical Magazine*, 4, pp. 329–33.

1835, 'Remarks on the Heavy Swell along some of the West-India Islands, commonly called "Ground" or "North Sea"; and on the Set and Velocity of the Tides, and the Effects produced by their Transporting Power, among the Virgin Islands', *Journal of the Royal Geographical Society*, 5, pp. 23–38.

1836, 'Report of an Expedition into the Interior of British Guayana, in 1835–36', *Journal of the Royal Geographical Society*, 6, pp. 224–84.

1837, 'Diary of an Ascent of the River Berbice, in British Guayana, in 1836–7', *Journal of the Royal Geographical Society*, 7, pp. 302–50.

1837, 'Diary of an Ascent of the River Corentyn in British Guayana, in October, 1836', *Journal of the Royal Geographical Society*, 7, pp. 285–301.

1837, 'On the Identity of Three supposed Genera of Orchideous Epiphytes. In a Letter to A. B. Lambert, Esq., V.P.L.S.', *The Transactions of the Linnean Society of London*, 17, pp. 551–2.

1838, 'On the Ant Tree of Guiana (*Triplaris americana*)', *Annals of Natural History; or, Magazine of Zoology, Botany, and Geology*, 1, pp. 264–7.

1838, 'Information respecting Botanical Travellers', *Annals of Natural History; or, Magazine of Zoology, Botany, and Geology*, 1, pp. 63–8

1838, 'Ueber die giftigen Wirkungen des Manschinell-Baums', *Linnaea*, 12, pp. 248–54.

1839, 'On the Habits of the King of Vultures', *Annals of Natural History; or, Magazine of Zoology, Botany, and Geology*, 2, pp. 255–60.

1839, 'Remarks on the Great Ant-bear (*Myrmecophaga jubata*)', *Proceedings of the Zoological Society of London*, 7, pp. 21–6.

1840, *Catalogue of Objects in Illustration of Ethnography & Natural History, composing the Guiana Exhibition*, London, E. & J. Thomas.

1840, *A Description of British Guiana, Geographical and Statistical: exhibiting its Resources and Capabilities, together with the Present and Future Condition and Prospects of the Colony*, London, Simpkin, Marshall & Co.

1840, 'Description of the Snake-nut Tree of Guiana', *Annals of Natural History or, Magazine of Zoology, Botany and Geology*, 5, pp. 202–4.

1840, 'The History and Habits of Crotophaga', *Annals of Natural History; or, Magazine of Zoology, Botany, and Geology*, 4, pp. 318–22.

1840, 'Information respecting Botanical Travellers', *Annals of Natural History; or, Magazine of Zoology, Botany, and Geology*, 4, pp. 194–9, 262–6, 322–8, 429–34.

1840, 'Information respecting Botanical Travellers', *Annals of Natural History; or, Magazine of Zoology, Botany, and Geology*, 5, pp. 29–35, 282–8, 343–8, 399–405.

1840, 'Observations on the Migrations of Certain Butterflies in British Guiana', *Journal of the Proceedings of the Entomological Society of London commencing January 6, 1840*, pp. 15–16 (in *The Transactions of the Entomological Society of London*, 4, 1846).

1840, 'Remarks on the Great Ant-bear (Myrmecophaga jubata)', *Annals of Natural History; or, Magazine of Zoology, Botany, and Geology*, 4, pp. 202–7.

1840, 'Über die problematische Existenz der Laguna Parime oder Laguna del Dorado', *Monatsberiche über die Verhandlungen der Gesellschaft für Erdkunde zu Berlin*, 1, pp. 52–5.

1841, 'Description of the Mora Tree', *The Transactions of the Linnean Society of London*, 18, pp. 207–11.

1841, *Twelve Views in the Interior of Guiana: from Drawings executed by Mr. Charles Bentley, after Sketches taken during the Expedition carried on in the Years 1835 to 1839, under the Direction of the Royal Geographical Society of London, and aided by Her Majesty's Government*, London, Ackermann & Co.

1841, *Fishes of Guiana*, Part 1, in *The Naturalist's Library* (conducted by Sir William Jardine), Vol. 32 (Ichthyology, Vol. III), Edinburgh, W. H. Lizars.

1841, 'Journey from Esmeralda, on the Orinoco, to San Carlos and Moura on the Rio Negro, and thence by Fort San Joaquim, to Demerara, in the spring of 1839', *Journal of the Royal Geographical Society*, 10, pp. 248–67.

1841, 'Journey from Fort San Joaquim, on the Rio Branco to Roraima, and thence by the River Parima and Merewari to Esmeralda, on the Orinoco, in 1838–9', *Journal of the Royal Geographical Society*, 10, pp. 191–247.

1841, 'On the Urari, the Arrow Poison of the Indians of Guiana; with a Description of the Plant from which it is extracted', *The Annals and Magazine of Natural History, including Zoology, Botany, and Geology*, 7, pp. 407–27.

1841, 'Report of the Third Expedition into the Interior of Guayana, comprising the Journey to the Sources of the Essequibo, to the Carumá Mountains, and to Fort San Joaquim, on the Rio Branco, in 1837–8', *Journal of the Royal Geographical Society*, 10, pp. 159–90.

1841, *R. H. Schomburgk's Reisen in Guiana und am Orinoko während der Jahre 1835–1839. Nach seinen Berichten und Mittheilungen an die Geographische Gesellschaft in London herausgegeben von O. A. Schomburgk*, Leipzig, George Wizard.

1841, 'Some Account of the Curata, a Grass of the Tribe of Bambuseae, of the Culm of which the Indians of Guiana prepare their Sarbacans or Blowpipes', *The Transactions of the Linnean Society of London*, 18, pp. 557–62.
1842, 'Excursion up the Barima and Cuyuni Rivers, in British Guiana, in 1841', *Journal of the Royal Geographical Society*', 12, pp. 178–96.
1842, 'Expedition to the Lower Parts of the Barima and Guiania Rivers, in British Guiana', *Journal of the Royal Geographical Society*, 12, pp. 169–78.
1843, 'Bericht über seinen Ausflug nach dem Canuku-Gebirge', *Monatsberichte über die Verhandlungen der Gesellschaft für Erdkunde zu Berlin*, 4, pp. 127–38.
1843, *Fishes of Guiana*, Part 2, in *The Naturalist's Library* (conducted by Sir William Jardine), Vol. 38 (Ichthyology, Vol. V), Edinburgh, W. H. Lizars.
1843, 'Visit to the Sources of the Takutu, in British Guiana, in the Year 1842', *Journal of the Royal Geographical Society*, 13, pp. 18–75.
1844, 'Description of a New Species of *Calycophyllum* from British Guiana', *London Journal of Botany*, 3, pp. 621–3.
1844, 'On Two New Species of the Family Laurineae from the Forests of Guiana', *London Journal of Botany*, 3, pp. 624–31.
1844, 'Über seine letzte Reise in Guyana vom Juni bis zum October 1843', *Monatsberichte über die Verhandlungen der Gesellschaft für Erdkunde zu Berlin*, new ser. 1, pp. 195–204.
1845, *Die Barbacenia Alexandrinae und Alexandra Imperatricis, entdeckt und beschrieben*, Braunschweig, Von Friedrich Vieweg & Sohn.
1845, 'Characters of Two New Plants discovered in British Guiana', *London Journal of Botany*, 4, pp. 12–13.
1845, 'A Description of the Murichi, or Ita palm of Guiana (*Mauritia flexuosa, L.*) and its Uses by the Aborigines', *Simmonds's Colonial Magazine*, 6, pp. 43–9.
1845, 'A Description of *Ophiocaryon paradoxum*, or the Snake Nut Tree of Guiana', *London Journal of Botany*, 4, pp. 375–8.
1845, 'Herrn Rob. Schomburgk's meteorologische Mittheilungen aus Guiana', *Monatsberichte über die Verhandlungen der Gesellschaft für Erdkunde zu Berlin*, new ser. 2, pp. 75–87. [From letter to von Humboldt dated Georgetown, 4 April 1844.]
1845, 'Herrn Rob. H. Schomburgk's neue Mittheilungen über seine Grenzexpedition in Guiana, in den Jahren 1841–1843', *Monatsberichte über die Verhandlungen der Gesellschaft für Erdkunde zu Berlin*, new ser. 2, pp. 66–75. [From letters to von Humboldt dated Georgetown, 28 January, 30 January and 4 April 1844.]
1845, 'Journal of an Expedition from Pirara to the Upper Corentyne, and from thence to Demerara', *Journal of the Royal Geographical Society*, 15, pp. 1–104.
1845, 'On the Lake Parima (The "El Dorado" of Sir Walter Raleigh) and the Geography of Guiana', *Simmonds's Colonial Magazine*, 5, pp. 381–91.
1845, *Die Rapatea Friderici Augusti, und Saxo-Fridericia Regalis, entdeckt und beschrieben*, Braunschweig, Von Friedrich Vieweg und Sohn.
1845, 'Remarks on the Geology of British Guiana', *The Quarterly Journal of the Geological Society of London*, 1, pp. 298–300.
1846, 'Remarks on the Cultivation of the Cochineal Insect in Barbados', *Simmonds's Colonial Magazine*, 9, pp. 211–14.
1846, 'A Visit to Turner's Hall Wood, Barbados, in 1846', *Simmonds's Colonial Magazine*, 8, pp. 482–8.

1847, 'Abhandlung über Barbadoes mit, worin die geographischen, statistischen und meteorologischen Verhältnisse der Insel dargestellt waren', *Monatsberichte über die Verhandlungen der Gesellschaft für Erdkunde zu Berlin*, new ser. 4, pp. 62–75.

1847, 'Beschreibung dreier neuer Pflanzen aus dem Flussgebiete des Carimani oder Caramang, eines Zuflusses des Mazaruni', *Linnaea*, 20, pp. 751–60.

1847, 'Description of some Grasses and Sedges from the East Coast of Demerara, with Remarks on the Geographical Distribution of the Species', *The Annals and Magazine of Natural History, including Zoology, Botany, and Geology*, 20, pp. 396–409.

1847, 'The Microscopical Siliceous Polycystina of Barbados, and their Relation to Existing Animals, as described in a Lecture of Professor Ehrenberg of Berlin, delivered before the Royal Academy of Sciences on 11th February 1847', *The Annals and Magazine of Natural History, including Zoology, Botany, and Geology*, 20, pp. 115–27.

1847, ' Pictures of Barbados', *Bentley's Miscellany*, 22, pp. 286–92.

1847, 'Rambles in Search of the Picturesque in Barbados', *Bentley's Miscellany*, 22, pp. 443–9.

1847, ' Steam-boat Voyage to Barbados', *Bentley's Miscellany*, 22, pp. 30–41.

1847, West Indian Fruits and Esculents that may be advantageously introduced into Cultivation in England', *The Journal of the Horticultural Society of London*, 2, pp. 148–55.

1848, 'Contributions to the Commercial Statistics of the Republics in South America. I. The Republics of Venezuela, New Granada, and Ecuador, with some Preceding Remarks on a Communication between the Atlantic and Pacific Oceans', *Simmonds's Colonial Magazine*, 13, pp. 234–60.

1848, 'Contributions to the Commercial Statistics of the Republics in South America. II. The Republic of Chile', *Simmonds's Colonial Magazine*, 14, pp. 40–35.

1848, 'Contributions to the Philological Ethnography of South America', *Proceedings of the Philological Society*, 3, pp. 228–37.

1848, 'Description of some New Fossil Shells from Bissex Hill and Springfield in Barbados', *The Annals and Magazine of Natural History, including Zoology, Botany, and Geology*, 2nd ser. 1, pp. 347–9. [Reprinted from *History of Barbados*.]

1848, 'A Description of some New Species of Fishes from the Sea surrounding the Island of Barbados', *The Annals and Magazine of Natural History, including Zoology, Botany, and Geology*, 2nd ser. 2, pp. 11–20. [Reprinted from *History of Barbados*.]

1848, (Ed.) *The Discovery of the Large, Rich, and Beautiful Empire of Guiana, with a Relation of the Great and Golden City of Manoa (which the Spaniards call El Dorado), etc. Performed in the Year 1595, by Sir Walter Ralegh, Knt.*, London, The Hakluyt Society, 1st ser. 3.

1848, *The History of Barbados comprising a Geographical and Statistical Description of the Island; a Sketch of the Historical Events since the Settlement; and an Account of its Geology and Natural Productions*, London, Longman, Brown, Green and Longmans.

1848, 'On the Manufacture of Beet-root Sugar in the German Custom-house League or Zollverein', *Simmonds's Colonial Magazine*, 14, pp. 134–40.

1848, 'On the Natives of Guiana', *Journal of the Ethnological Society of London*, 1, pp. 253–76.

1848, 'On a Uniform System to reduce Unwritten Languages to Alphabetical Writing in Roman Characters', *Simmonds's Colonial Magazine*, 15, pp. 77–81.

1848, ' Remarks to accompany a Comparative Vocabulary of Eighteen Languages and

Dialects of Indian Tribes inhabiting Guiana', *Simmonds's Colonial Magazine*, 15, pp. 46–64.
1849, *Travels of his Royal Highness Prince Adalbert of Prussia, in the South of Europe and in Brazil, with a Voyage up the Amazon and the Xingu*, translated by Sir Robert H. Schomburgk and J. E. Taylor, London, David Boyne.
1850, 'On the Currents and Tides of the Mona Passage', *The Nautical Magazine*, 19, pp. 585–8.
1850, 'A Vocabulary of the Maiongkong Language', *Proceedings of the Philological Society*, 4, pp. 217–22.
1852, 'On the Currents and Tides of the Southern Coast of Santo Domingo', *The Nautical Magazine*, 21, pp. 289–92.
1852, 'On the Geographical Position of the City of Santo Domingo', *The Nautical Magazine*, 21, pp. 412–18.
1852, 'Visit to the Valley of Constanza in the Cibao Mountains of the Island of Santo Domingo, and to an Indian Burial-ground in its Vicinity', *The Athenaeum*, 1291, pp. 797–9.
1853, 'Effect of a Tropical Climate upon Plants of a Temperate Zone', *The Journal of the Horticultural Society of London*, 8, pp. 32–6.
1853, 'Harbour of San Domingo', *The Nautical Magazine*, 22, p. 276.
1853, 'The Loadstone Mountain at Santo Domingo'. *The Athenaeum*, 1338, pp. 738–40.
1853, 'The Peninsula and Bay of Samaná, in the Dominican Republic', *Journal of the Royal Geographical Society*, 23, pp. 264–83.
1853, 'Remarks on the Principal Ports and Anchoring Places along the Coast of the Dominican Republic', *The Nautical Magazine*, 22, pp. 202–7, 225–32, 304–13, 340–49.
1854, 'Ethnological Researches in Santo Domingo', *Journal of the Ethnological Society of London*, 3, pp. 115–22.
1855, 'Hurricane at Santo Domingo', *The Athenaeum*, 1462, pp. 1273–4.
1855, 'On the Forest-trees of British Guiana and their Uses in Naval and Civil Architecture', *Proceedings of the Linnean Society of London (1845–55)*, 2, pp. 158–65
1857, 'The Urari, or Arrow Poison of the Indians of Guiana', *Pharmaceutical Journal and Transactions*, 16, pp. 500–507.
1858, 'Description of a Remarkable Spike or Bunch of Fruits of the Fig Banana (*Musa sapientum*), *var.*', *Journal of the Proceedings of the Linnean Society (Botany)*, 2, pp. 130–32.
1859, 'Autobiographie', *Leopoldina*, 1, pp. 34–9.
1861, 'Boat Excursion from Bangkok, in Siam, to the Pechaburri, on the Western Shore of the Gulf of Siam', *Journal of the Royal Geographical Society*, 31, pp. 302–21.
1861, 'The Vegetable Products of Siam', *The Technologist*, 1, pp. 355–62.
1862, 'Siamese Products', *The Technologist*, 2, pp. 444–50.
1862, 'The Vegetable Products of Siam', *Pharmaceutical Journal and Transactions*, 2nd ser. 3, pp. 123–8.
1863, 'A Visit to Xiengmai, the Principal City of the Laos or Shan States', *Journal of the Asiatic Society of Bengal*, 32, pp. 387–99.
1864, 'The Paper-tree of Siam (Ton-khai), *Trophis aspera*, W.', *The Technologist*, 4, pp. 337–9.
1931, *Robert Hermann Schomburgk's Travels in Guiana and on the Orinoco during the Years 1835–1839*, translated by W. E. Roth, Georgetown, "The Argosy" Company.

2004, 'Robert Herman Schomburgk – An Autobiography', in *Sir Robert H. Schomburgk. Essays in Honour of an Explorer and Natural Scientist*, ed. James Rose, Georgetown, University of Guyana. [Translation by B. J. H. ter Welle of 1859, 'Autobiographie', *Leopoldina*, 1, pp. 34–9.]

Part 2: Works Cited

Acuña, Christoval de, *Nuevo descubrimiento del Gran Rio de las Amazones*, Madrid, en la Imprenta del Reyno, 1641.

Adair, James, *The History of the American Indians; particularly those Nations adjoining to the Missisppi, East and West Florida, Georgia, South and North Carolina, and Virginia*, London, Edward and Charles Dilly, 1775.

Alexander, J. E., *Transatlantic Sketches, comprising Visits to the most interesting Scenes in North and South America and the West Indies with Notes on Negro slavery and Canadian Emigration*, London, Richard Bentley, 1833.

Allsopp, Richard, *Dictionary of Caribbean English Usage*, Oxford, Oxford University Press, 1996.

Anon., 'Journal of an Excursion from St. Thome de Angostura, in Spanish Guayana, to the Capuchin Missions of the Caroni', *The Quarterly Journal of Science, Literature, and the Arts*, 9, 1820, pp. 1–32.

Anon., 'Journal of a Trip from St. Thome de Angostura, in Spanish Guayana, to the Capuchin Missions of the Caroni', *The Quarterly Journal of Science, Literature, and the Arts*, 8, 1820, pp. 260–87.

Aublet, Fusée, *Histoire des plantes de Guiane Françoise: rangées suivant la méthode sexuelle, avec plusieurs mémoires sur différens objets intéressans, relatifs à la culture et au commerce de la Guiane Françoise, et une notice des plantes de l'Isle-de-France*, 4 vols, London, Paris, Pierre-François Didot jeune, 1775.

d'Azara, Félix, *Apuntamientos para la historia natural de los Páxaros de Paragüay y Rio de la Plata*, 3 vols, Madrid, Ibarra, 1802–5.

d'Azara, Félix, *Essais sur l'histoire naturelle des quadrupedes de la Province du Paraguay*, 2 vols, Paris, Charles Pougens, 1801.

Barata, Francisco José Rodrigues, 'Diario da viagem que fez á Colonia Hollandeza de Surinam o porta bandeira da Setima compania do regimento da cidade do Pará, pelos sertões e rios d'este estado, em diligencia do Real Serviço', *Revista trimensal de historia e geographia*, 8, 1846, pp. 1–53, 157–204.

Barrère, Pierre, *Nouvelle relation de la France Equinoxiale*, Paris, Chez Piget, Damonneville, Durand, 1743.

Barrington Brown, C., *Canoe and Camp Life in British Guiana*, London, Edward Stanford, 1876.

Barry, John V., *Alexander Maconochie of Norfolk Island: A Study of a Pioneer in Penal Reform*, Melbourne, Oxford University Press, 1958.

Beebe, William, G. Inness Hartley and Paul G. Howes, *Tropical Wild Life in British Guiana*, New York, New York Zoological Society, 1917.

Bennett, John P., *An Arawak–English Dictionary*, Georgetown, Walter Roth Museum, 1989.

Bentham, George, 'Contributions towards a Flora of South America – Enumeration of

Plants collected by Mr Schomburgk, British Guiana', *Annals of Natural History; or, Magazine of Zoology, Botany, and Geology*, 2, 1839, pp. 105–11, 441–51.

Bentham, George, 'Contributions towards a Flora of South America – Enumeration of Plants collected by Mr Schomburgk, British Guiana', *Annals of Natural History; or, Magazine of Zoology, Botany, and Geology*, 3, 1839, pp. 427–38.

Bentham, George, 'Contributions towards a Flora of South America – Enumeration of Plants collected by Mr Schomburgk in British Guiana', *Journal of Botany*, 2, 1840, pp. 39–103, 127–46, 210–33, 286–324.

Bentham, George, 'Contributions towards a Flora of South America – Enumeration of Plants collected by Mr Schomburgk in British Guiana', *Journal of Botany*, 3, 1841, pp. 212–50.

Bentham, George, 'Contributions towards a Flora of South America – Enumeration of Plants collected by Mr Schomburgk in British Guiana', *Journal of Botany*, 4, 1842, pp. 99–133, 321–23.

Bentham, George, 'Contributions towards a Flora of South America – Enumeration of Plants collected by Mr Schomburgk in British Guiana, *London Journal of Botany*, 1, 1842, pp. 193–203.

Bentham, George, 'Contributions towards a Flora of South America – Enumeration of Plants collected by Mr Schomburgk in British Guiana, *London Journal of Botany*, 2, 1843, pp. 42–52, 359–78, 670–74.

Bentham, George, 'Contributions towards a Flora of South America – Enumeration of Plants collected by Sir Robert Schomburgk in British Guiana', *London Journal of Botany*, 4, 1845, pp. 622–37.

Bentham, George, 'Contributions towards a Flora of South America – Enumeration of Plants collected by Sir Robert Schomburgk in British Guiana', *London Journal of Botany*, 5, 1846, pp. 351–65.

Bentham, George, 'Contributions towards a Flora of South America – Enumeration of Plants collected in British, Dutch, and French Guiana by Sir Robert and Richard Schomburgk, Dr Hostmann, M. Leprieur, and others', *London Journal of Botany*, 7, 1848, pp. 116–37.

Bentham, George, 'On the *Heliamphora nutans*, a new Pitcher-plant from British Guiana', *The Transactions of the Linnean Society of London*, 18, 1841, pp. 429–33.

Bentham, George, *Labiatarum Genera et Species: or, A description of the Genera and Species of Plants of the Order Labiatae*, London, James Ridgway & Sons, 1832–6.

Bernau, John, *Missionary Labours in British Guiana*, London, Edward Stanford, 1847.

Blackburn, Julia, *Charles Waterton: 1782–1865*, London, Bodley Head, 1989.

Bos, G., 'Atorai, Trio, Tunayana and Wai Wai in early Eighteenth Century Records', *Folk*, 27, 1985, pp. 5–15.

Bos, G., *Some Recoveries in Guiana Indian Ethnohistory*, Amsterdam, VU Uitgeverij, 1998.

Bougainville, Louis Antoine de, *Voyage autour du monde, par la frégate du Roi La Boudeuse, et la flûte L'Étoile; en 1766, 1767, 1768 et 1769*, Paris, 1771.

Bowring, John, *The Kingdom and People of Siam; with a Narrative of the Mission to that Country in 1855*, 2 vols, London, John W. Parker and Son, 1857.

Buffon, Georges Louis Leclerc, Comte de, *Oeuvres complètes. Histoire des animaux quadrupèdes*, 10 vols, Paris, 1775–9.

Burnett, D. Graham, *Masters of All They Surveyed. Exploration, Geography, and a British El Dorado*, Chicago and London, University of Chicago Press, 2000.

Butt Colson, Audrey, 'Itoto (Kanaima*) as Death and Anti-structure' in *Beyond the Visible and the Material. The Amerindianization of Society in the Work of Peter Rivière*, ed. Laura M. Rival and Neil L. Whitehead, Oxford, Oxford University Press, 2001, pp. 221–33.

Butt Colson, Audrey, 'The Spatial Component in the Political Structure of the Carib Speakers of the Guiana Highlands: Kapon and Pemon', *Antropologica*, 59–62, 1983–4, pp. 73–124.

Candolle, Augustin Pyramus de, *Prodromus systematis naturalis regni vegetabilis*, 18 vols, Paris, Treuttel & Würtz, 1824–73.

Carmichael, A. C., *Domestic Manners and Social Conditions of the White, Coloured, and Negro Population of the West Indies*, 2 vols, London, Whittaker, Treacher & Co., 1833.

Caulin, Antonio, *Historia coro-graphica natural y evangelista de la Nueva Andalucia, provincias de Cumaná, Guyana, y vertíentes de Río Orínoco*, Madrid, Dada á luz de Orden, y á Expen^s de S. M., 1779.

Clay, John, *Maconochie's Experiment*, London, John Murray, 2001.

Corrêa, M. P., *Dicionário das plantas úteis do Brasil e das exóticas cultivadas*, 6 vols, Rio de Janeiro, Imprensa Nacional, 1926–78.

Coudreau, Henri, *La France équinoxiale*, 2 vols, Paris, Challamel Ainé, 1886–7.

Dam, J. A. C. van., 'The Guyanan Plant Collections of Robert and Richard Schomburgk', in *Flora of the Guianas*, ed. M. J. Jansen-Jacobs, Kew, Royal Botanic Gardens, 2002.

Daszkiewicz, Piotr and Michel Jegu, 'Correspondence between Adolphe Brongniart and Robert Schomburgk: Trading Natural History Collections for Honours', *Archives of Natural History*, 29, 2002, pp. 333–6.

De Goeje, C. H., *Beiträge zur völkerkund von Surinam*, Leiden, E. J. Brill, 1908.

De Goeje, C. H., *Bijdrage tot de ethnographie der Surinaamsche Indianen*, Leiden, E. J. Brill, 1906.

De La Beche, H. T., *Researches in Theoretical Geology*, London, Charles Knight, 1834.

Desmarest, Anselme Gaëtan, *Mammalogie, ou description des espèces mammifères*, 2 pts, Paris, Veuve Agasse, 1820–22.

Dressler, Robert L., *The Orchids. Natural History and Classification*, Cambridge, Mass., Harvard University Press, 1981.

Dubelaar, C. N., *South American and Caribbean Petroglyphs*, Koninklijk Instituut voor Taal-, Land- en Volkenkunde, Caribbean Series 3, Dordrecht and Riverton, Foris Publications, 1986.

Dunsterville, G. C. K. and Leslie A. Garay, *Venezuela Orchids Illustrated*, 6 vols, London, Andre Deutsche, 1959–76.

Endlicher, Stephen Ladislaus, *Genera plantarum secundum ordines naturales disposita*, 1 vol. and 4 supplements, Vindobonae [Vienna], Apud Fr. Beck Universitatis Bibliopolam, 1836–40.

Farabee, William C., 'A Pioneer in Amazonia: The Narrative of a Journey from Manaos to Georgetown', *Bulletin of the Geographical Society of Philadelphia*, 15, 1917, pp. 57–103.

Farabee, William C., *The Central Caribs*, Philadelphia, University Museum, 1967 [1924].

Farage, Nádia, *As muralhas dos sertões. Os povos indígenas no rio Branco e a colonização*, Rio de Janeiro, Paz e Terra, ANPOCS, 1991.

Farrar, T., *Notes on the History of the Church in Guiana*, British Guiana, Wm. Macdonald, "Gazette" Office, Berbice, [1892].

Fitch, W. and W. J. Hooker, *Victoria Regia; or Illustrations of the Royal Water-lily, in a Series of Figures chiefly made from Specimens Flowering at Syon and at Kew*, London, Reeve and Benham, 1851.

Fleming, Fergus, *Barrow's Boys*, London, Granta Books, 1998.

Freiberg, Marcos, *Turtles of South America*, Neptune, New Jersey, T. F. H. Publications, 1981.

Frikel, Protásio, 'Das Problem der Piankotó-Tiriyó', in *Beiträge zur Völkerkunde Südamerikas*, ed. Hans Becher, Hanover, Kommissionsverlag Münstermann-Druck, 1964, pp. 97–104.

Frost, Darrel R., ed., *Amphibian Species of the World*, Lawrence, Kansas, Allen Press and Association of Systematics Collections, 1985.

Froude, James Anthony, 'England's Forgotten Worthies', *The Westminster Review*, new ser. 2, July–October 1852, pp. 32–67.

Goodall, Edward, 'The Diary of Edward Goodall Esq. during his Sojourn in Georgetown from 28th July to 11th December 1841', *Journal of the British Guiana Museum and Zoo*, 35, 1962, pp. 39–53.

Goodall, Edward, 'Edward Goodall's Diary 23rd December 1841–13th June 1842 (on the Essequibo & Rupununi Rivers)', *Journal of the British Guiana Museum and Zoo*, 36, 1962, pp. 47–64.

Goodall, Edward, *Sketches of Amerindian Tribes 1841–43*, London, British Museum Publications, 1977.

Gorinsky, Conrad, 'The Isolation and Characterisation of Biologically Active Constituents from Cissampelos ovalifolia, D.C. and Clibadium sylvestre (Aubl.) Baill.', unpublished PhD thesis, University of London, 1974–5.

Grenand, P., C. Moretti and H. Jacquemin, *Pharmacopées traditionnelles en Guyane, Créoles, Palikur, Wayãpi*, Paris, ORSTOM, 1987.

Gumilla, Joseph, *Historia natural, civil y geografica de las naciones situadas en las riveras del Rio Orinoco*, Barcelona, Carlos Gibert y Tutó, 1791.

Hancock, John, *Observations on the Climate, Soil, and Productions of British Guiana, and on the Advantages of Emigration to, and Colonizing the Interior of that Country*, London, James Fraser; John Hatchard & Son; and George Mann, 1835.

Hart, H. G., *New Army List*, London, John Murray, 1844.

Hartsinck, J. J., *Beschryving van Guiana of de Wildekust in Zuid-America*, 2 vols, Amsterdam, Gerrit Tielenburg, 1770.

Hemming, John, 'How Brazil acquired Roraima', *The Hispanic American Historical Review*, 70:2, 1990, pp. 295–325.

Hemming, John, *The Search for El Dorado*, London, Joseph, 1978.

Hilhouse, William, *Indian Notices: or Sketches of the Habits, Characters, Languages, Superstitions, Soil, and Climate of the Several Nations; with Remarks on their Capacity for Colonization, Present Government and Suggestions for Future Improvements and Civilization. Also, the Icthyology of the Fresh Waters of the Interior*. Georgetown, National Commission for Research Materials, 1978 [1825].

Hilhouse, William, 'Journal of an Expedition up the River Cuyuny, in British Guayana, in March, 1837', *Journal of the Royal Geographical Society*, 7, 1837, pp. 446–54.

Hilhouse, William, 'Journal of a Voyage up the Massaroony in 1831', *Journal of the Royal Geographical Society*, 4, 1834, pp. 25–40.

The History of the Collections contained in the Natural History Departments of the British Museum, 2 vols, London, Trustees of the British Museum, 1904–6.

Humboldt, Alexander von, 'Sur quelques points importans de la géographie de la Guyane', *Nouvelles annales des voyages et des sciences géographique*, 74 (14, 3rd ser.), pp. 136–80, 1837.

Humboldt, Alexander von, *Voyage aux régions équinoxiales du Nouveau Continent, fait en 1799, 1800, 1801, 1803 et 1804*, 13 vols, Paris, Librairée Grecque-Latine-Allemande, 1816–31.

Humboldt, Alexander von and Aimé Bonpland, *Personal Narrative of Travels to the Equinoctial Regions of the New Continent during the Years 1799–1804*, translated by Helen Maria Williams, 7 vols, London, Longman, Hursts, Rees, Orme & Brown; J. Murray; and H. Colburn, 1814–29.

Humboldt, Alexander von and Aimé Bonpland, *Personal Narrative of Travels to the Equinoctial Regions of the New Continent during the Years 1799–1804*, translated by Thomasina Ross, 3 vols, London, George Bell & Sons, 1907.

Im Thurn, Everard, 'An Editor's Prologue', *Timehri being the Journal of the Royal Agricultural and Commercial Society of British Guiana*, 1, 1882, pp. 1–6.

Kasper, Johann, *Life of John Kaspar Lavater*, London, Religious Tract Society, 1849.

Keay, John, *The Great Arc. The Dramatic Tale of how India was Mapped and Everest was Named*, London, Harper Collins, 2000.

Keyes, Charles, F., *Thailand. Buddhist Kingdom as Modern Nation-State*, Boulder, Col., Westview Press, 1989.

Koch-Grünberg, Theodor, *Del Roraima al Orinoco*, 3 vols, Caracas, Ernesto Armitano Editor, 1981–2 [1917–24].

Kunth, Karl S., *Enumeratio plantarum omnium hucusque cognitarum, secundum familias naturales disposita, adjectis characteribus, differentiis et synonymis*, 4 vols, Stuttgart and Tubingen, J. G. Cottae, 1833–43

Kunth, Karl S., *Handbuch der Botanik*, Berlin, Duncker und Humblot, 1831.

La Condamine, Charles Marie de, *Relation abrégée d'un voyage fait dans l'interieur de l'Amérique Méridionale*, Paris, chez la Veuve Pissot, 1745.

Lawson, George, *The Royal Water-lily of South America, and the Water-lilies of our own Land: their History and Cultivation*, Edinburgh, James Hogg, 1851.

Leonowens, Anna Harriette, *The English Governess at the Siamese Court: being Recollections of Six Years in the Royal Palace at Bangkok*, London, Trübner & Co., 1870.

Lindley, John, *The Genera and Species of Orchidaceous Plants*, 6 pts, London, Ridgways, 1830–40.

Lindley, John, *An Introduction to the Natural System of Botany: or, a Systematic View of the Organisation, Natural Affinities, and Geographical Distribution, of the Whole Vegetable Kingdom*, London, Longman, Rees, Orme, Brown & Green, 1830.

Lindley, John, *A Natural System of Botany: or, a Systematic View of the Organisation, Natural Affinities, and Geographical Distribution, of the Whole Vegetable Kingdom*, London, James Ridgway & Sons, 1838.

Lindley, John, *Victoria regia*, London, Shakespeare Press, 1837.

Loukotka, Čestmír, *Classification of South American Indian Languages*, Los Angeles, Latin American Center, University of Los Angeles, 1968.

Mabberley, D. J., *The Plant-book. A Portable Dictionary of the Vascular Plants*, Cambridge, Cambridge University Press, 1997.

Macdonald, David, ed., *The New Encyclopedia of Mammals*, Oxford and New York, Oxford University Press, 2001.

Magaña, Edmundo, *Contribuciones al estudio de la mitología y astrononomía de los indios de las Guayanas*, Amsterdam, CEDLA, 1987.

Markham, Clements R., 'New Discovery of the Great River of the Amazons', in *Expeditions into the Valley of the Amazons, 1539, 1540, 1639*, ed. Clements R. Markham, London, Hakluyt Society, 1st ser. 24, 1859, pp. 41–142.

Martius, Carl F. P. von, *Historia naturalis palmarum*, 3 vols, Leipzig, Frid. Fleischer, 1823–50.

Maximilian, Prinzen zu Wied, *Beiträge zur Naturegeschichte von Brasilien*, 5 vols, Weimar, Im Verlage des Landes-Industrie-Comptoirs, 1825–33.

Menezes, M. N., *British Policy towards the Amerindians in British Guiana 1803–1873*, Oxford, Clarendon Press, 1977.

Menezes, M. N., 'Introduction' to Goodall, *Sketches of Amerindian Tribes 1841–1843*, pp. 11–30.

Mennega, E. A., W. C. M. Tammens-de Rooij and M. J. Jansen-Jacobs, eds, *Check-list of Woody Plants of Guyana: based on D. B. Fanshawe's "Check-list of the Indigenous Woody Plants of British Guiana"*, Ede, The Tropenbos Foundation, 1988.

Moore, D. T., J. C. Thackray, and D. L. Morgan, 'A Short History of the Geological Society of London, 1807–1911, with a Catalogue of the British and Irish Accessions, and Notes on Surviving Collections', *The Bulletin of the British Museum of Natural History (Historical Series)*, 19, 1991, pp. 51–160.

Mowat, Linda, 'Four Paintings by Edward A. Goodall at the Pitt Rivers Museum, Oxford', *Journal of the History of Collections*, 5, 1993, pp. 223–9.

Murchison, Roderick, 'Address to the Royal Geographical Society of London', *Journal of the Royal Geographical Society*, 14, 1844, pp. xiv–cxxviii.

Murphey, Robert M., 'Mammalia americae australe. A Table of Taxonomic and Vernacular Names', *Ciencia Interamericana*, 17, 1–4, 1976, pp. 16–32, 18–30, 26–35.

Nees von Esenbeck, C. G. D., 'Cyperaceae a Schomburgkio in Guiana Anglica collectae, ex Herbario Lindleyano', *Journal of Botany*, 2, 1840, pp. 393–9.

Nelson, Joseph S., *Fishes of the World*, New York, John Wiley, 1984.

O'Leary, Timothy J., *Ethnographic Bibliography of South America*, New Haven, Human Relations Area Files, 1963.

O'Toole, Christopher, ed., *The Encyclopedia of Insects*, Abingdon, Andromeda Oxford, 1995.

Paoletti, Maurizio Guido, Darna L. Dufour, Hugo Cerda, Franz Torres, Laura Pizzoferrato and David Pimentel, 'The Importance of Leaf- and Litter-feeding Invertebrates as Sources of Animal Protein for the Amazonian Indians', *Proceedings of the Royal Society, Biological Sciences*, 267, 2000, pp. 2247–52.

Parliamentary Papers, 'British Guiana', *Parliamentary Accounts and Papers*, Vol. XXXIV, Ordered, by the House of Commons, to be printed 11 May 1840, pp. 315–31.

Payne, Pauline, 'Dr Richard Schomburgk and Adelaide Botanic Gardens, 1865–1891', unpublished PhD thesis, University of Adelaide, 1992.

Perez, Carlos Federico, 'Sir Robert Schomburgk y la invasión haitiana de 1855–56', *Aula. Revista General de la Universidad Nacional Pedro Henriquez Ureña*, 2, 4, 1973, pp. 11–29.

Peters, James Lee, *Check-list of Birds of the World*, 16 vols, Cambridge, Mass., Harvard University Press, 1931–87.

Prance, Ghillean T., and Jorge R. Arias, 'A Study of the Floral Biology of *Victoria amazonica* (Poepp.) Sowerby (Nymphaeaceae)', *Acta Amazonica*, 5, 1975, pp. 109–39.

Price, Richard, *Alabi's World*, Baltimore, Johns Hopkins University Press, 1990.

Price, Richard, *The Guiana Maroons: A Historical and Bibliographical Introduction*, Baltimore, Johns Hopkins University Press, 1976.

Questions des limites du Brésil et de la Guyane Anglaise a l'arbitrage de S.M. le Roi d'Italie. Atlas accompagnant le premier Mémoire du Brésil, Paris, Ducourtioux et Huillard, 1903–4.

Ralegh, Walter, *The Discoverie of the Large, Rich and Bewtiful Empyre of Guiana*, ed. Neil Whitehead, Manchester, Manchester University Press, 1997.

Ralegh, Walter, *Discovery of the Large, Rich, and Beautiful Empire of Guiana, with a Relation of the Great and Golden City of Manoa (which the Spaniards call El Dorado), etc. Performed in the Year 1595*, ed. Robert Schomburgk, London, The Hakluyt Society, 1st ser. 3, 1848.

Ribeiro de Sampaio, Francisco Xavier, *Diario de uma viagem que em visita, e correição das povações da Cap de S. José do Rio Negro fez o Ouvidor e Intendente Geral da mesma, Fco. Xavier R. Ribeiro de Sampaio, nos annos de 1774 e 1775*, Lisbon, Na Typografia da Academia, 1825.

Rivière, Peter, *Absent-minded Imperialism. Britain and the Expansion of Empire in Nineteenth-century Brazil*, London, Tauris Academic Studies, 1995.

Rivière, Peter, 'Confused Identities – the Schomburgk Brothers', *The Orchid Review*, 112, 2004, pp. 51–5.

Rivière, Peter, *An Ethnographic Survey of the Indians on the Divide of the Guianese and Amazonian River Systems*, unpublished BLitt thesis, University of Oxford, 1963.

Rivière, Peter, *The Forgotten Frontier. Ranchers of Northern Brazil*, New York, Holt, Rinehart and Winston, 1972.

Rivière, Peter, 'From Science to Imperialism: Robert Schomburgk's Humanitarianism', *Archives of Natural History*, 25, 1998, pp. 1–8.

Rivière, Peter, *Individual and Society in Guiana. A Comparative Study of Amerindian Social Organization*, Cambridge, Cambridge University Press, 1984.

Rivière, Peter, *Marriage among the Trio. A Principle of Social Organisation*, Oxford, Clarendon Press, 1969.

Rodríguez Demorizi, Emilio, 'Schomburgk, primer cónsul de Inglaterra en Santo Domingo', *Boletin del Archivo General de la Nación*, 10, 50–51, 1947, pp. 3–8.

Rodway, James, *History of British Guiana, from the Year 1668 to the Present Time*, 3 vols, Georgetown, J. Thomson, 1891–4.

Rose, James, ed., *Sir Robert H. Schomburgk. Essays in Honour of an Explorer and Natural Scientist*, Georgetown, University of Guyana and German Technical Cooperation, 2004.

Roth, Vincent, *Notes and Observations on Fish Life in British Guiana*, Georgetown, *The Daily Chronicle*, 1943.

Roth, W. E., 'An Introductory Study of the Arts, Crafts, and Customs of the Guiana Indians', *38th Annual Report of the Bureau of American Ethnology*, 1916–17, pp. 25–745.

Saint-Hilaire, Auguste de, *Plantes usuelles des Brasiliens*, Paris, Grimbert, 1824.

Saint-Hilaire, Auguste de, *Voyage dans les Provinces de Rio de Janeiro et de Minas Geraes*, 2 vols, Paris, Grimbert et Dorez, 1830.

Schomburgk, Richard, *Reisen in Britisch-Guiana in den Jahren 1840–1844*, 3 vols, Leipzig, Verlagsbuchhandlung von J. J. Webber, 1847–8.

Schomburgk, Richard, *Travels in British Guiana 1840–1844*, 2 vols, translated and edited by W. E. Roth, Georgetown, *Daily Chronicle*, 1922.

Smyth, Lieutenant, 'Account of the Rivers Amazon and Negro, from Recent Observations', *Journal of the Royal Geographical Society*, 6, 1836, pp. 11–23.

Smyth, W. and F. Lowe, *Narrative of a Journey from Lima to Para*, London, John Murray, 1836.

Snyder, Dorothy E., *The Birds of Guyana*, Salem, Mass., Peabody Museum, 1966.

Southey, Robert, *History of Brazil*, 3 vols, London, Longman, Hurst, Rees, Orme & Brown, 1810–19.

Spix, Johann Baptist von, and Karl Friedrich Philipp von Martius, *Reise in Brasilien auf Befehl Sr. Majestät Maximilian Joseph I, Königs von Baiern in den Jahren 1817 bis 1820*, 3 vols, Munich, Lindauar, 1823–31.

Sprengel, Kurt P. J., *Caroli Linnaei … systema vegetabilium*, 5 vols, Gottingae: sumtibus Librariae Dieterichianae, 1825–8.

Spruce, Richard, *Notes of a Botanist on the Amazon and Andes*, 2 vols, London, Macmillan, 1908.

Stanier, Peter, *Cornwall's Geological Heritage*, Truro, Twelveheads Press, 1990.

Stearn, William T., ed., *John Lindley 1799–1856. Gardener, Botanist and Pioneer Orchidologist*, Woodbridge, Antique Collectors' Club and Royal Horticultural Society, 1998.

Stephens, Thomas M., *The Dictionary of Latin American Racial and Ethnic Terminology*, Gainesville, University Press of Florida, 1999.

Swainson, William, *Animals in Menageries* (*The Cabinet Cyclopaedia*), London, Longman, Orme, Brown, Green and Longmans, 1838.

Swainson, William, *On the Natural History and Classification of Birds*, London, Longman, Rees, Orme, Brown, Green & Longman, 1836–7.

Tardy de Montravel, L. M. F., 'Exploration du fleuve des Amazones, depuis Para jusqu'a Obidos', *Revue coloniale*, 3, 1844, pp. 457–512.

Tello, Jaime, *Mamiferos de Venezuela*, Caracas, Fundación La Salle de Ciencias Naturales, 1979.

Van Heuvel, Jacob A., *El Dorado; being a Narrative of the Circumstances which gave rise to Reports in the Sixteenth Century, of the Existence of a Rich and Splendid City in South America, to which that Name was given, and which led to many Enterprises in search of it; including a Defence of Sir Walter Raleigh, in regard to the Relations made by him respecting it, etc.*, New York, J. Winchester, New World Press, 1844.

Vellard, Jean, *Histoire du curare. Les Poisons de chasse en Amérique du Sud*, Paris, Gallimard, 1965.

Vieillot, Louis Pierre, *Analyse d'une nouvelle ornithologie élémentaire*, Paris, Deterville, 1816.

Wallace, Alfred Russel, *A Narrative of Travels on the Amazon and Rio Negro*, London, New York and Melbourne, Ward, Lock & Co., 1890.

Waterton, Charles, *Wanderings in South America, the North-West of the United States, and the Antilles, in the years 1812, 1816, 1820 and 1824 with Original Instructions for the Perfect Preservation of Birds and for Cabinets of Natural History*, London, J. Mawman, 1825.

Weiss, C. S., *Uber das südliche Ende des Gebirgszuges von Brasilien in der Provinz S. Pedro do Sul und der Bande oriental, oder dem Staate vor Monte Video*, Berlin, 1840.

Whitehead, Neil, 'Kanaimà: Shamanism and Ritual Death in the Pakaraima Mountains, Guyana', in *Beyond the Visible and the Material. The Amerindianization of Society in the Work of Peter Rivière*, ed. Laura M. Rival and Neil L. Whitehead, Oxford, Oxford University Press, 2001, pp. 235–45.

Wilbert, Johannes, 'To Become a Maker of Canoes: An Essay in Warao Enculturation', in *Enculturation in Latin America. An Anthology*, ed. Johannes Wilbert, Los Angeles, University of California Press, 1976, pp. 303–58.

Wilbert, Johannes, *Mindful of Famine. Religious Climatology of the Warao Indians*, Cambridge, Mass., Harvard University, Center for the Study of World Religions, 1996.

Wilbert, Johannes, *Tobacco and Shamanism in South America*, New Haven and London, Yale University Press, 1987.

INDEX

Schomburgk was inconsistent in his spelling of names of both people and places. Where the spelling differences are major (e.g., with a different initial letter), the reader is referred to a standardized spelling; where they are minor, the variations are noted in parentheses after the standardized entry. Individuals and places that have only a single mention and that cannot be identified or located are not included in the index. Given that Schomburgk is best known as a botanist, more detailed entries are provided for botanical species, under the heading 'Plants', than for other natural species. Plants are listed under their scientific names as species, genus or family, depending on the level at which they can be identified. To find references to a plant with a vernacular name it is necessary first to consult Appendix 2, 'Glossary of Vernacular Plant Names'. References to 'Amphibia', 'Animals', 'Fishes', 'Insects', and 'Reptiles' are under those headings and simply by their popular names, e.g., 'Jaguar'. In all cases of natural species, references to their uses have also been included under this heading, e.g., under Birds: Parrots, will be found references to the use of parrot feathers as decoration.